Forming and Machining of Polymers, Ceramics, and Composites

Forming and Machining of Polymers, Ceramics, and Composites targets the two important manufacturing processes where plastic deformation is involved to give the required shape and size to the raw material. The main goal of the book is to represent the recent developments in the field of forming and machining of different non-metals, especially polymers, composites, and ceramics. Special focus is on the advancement of these processes to manufacture components from these non-metals.

- Presents exclusive material dedicated to forming and machining of non-metals, that is, polymers, ceramics, and composites
- Provides comprehensive coverage of all important topics related to non-metals processing
- Covers basics and current research in the field of forming and machining of non-metals
- Focuses on sustainability interventions and intelligent manufacturing techniques for quality and productivity in forming and machining of non-metals
- Discusses conventional and non-conventional machining and microfabrication aspects for fabrication and processing of non-metals

This book is aimed at graduate students and researchers in materials processing and machine design.

AF386499

Advanced Materials Processing and Manufacturing

Series Editor: Kapil Gupta

The CRC Press Series in Advanced Materials Processing and Manufacturing covers the complete spectrum of materials and manufacturing technology, including fundamental principles, theoretical background, and advancements. Considering the accelerated importance of advances for producing quality products for a wide range of applications, the titles in this series reflect the state-of-the-art in understanding and engineering the materials processing and manufacturing operations. Technological advancements for enhancement of product quality, process productivity, and sustainability, are on special focus including processing for all materials and novel processes. This series aims to foster knowledge enrichment on conventional and modern machining processes. Micro-manufacturing technologies such as micro-machining, micro-forming, and micro-joining, and Hybrid manufacturing, additive manufacturing, near net shape manufacturing, and ultra-precision finishing techniques are also covered.

Thin-Films for Machining Difficult-to-Cut Materials
Challenges, Applications, and Future Prospects
Ch Sateesh Kumar and Filipe Daniel Fernandes

Advanced Materials Processing and Manufacturing
Research, Technology, and Applications
Amogelang Sylvester Bolokang and Maria Ntsoaki Mathabathe

Advanced Joining Technologies
Edited by Manjaiah M, Shivraman Thapliyal and Adepu Kumar

Nanofinishing of Materials for Advanced Industrial Applications
Edited by Faiz Iqbal, Dilshad Ahmad Khan and Zafar Alam

Environmentally Benign Machining
Edited by Şenol Bayraktar and Sunil Pathak

Post-Processing of Parts and Components Fabricated by Fused Deposition Modeling
Techniques and Advancements
Edited by Vinayak R. Malik, Vivek Tiwary and Arunkumar P.

Forming and Machining of Polymers, Ceramics, and Composites
Edited by Matruprasad Rout and Kishore Debnath

For more information about this series, please visit: www.routledge.com/Advanced-Materials-Processing-and-Manufacturing/book-series/CRCAMPM

Forming and Machining of Polymers, Ceramics, and Composites

Edited by
Matruprasad Rout
Kishore Debnath

CRC Press
Taylor & Francis Group
Boca Raton London New York

CRC Press is an imprint of the
Taylor & Francis Group, an **informa** business

Designed cover image: Adobe Stock

First edition published 2025
by CRC Press
2385 NW Executive Center Drive, Suite 320, Boca Raton FL 33431

and by CRC Press
4 Park Square, Milton Park, Abingdon, Oxon, OX14 4RN

CRC Press is an imprint of Taylor & Francis Group, LLC

ISBN: 978-1-032-52790-1 (hbk)
ISBN: 978-1-032-66536-8 (pbk)
ISBN: 978-1-032-66537-5 (ebk)

DOI: 10.1201/9781032665375

Typeset in Times
by SPi Technologies India Pvt Ltd (Straive)

Contents

About the Editors

Matruprasad Rout is currently working as an Assistant Professor in the Department of Production Engineering, National Institute of Technology Tiruchirappalli, India. He completed his PhD from the Indian Institute of Technology (IIT) Kharagpur, India, in 2018. His research interests include metal forming, mechanical metallurgy, and material characterization. He has published more than 25 research papers in peer-reviewed international journals and conference proceedings. He has also published seven book chapters with publishers of international repute.

Kishore Debnath completed his bachelor's and master's degrees in mechanical engineering from National Institute of Technology (NIT) Agartala and National Institute of Technology Rourkela, respectively. He completed his PhD from IIT Roorkee in 2015 in the field of manufacturing of composite materials. He started his career as an assistant professor in the Department of Mechanical Engineering of National Institute of Technology Meghalaya in 2015. He has published more than 70 manuscripts in peer-reviewed journals. He has also contributed several chapters in different books published by internationally renowned publishers. His current research is focused on the area of conceptualization and development of biodegradable composites and cost-effective manufacturing methods. He was honored with the prestigious 'Excellent Research Contribution Award' of NIT Meghalaya in 2019 and 2023.

Contributors

Adefemi Adeodu
Department of Mechanical Engineering
University of South Africa
Florida, South Africa

Jasti Anurag
Composite Processing Laboratory
Department of Mechanical Engineering
National Institute of Technology
    Rourkela
Odisha, India

Sivanandam Aravindan
Department of Mechanical Engineering
Indian Institute of Technology Delhi
New Delhi, India

Vivek Bahuguna
Department of Mechanical Engineering
    & Design Innovation Center
National Institute of Technology
    Uttarakhand
Srinagar, India

Pramendra Kumar Bajpai
Netaji Subhas University of Technology
Delhi, India

Gobinda Chandra Behera
Department of Mechanical Engineering
Indian Institute of Technology
    Kharagpur
Kharagpur, India

Shaurya Bhatt
Department of Mechanical Engineering &
    Design Innovation Center
National Institute of Technology
    Uttarakhand
Srinagar, India

Alfa Bisoi
Department of Mechanical Engineering
National Institute of Technology
    Silchar
Silchar, India

Sandhyarani Biswas
Composite Processing Laboratory
Department of Mechanical Engineering
National Institute of Technology
    Rourkela
Odisha, India

Dipankar Chetia
Department of Mechanical Engineering
National Institute of Technology Silchar
Silchar, India

Ilesanmi Daniyan
Department of Industrial Engineering
Tshwane University of Technology
Pretoria, South Africa

Sankha Deb
Department of Mechanical Engineering
Indian Institute of Technology
    Kharagpur
Kharagpur, India

Sandeep Gairola
Centre of Excellence in Disaster
    Mitigation and Management
Indian Institute of Technology Roorkee
Roorkee, India

Danish Handa
Department of Aerospace Engineering
Indian Institute of Space Science and
    Technology
India

A. Kannan
Department of Mechanical Engineering
National Institute of Technology
 Puducherry
Karaikal, Puducherry, India

Simanchal Kar
Department of Mechanical Engineering
National Institute of Technology Silchar
Silchar, India

**Mukondeleli Grace Kana-Kana
 Katumba**
Department of Industrial Engineering
Tshwane University of Technology
Pretoria, South Africa

Deepak Kaushik
Department of Mechanical and
 Industrial Engineering
Indian Institute of Technology Roorkee
Roorkee, India

Nakka Anil Kumar
Department of Production Engineering
National Institute of Technology
 Tiruchirappalli
Tiruchirappalli, India

Pintu Kumar
Department of Production Engineering
National Institute of Technology
 Tiruchirappalli
Tiruchirappalli, India

Rajesh Kumar
Department of Mechanical Engineering
 & Design Innovation Center
National Institute of Technology
 Uttarakhand
Srinagar, India

Rendani Maladzhi
Department of Mechanical Engineering
University of South Africa
Florida, South Africa

V. Murugabalaji
Department of Production Engineering
National Institute of Technology
 Tiruchirappalli
Tiruchirappalli, India

Rounak Pal
Department of Mechanical Engineering
National Institute of Technology
 Silchar
Silchar, India

Deepak Patil
Department of Production Engineering
National Institute of Technology
 Tiruchirappalli
Tiruchirappalli, India

Mayank Pokhriyal
Department of Mechanical Engineering
 & Design Innovation Center
National Institute of Technology
 Uttarakhand
Srinagar, India

Pawan Kumar Rakesh
Department of Mechanical Engineering
 & Design Innovation Center
National Institute of Technology
 Uttarakhand
Srinagar, India

Khushi Ram
Indian Institute of Technology
Delhi, India

Suman Saha
Department of Mechanical Engineering
Indian Institute of Technology Dhanbad
Dhanbad, India

V. Satheeshkumar
Department of Production Engineering
National Institute of Technology
 Tiruchirappalli
Tiruchirappalli, India

Hitesh Sharma
Department of Mechanical Engineering
National Institute of Technology
    Uttarakhand
Srinagar, India

Tobilola Sholanke
Department of Mechanical and
    Mechatronics Engineering
Afe Babalola University
Ado-Ekiti, Nigeria

Inderdeep Singh
Department of Mechanical and
    Industrial Engineering
Indian Institute of Technology Roorkee
Roorkee, India

V. S. Sooraj
Department of Aerospace Engineering
Indian Institute of Space Science and
    Technology
India

B. N. Sreeharan
Department of Mechanical
    Engineering
Kumaraguru College of Technology
Coimbatore, India

Binaz Varikkadinmel
Department of Mechanical and
    Industrial Engineering
Indian Institute of Technology Roorkee
Roorkee, India

Preface

Materials like polymers, ceramics, and composites have been gaining importance in recent years as they find their application in almost all unique and strategic applications more economically. They possess excellent mechanical, chemical, thermal, and optical properties, to name a few. The manufacturing of products using these materials depends on processing them to undergo plastic deformation and machining them to the appropriate shape and size. Over the years, there have been developments in the field of polymers, ceramics and composites and so in the processing to produce components from these newly developed materials. This book presents two important manufacturing processes, namely machining and forming where plastic deformation is involved to give the required shape and size to the raw material. The book is intended for researchers, engineers, and practitioners in the fields of manufacturing engineering, mechanical engineering, production engineering, metallurgy, and materials engineering. The chapters in this book provide reviews and detailed descriptions of the processes, as well as a few case studies demonstrating the effectiveness of the processes. The book provides fourteen chapters and the chapter-wise summary is as follows:

Chapter 1 provides the reader with basic knowledge about the hot embossing process for creating micro-nano structures in polymer materials. The chapter also discusses the process parameters and their impact on the hot embossing process.

Chapter 2 provides a brief idea about the processes, extrusion and equal channel angular pressing. It demonstrates a method for fusing two acrylonitrile–butadiene–styrene (polyamide) polymers using a Y-shape die channel.

Chapter 3 is focused on the forming of tailor-made sandwich blanks. This chapter considers various aspects of sandwich blanks (SBs), for example the fabrication methods, the mechanical response, and the formability of SBs, factors influencing the formability of SBs and the deformation pattern in the forming of SBs, among others, and provides a comprehensive review.

Chapter 4 explores the hot isostatic pressing (HIP) for ceramics with an emphasis on the preparation of transparent ceramics with optimal product quality. The microstructural characteristics achieved for various ceramics after HIP are also discussed.

Chapter 5 presents a detailed review of the recent developments in the machining of polymer composites and provides an overview of the challenges and future scope in the machining of polymer-based composites. The chapter also discusses the influence of different factors on the performance of machined parts and various approaches along with the recent optimization techniques used for machining polymer composites.

Chapter 6 discusses the latest developments in machining techniques that can be used to achieve high surface quality and precise machining of polymer matrix composites (PMCs). Processes used for machining PMCs like electrical discharge machining, laser cutting, abrasive water jet machining, and hybrid machining are presented.

Chapter 7 introduces a new methodology, namely eccentric sleeve grinding (ESG) for abrasive-based surface machining of fibre-reinforced polymer composites. The chapter discusses various aspects of ESG, including kinematics, mechanics, process variable selection, surface generation, thermal performance, and more and includes specific case studies to illustrate the merits of ESG and its comparison with conventional surface grinding.

Chapter 8 represents the drilling characteristics of fibre-reinforced polylactic acid composites with a dedicated case study on the drilling characteristics of *Himalaya-calamus falconeri* fibre–based composites with two different drill geometries, namely the twist drill and the JO drill.

Chapter 9 emphasizes two nonconventional machining processes for fibre-reinforced polymer composites, namely water jet machining (WJM) and abrasive water jet machining (AWJM). The chapter discusses the effect of the process parameters of the previously mentioned processes on surface roughness, dimensional accuracy of machined parts, and damage to the fibre and matrix of the polymer composites.

Chapter 10 delivers a thorough examination of the application of AWJM in sustainable composites, elucidating its advantages, limitations, and potential for future applications, focusing on the hole-making process, accompanied by a detailed case study.

Chapter 11 is based on the machine learning approaches that include artificial neural network (ANN) modelling for the drilling of glass fibre-reinforced plastic laminates. The chapter highlights the implementation of ANNs, through a dedicated case study, to evaluate the impact of varying input parameters on drilling-induced damage, allowing for the identification of optimal settings that minimize damage and maximize efficiency.

Chapter 12 provides an overview of the process optimization for PMCs and discusses the same through a case study on process optimization for the mechanical and microstructural behaviour of aluminium-reinforced epoxy composites.

Chapter 13 presents a detailed overview of the implementation of intelligent approaches like ANNs, fuzzy expert systems, and hybrid neuro-fuzzy and meta-heuristic optimization algorithms during the machining of engineering composites. The chapter focuses on the implementation of these strategies towards controlling the machining output such as surface integrity, tool wear, tool life, material removal rate, productivity, and feature accuracy.

Chapter 14 discusses the grinding behaviour of the ceramic surface coating obtained through plasma spraying. The chapter discusses the various parameters to access the ceramic coating and presents a dedicated case study on accessing the grindability of alumina coating obtained through thermal spraying.

We want to thank all of the authors who took the time and effort to contribute chapters to this book, as well as everyone else who helped us publish it. We think that this book will be useful to individuals interested in the field of machining and forming processes for polymers, ceramics, and composites.

1 Polymer Flow Behavior in Micro-Nano Molds

Recent Advances in Modeling and Fabrication Processes

Deepak Patil and Sivanandam Aravindan

1.1 INTRODUCTION TO MICRO-NANO FORMING OF POLYMERS

Today, cost-effective component replication in polymers is critical for the mass production of a wide range of products, particularly in the consumer market. The replication techniques, particularly the injection molding process, are highly automated and optimized for replicating goods with a broad range of forms and dimensions. With the advancement of microsystem technology and, more recently, nanotechnology, replication was set to face a new challenge: the replication of structures in the micro or even nano range. The fabrication costs of micro and nanostructures, as well as the complexity of the fabrication methods, highlight the demand for cost-effective replication. The established replication technologies that have been optimised for structures in macroscopic dimensions cannot be used for structures in microscopic dimensions without modification.

The technological requirements are different because a decreasing structure size is characterized by an increasing relation between the surface and volume of a structure. Furthermore, the structure sizes require replication processes and technologies based on high-precision manufacturing and precise control, which makes it necessary to adapt established techniques or to develop new molding techniques. Today, hot embossing is, besides the micro injection molding process, one of the common replication technologies for replicating microstructures in polymers. With the upcoming nanoimprint technologies, the importance of the embossing technique increases. The thermal nanoimprint and the hot embossing processes are especially characterized by similar process steps. This chapter gives the reader a current development on the different aspects of hot embossing. The versatility of hot embossing will be illustrated by applications where the process plays an important role in the fabrication line. Today, microstructures are

DOI: 10.1201/9781032665375-1

part of our life. In many applications of everyday use, and especially in sophisticated applications, microstructures are integrated and fulfill, often invisibly, essential tasks in these applications. The size of these structures is, concerning the name "micro" structures, in the dimensions of micrometers. The range of these structures includes the range of several hundred micrometers down to the submicron range, at least a large bandwidth. But not in each of the three dimensions do these structures have to be characterized in these ranges. Often only in two dimensions structures are characterized by a resolution in the micron range; in the third geometrical dimension the size of millimeters or even centimeters are typical, for example, microchannels with a cross section of $50 \times 50 \ \mu m^2$ and a length up to several centimeters. The fabrication of these structures is part of the science of microstructure technology, and the further assembling into a microsystem, the science of microsystem technology [1]. The fabrication of microstructures can be done by several processes like mechanical machining or lithographic processes, which are cost- and time-intensive processes. To obtain a further distribution of structures and the corresponding microsystem, replication from master structures is recommended and a precondition for its use in commercial applications. In macroscopic dimensions, replication processes and their technologies are already established. Here, especially, the process of injection molding of thermoplastic polymers is responsible for the mass production of a wide range of structures. Replicating structures in microscopic dimensions requires other preconditions and has developed into independent processes with their technologies. Nevertheless, structures in the micron range can also be replicated with the established macroscopic replication processes and technologies, but with a decrease in structure size and molding area, the macroscopic-oriented processes come to their limits. Here, modified technologies and adapted processes close the gap and pave the way for the mass production of microstructures and mechanical microsystems.

Besides the process of micro injection molding or thermoforming, the process of hot embossing is one of the established technologies [2, 3]. Hot embossing is like the other micro-replication processes named earlier, a process that replicates a microstructured master, a so-called mold insert, in the polymer. Polymers are cheap and are available in different modifications with a wide range of properties. Therefore, microstructures with a large bandwidth of properties can be replicated, suitable for the application for which they are needed.

1.2 HOT EMBOSSING: DEVELOPMENT OF FABRICATION PROCESS AND PROCESS ANALYSIS

1.2.1 Principle of Hot Embossing

The hot embossing process is carried out in four steps; heating of the polymer film to molding temperature, isothermal molding, cooling of the molded polymer film to room temperature with the constant force being maintained, and removing the

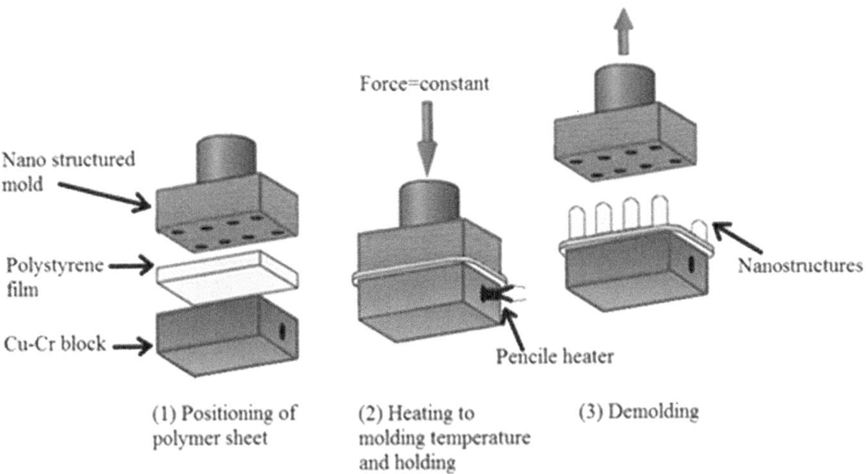

FIGURE 1.1 Schematic of the hot embossing process: positioning of polymer film, heating, and demolding.

component by withdrawing the mold. Figure 1.1 demonstrates the working principle of the hot embossing process.

The polystyrene film was aligned and kept in between the silicon mold and copper block. The total surface area (4×4 cm^2) of the polymer film is greater than the mold size ($8 \times 8.3 \times 0.7$ mm^3) to imprint nanostructures at least on a 3.2×3.2 cm^2 area. The copper block was heated with the help of a pencil heater up to polymer molding temperature. The instant temperature was sensed by the thermocouple and can be set to a limited value using a thermostat. The parameters (molding temperate: 100°C, embossing force: 20 N, holding time: 5 min) used were kept constant for the embossing of both types of nanostructures.

1.2.2 COMPONENTS OF HOT EMBOSSING PROCESS

Independent from the manifold of the embossing processes described above, hot embossing is mainly based on the thermal forming of thermoplastic polymers by using an embossing technique. The details of the technique depend on the hot embossing machine and may vary. In general, however, the components shown in Figure 1.2 are essential for hot embossing. A microstructured mold has to be replicated in the tool to integrate the microstructured mold. This tool consists of the heating and cooling unit and if needed, an alignment system for double-sided or positioned molding the molding machine, in which the tool will be integrated. The machine consists of a press unit with a frame of high stiffness. This unit has to generate the molding force and the relative movement between mold and polymer, a vacuum chamber in which the tool and the mold can be integrated.

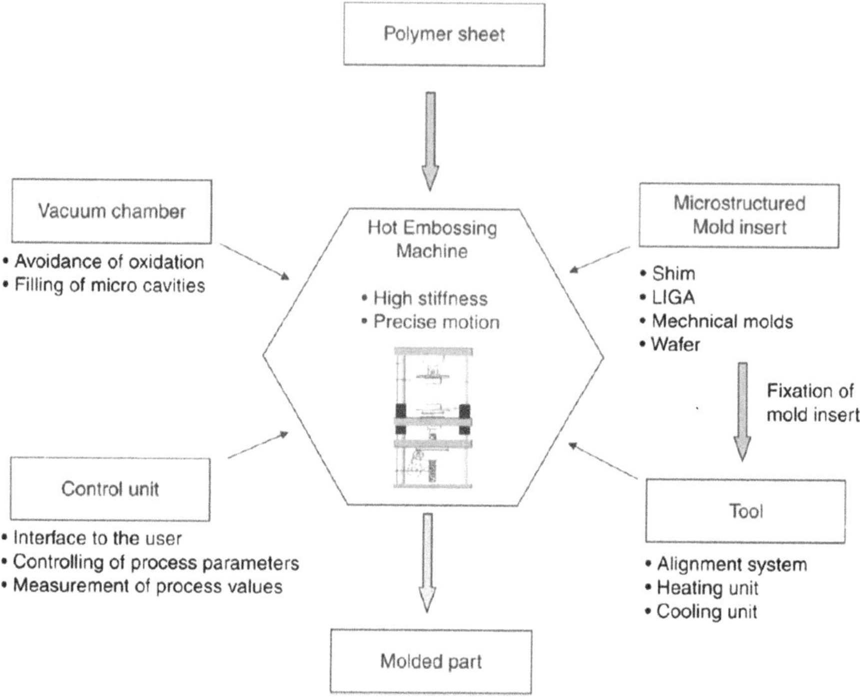

FIGURE 1.2   Schematic view of the components needed for successful hot embossing: machine, control unit, tool, mold insert, and a polymer film or sheet.

Reproduced with permission from Ref. [1], Copyright @ Elsevier, Hot Embossing by Matthias Worgull.

1.2.3   PROCESS AND MATERIAL PARAMETERS

1.2.3.1   Material Modelling

The actual hot embossing method is highly complex since temperature-dependent transitions between states that resemble solids or fluids are required. Because the behavior of the polymer (here, e.g., poly(methyl methacrylate), or PMMA) similarly switches from an elastic to a plastic range, it is difficult to theoretically foresee or estimate the amount of deformation generated by the procedure in advance. To investigate the embossing phenomenon explanation of the behavior of PMMA materials in terms of strain softening, work hardening, and temperature softening, the arbitrary Lagrangian–Eulerian (ALE) remesh approach and the finite element method (FEM).

It was assumed that the flow stress of a material is impacted by stress, strain, temperature, and strain rate. A new constitutive model for PMMA was created to better comprehend work hardening, strain and temperature softening, and strain rate hardening. This modeling was then used to do finite element analysis in ABAQUS. To begin the analysis, it was necessary to properly quantify flow stress for a wide variety of stresses, strain rates, and temperatures. The flow stress grows along with the strain

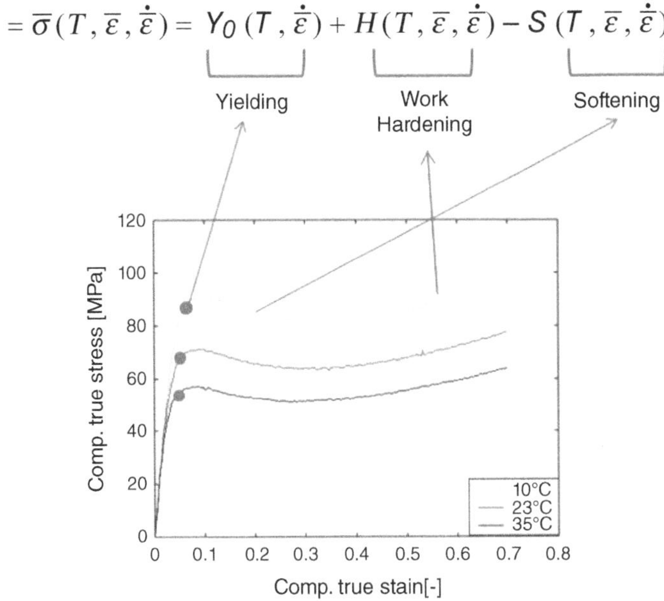

FIGURE 1.3 Stress-strain relation of poly(methyl methacrylate) (PMMA) in the low-temperature region.

Reproduced with permission from Ref. [4], Copyright @ MDPI.

rate, but it falls as the temperature rises. Work hardening and softening are supposed to be functions of strain, strain rate, and temperature, respectively, whereas the yield point is assumed to be a function of both strain rate and temperature. Figure 1.3 depicts the stress–strain relationship of PMMA at the low-temperature region.

First, to do the study, it was important to precisely define the flow stress for a wide range of temperatures, strain, and strain rate. PMMA has a distinct tendency in the flow stress curve from metals like steel or aluminum. For instance, PMMA exhibits softening in a specific region with increasing strain, which is different from the softening characteristic as per temperature. By comparison, metal exhibits work hardening at an increasing strain rate. A two-dimensional (2D) model used in the analysis is given in Figure 1.4.

By assuming symmetry, analysis was done for just one pattern. A temperature-controlled boundary condition was applied for heat transfer analysis, and the top stamp was treated as a rigid body. The derived constitutive model was applied to the finite element (FE) analysis using a user subroutine. A study of the heat transfer from the die to PMMA material and inside the PMMA was performed along with the forming analysis. The deformed shape was accurately predicted through this research.

In the future, it could be employed as a design tool. In the micro-hot embossing (HE) operation, the polymer has several benefits over a large variety of other crystalline, glass, or a quartz-like geometries (rounded, rectangular, high aspect ratios), material properties, ease of fabrication, and cheap cost [5]. Since its physical behavior

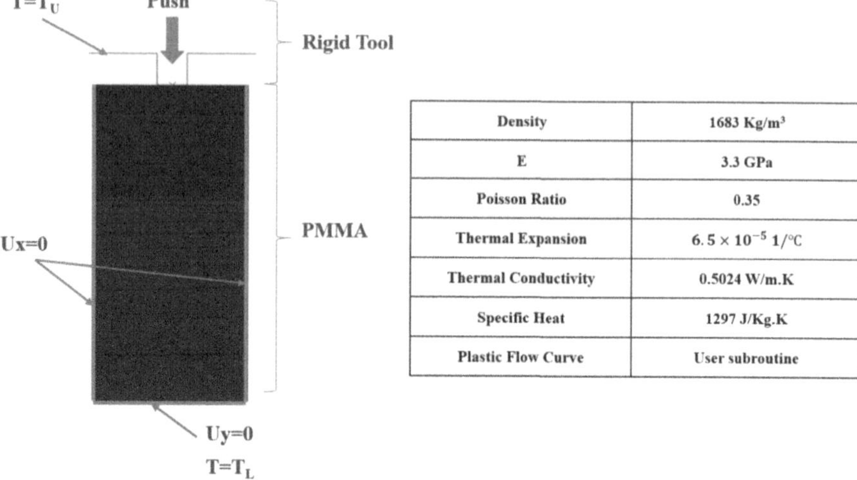

Density	1683 Kg/m³
E	3.3 GPa
Poisson Ratio	0.35
Thermal Expansion	6.5×10^{-5} 1/°C
Thermal Conductivity	0.5024 W/m.K
Specific Heat	1297 J/Kg.K
Plastic Flow Curve	User subroutine

FIGURE 1.4   Analysis model and material properties.

Reproduced with permission from Ref. [4], Copyright @ MDPI.

is substantially influenced by heating, the polymeric material is ideally suited for micro-HE. A thermoplastic substrate is preheated above the transmission glass temperature (Tg) during the compression stage and allowed to cool down below Tg in a demolding stage during the micro-HE operation. The physical behavior of the thermoplastic polymers over the complete micro-HE temperature range must be thoroughly explored to optimize this process's parameters and increase microstructure-replicating precision [6]. Different models are available, which are discussed in the following subsections.

1.2.3.2   Models of Viscoelasticity Based on Physical Principles

The elastic reaction is represented by the spring, while the viscous response is represented by the dashpot in the physically based model. Graphics objects now have new levels of representation due to physically based modeling. In addition to geometry, various physical quantities such as forces, velocities, torques, kinetic and potential energy, accelerations, heat, and others influence the creation and development of viscoelastic models (or time-dependent response). The behavior rules for the spring and the dashpot system are represented as Equations 1.1 and 1.2. [7–9]

$$\sigma = E\varepsilon \ldots \tag{1.1}$$

$$\sigma = \eta \frac{d\varepsilon}{dt} \ldots \tag{1.2}$$

1.2.3.3   Models with Viscoelastic and Viscoplastic Properties

The viscoelastic and viscoplastic characteristics of polymer composites have been extensively explored in the literature [10–14]. Whenever the strain is recovered and is thermodynamically reproducible after deformation, the term elasticity is employed [15]. When a strain is impossible to recover from thermodynamically irreversible and is thought to rely exclusively on the size of localized stress/strain, the word *plastic* is employed. The word *visco* is used when the strain varies over time [16]. During the micro-HE procedure, the behavior of the polymer is temperature and time-dependent, combining elastic, viscoelastic, and viscoplastic behaviors. There are several types of models for micro-HE materials.

1.3 REPLICATION ACCURACY IN THE MICRO-HE PROCEDURE

The filling stage is the most crucial phase in the micro-nano hot embossing since it dictates the replication accuracy of the final products. See the example in Table 1.1, where there is a slight change in the height of replicated nanostructures. The possible reason could be polymer shrikange in nano-cavity [17]. However, the exact reason for not completely filling the cavity is vital. That's where the simulation can forecast the polymer deformation throughout the filling of the mold cavity, which decreases the chances of defects in the final product. In the simulation of the micro-HE replication,

---

TABLE 1.1

Dimensions of Fabricated Nanostructures Compared with Structures Present on a Silicon Mold

Types of Structures	Nanostructured Silicon Mold (Avg. Value Dimensions)	Nanostructured on Polystyrene Film (Measured Avg. Dimensions)
Pillar 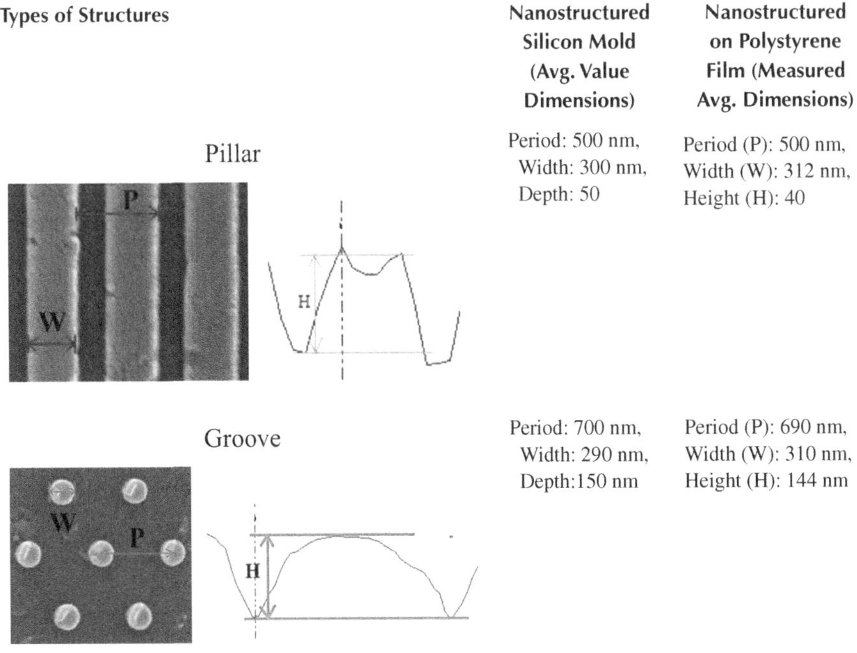	Period: 500 nm, Width: 300 nm, Depth: 50	Period (P): 500 nm, Width (W): 312 nm, Height (H): 40
Groove	Period: 700 nm, Width: 290 nm, Depth:150 nm	Period (P): 690 nm, Width (W): 310 nm, Height (H): 144 nm

Source: (Reproduced with permission from Ref. [17], Copyright @ IET Library)

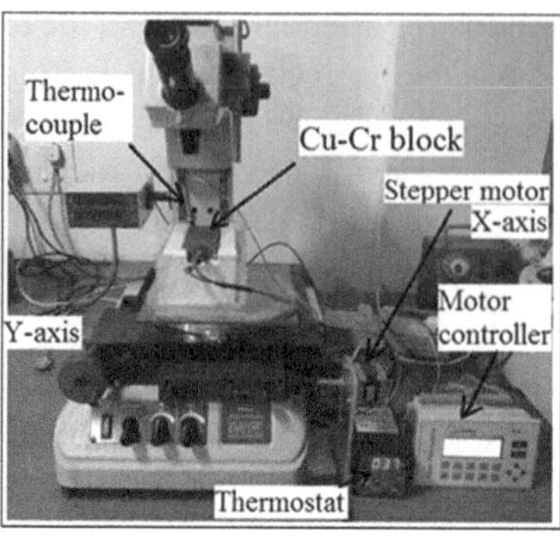

FIGURE 1.5   In-housed developed hot embossing setup.

Reproduced with permission from ref. [17], Copyright @ IET Library.

accuracy has been noticed for the location of the mold die cavity by compressing on a polymer substrate with pressure. Where the reservoir and channel are filled, a microchannel depth not more than 5% of the mold die cavity is designated as exact replication.

Huet et al. simulated a low strain rate to reduce thermal stress and study deformation behavior during micro hot embossing [18, 19]. The simulation model of a micro-HE glass-micro-prism array is depicted in Figure 1.5. It has been found that there is only a 7% deviation between theoretical and experimental forming. With a 630 °C embossing temperature, 0.335-N embossing force, and a 12-minute holding time, the maximum experimental forming rate was 66.56%. Based on the simulation results, molten glass formed a trapezoidal micro-prism array in the core cavity after flowing in a vortex-like fashion. Using the perfect hot embossing temperature and force resulted in a low strain rate, which, in turn, meant the online cooling process was not necessary, resulting in a faster hot embossing process. Hot melt glass is stuck to the die core microgroove surfaces when the embossing temperature and force are high, and optical glass substrate cracks, probably due to uneven thermal stress. Strain rate changes and deformation changes concerning embossing time during micro hot embossing are depicted in Figures 1.6a and 1.6b.

Apart from the influencing factors of the embossing machine and the tool, the influence of the microstructured mold insert is essential for the quality of the molded part. Most effective are undercuts of microcavities. These undercuts result in high demolding forces and increase the risk of damage to structures during demolding. The roughness of the mold surface is another factor responsible for high friction and demolding forces. It may cause the risk of damaging the structures to increase during demolding. Using mold coatings and release agents can reduce friction, thereby reducing the risk of damaging the structures during demolding. The flatness of the mold is essential for molding structures on a thin residual layer. There have been many advances in fabricating micro-nano dies/molds for polymer forming. Micro-electric discharge machining, electrochemical

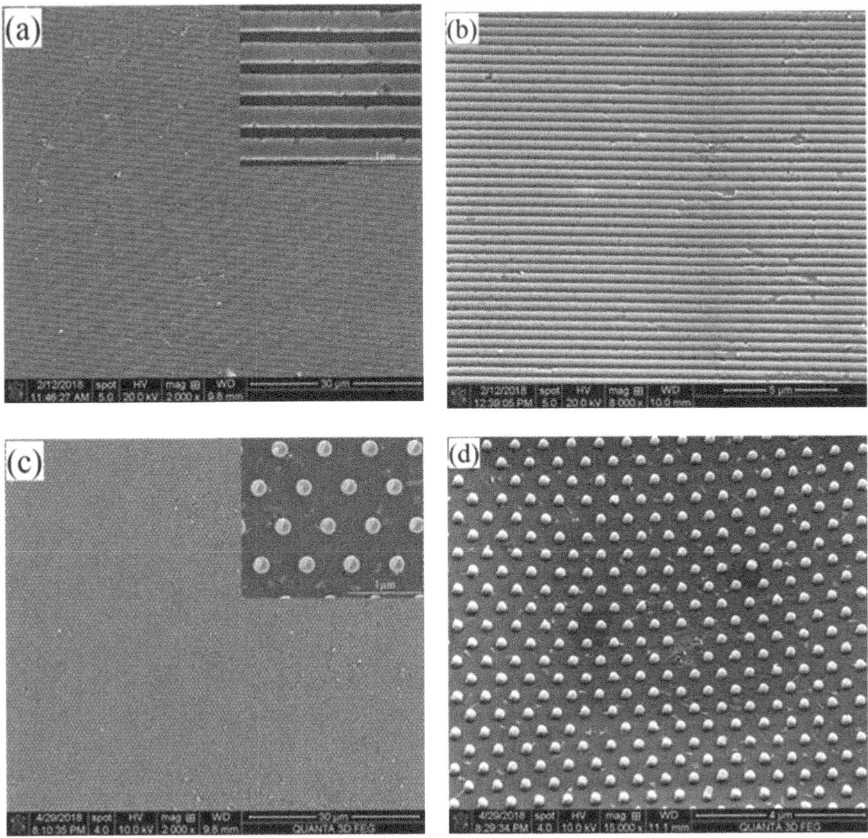

FIGURE 1.6 FE-SEM images of nanostructured polystyrene film (a) overhead image of grooves (b) Tilted (at 42°) image of grooves (c) overhead image of nanopillars (d) Tilted (at 42°) image of nanopillars.

Reproduced with permission from ref. [17], Copyright @ IET Library.

machining, laser textured molds, reactive ion etching, and focus ion beam machining are the popular techniques used for making reusable molds.

1.4 APPLICATIONS OF MICRO-NANOSTRUCTURED POLYMER SURFACES WITH A CASE STUDY

In hot embossing technique, three types of configurations exist: plate-to-plate embossing, roll-to-plate embossing, and roll-to-roll embossing. In this case study, the plate-to-plate in-house hot embossing setup is discussed. It had two sets of manual wheels that could move the stage in two orthogonal directions with a resolution of 1 μm. The distance traveled by the stage in both directions was displayed in real time through a digital display fixed on the stage. The setup was modified by automating the X-axis of the stage using a stepper motor. Stepper motor needs an

AC-to-DC converter, stepper motor controller, and stepper motor driver for its function. The stepper motor used for this system provides 2.5 µm/step; that is, the minimum distance moved by one step of the motor is 2.5 µm. Hence, the movement in the transverse direction was provided through the stepper motor assembly. For the development of setup, the apparatus required is a copper–chromium block of dimensions 100 × 40 × 50 mm, pencil heater, thermostat, thermocouple, and ceramic as an insulator. The copper–chromium block has good electrical conductivity, high strength, and good corrosion resistance (Figure 1.5). Due to the presence of chromium, the corrosion resistance property of copper–chromium block is good compared to a pure copper block. A pencil heater has a high watt density that makes it capable of generating high heat flux in less space without compromising its life span. A thermostat is often the main control unit for heating, by setting the target temperature. Ceramics can withstand a temperature up to 1600°C. Therefore, the insulation of ceramic was used and wrapped around the copper–chromium block.

1.4.1 Fabrication of Nanostructures on Polystyrene Film

The polystyrene film was aligned and kept in between the silicon mold and copper block. The total surface area (4 × 4 cm^2) of the polymer film is greater than the mold size (8 × 8.3 × 0.7 mm^3) to imprint nanostructures at least on a 3.2 × 3.2 cm^2 area. The copper block was heated with the help of a pencil heater up to polymer molding temperature. The instant temperature was sensed by the thermocouple and can be set to a limited value using a thermostat. The parameters (molding temperate: 100°C, embossing force: 20 N, holding time: 5 min) used were kept constant for the embossing of both types of nanostructures.

1.4.2 Characterization of the Structured Polymer Film

The photograph of a developed hot embossing setup is shown in Figure 1.6. The nano-grooves and pillar structures are fabricated on polystyrene film using this setup on 3.2 × 3.2 cm^2 area. The field emission-scanning electron microscope (FE-SEM) (Make: Quanta 3D FEG, FEI) images of nano-grooves and pillar structures were recorded after the DC sputter coater was coated with 7 nm of gold. The images are shown in Figure 1.6.

Figures 1.6a and 1.6c show the overhead FE-SEM images taken at lower magnification indicate the large area coverage of nanostructures whereas Figures 1.6b and 1.6d show the tilted view of nanostructured polystyrene surface at higher magnification to observe the detail ordered nano-grooves and nanopillars respectively. Atomic force microscopy (AFM) is used in order to know the height of nanostructures. The AFM (Make: Bruker, USA) images and a cross-sectional analysis of both nanostructures are shown in Figure 1.7.

The nanopillar structures are ordered in a hexagon pattern as can be seen from Figure 1.6c. The width and spacing between the grooves were measured using image-J analysis software and compared with the dimensions of the silicon mold used for the embossing. Table 1.1 shows the comparison of dimensions of fabricated nanostructures with dimensions of nanostructured silicon mold.

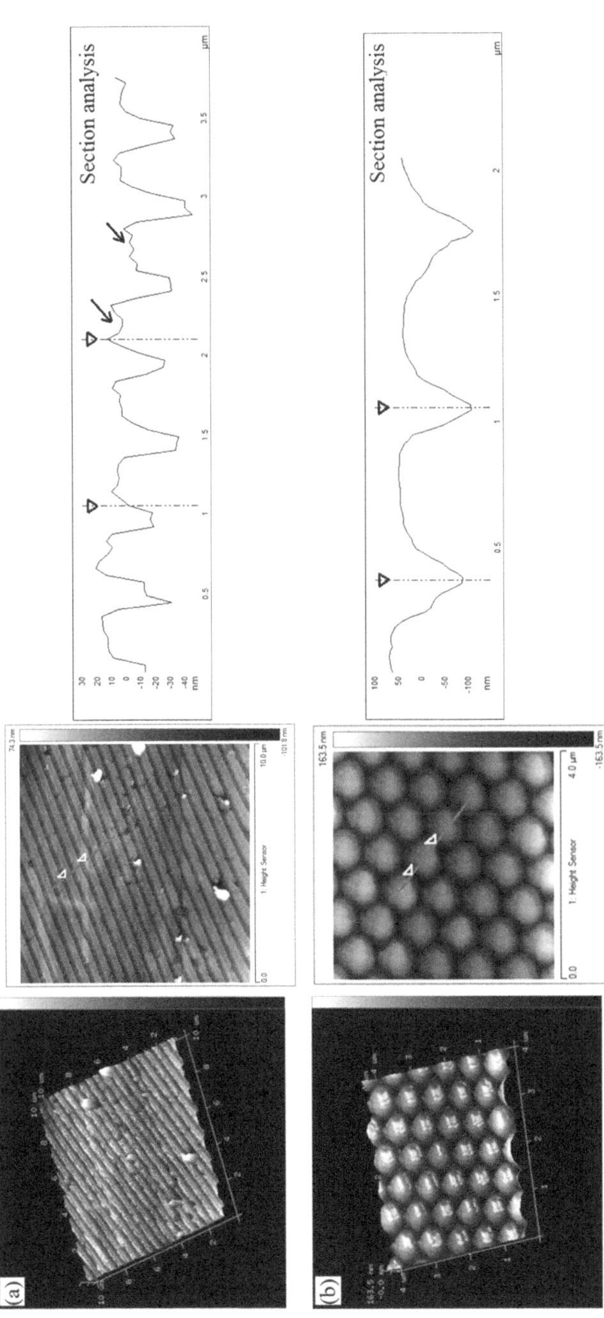

FIGURE 1.7 AFM images of nanostructured polystyrene film (a) grooves (arrow indicates uneven shrinkage of polystyrene film) (b) nanopillars. Reproduced with permission from ref. [17], Copyright @ IET Library.

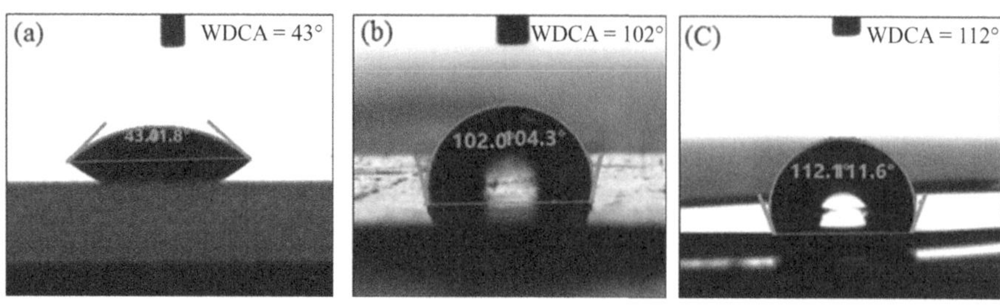

FIGURE 1.8   WDCA on (a) a plain polystyrene surface, (b) a nanogroove-structured surface, and (c) a nanopillar-structured surface.

Reproduced with permission from ref. [17], Copyright @ IET Library.

1.4.3 Application in Altering Surface Wettability

The water droplet contact angle (WDCA) was measured on plain and structured polystyrene surfaces using drop shape analyzer equipment (Make: DSA25, KRUSS, Germany). The WDCA greater than 90° is termed as the hydrophobic surface. Figure 1.8a shows the WDCA on a plain polystyrene surface (WDCA = 43°), which is hydrophilic. However, the WDCA on nanogroove structured surface is 102 ± 3° and 112° ± 2° on the nanopillar-structured surface. This transition from hydrophilic to the hydrophobic surface is due to the fabricated structures which enhance the value of surface roughness. Wenzel and Cassie-Baxter developed wettability models for structured surfaces. According to the Cassie-Baxter theory, the liquid droplet is supported by the structures present on the polystyrene surface, and the air gets entrapped between the structures and the liquid meniscus. Moreover, the wettability of the surface depends on the dimensions of the structured surface. During embossing, no chemical reagent is applied to the mold. Hence, the change in WDCA is only due to the presence of nanostructures on the polymer film. Hot embossing is a simple method of physical surface structuring capable of permanently altering the wettability surface from a hydrophilic to a hydrophobic surface.

1.4.4 Antibacterial Testing

The bacterial testing was carried out on plain and structured polystyrene surfaces using the plate counting method. The number of colonies indicates the number of live bacteria present on the respective polystyrene surface. A single colony originates from single live bacteria. The optical micrographs of colony-forming units are shown in Figure 1.9 The complete lawn of bacteria is observed on plain polystyrene surface as can be seen in Figures 1.9a and d whereas the countable colonies are observed on nanostructured polystyrene surface as can be seen from Figures 1.9b–e.

Thus, the developed setup enables the fabrication of nanostructures on a large area, which is very important for antibiofouling applications to minimize surface area for bacterial attachment and make the surface hydrophobic.

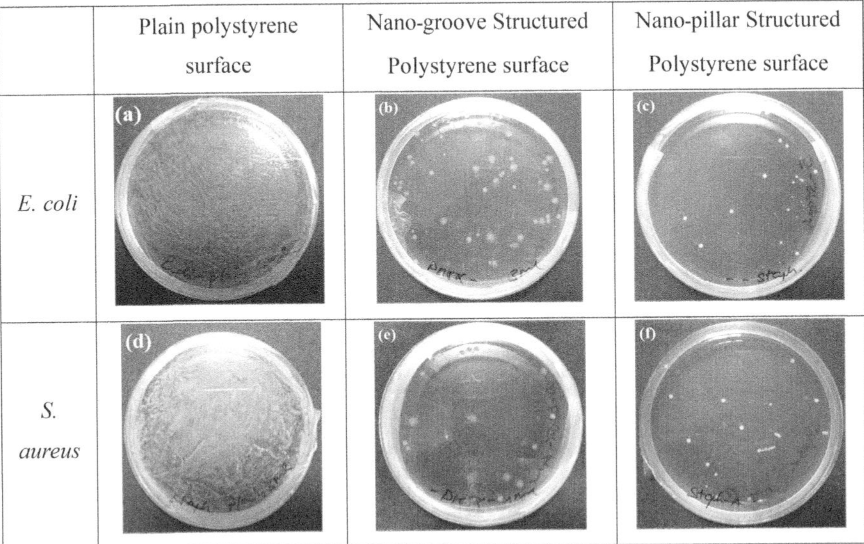

	Plain polystyrene surface	Nano-groove Structured Polystyrene surface	Nano-pillar Structured Polystyrene surface
E. coli	(a)	(b)	(c)
S. aureus	(d)	(e)	(f)

FIGURE 1.9   Optical micrographs of colonies formed in petri dishes after 24 h of incubation. Reproduced with permission from ref. [17], Copyright @ IET Library.

1.5 SUMMARY AND FUTURE DIRECTION

The applications mentioned earlier are just a few of many, but they highlight the importance of hot embossing and thermal nanoimprinting as established replication technologies. New applications will drive further advancements in hot embossing and nanoimprinting, particularly if large series are required. Automation and possibly standardization of, for example, molding formats will aid in establishing these technologies in industry. Cycle times must be lower in the future to achieve an effective heating and cooling system. Thus, optimized molding tools are essential for cost-effective molding. Combining hot embossing with well-known handling methods in the macroscopic arena can minimize process times, and hot embossing can be established as an industrial duplication technology. For cost-effective duplication, large molding areas are also necessary. These molding areas relate to the individually constrained production processes of the mold inserts. Molding tools with multiple mold inserts can be utilized to get around these restrictions. These features are primarily of interest for a cost-effective application of hot embossing. Hot embossing will continue to be a good choice for the first replications of prototypes, regardless of the potential for cost savings. In this instance, future replication may shift to structures with smaller structure sizes in the nano range, along with high aspect ratios, such as greater than 5 nm. The majority of work employed PMMA sheet as the working substrate, although relatively few researchers used polycarbonate and other materials. The majority of the work is done at the micron scale, although there is room for advancement to the nanoscale. Only simple microstructures have been created.

REFERENCES

[1] M. Worgull, *Hot embossing, Micro and Nano Technologies*, ISBN: 978-0-8155-1579-1, 2009.

[2] N. Bogdanski, H. Schulz, M. Wissen, H.-C. Scheer, J. Zajadacz, and K. Zimmer, 3D-hot embossing of undercut structures-an approach to microzippers. *Microelectronic Engineering*, 73–74, pp. 190–195, 2004.

[3] H. Chien and Z. P. Chen, Fabrication of light guiding plate of double-sided hot-embossing. *Materials and Science Forum*, 505–507, pp. 211–216, 2006.

[4] Y. Dand and J. B. Kim, Material modeling of PMMA film for hot embossing process, *Polymers*, 13, 2021.

[5] B. Hand and U. Heim, Hot embossing as a method for the fabrication of polymer high aspect ratio structures, *Sensors and Actuators A: Physical*, 83, pp. 130–135, 2000.

[6] J. Narasimhan and I. Papautsky, Polymer embossing tools for rapid prototyping of plastic microfluidic devices, *Journal of Micromechanics Microengineering*, 14(96–103), pp. 93–97, 2004.

[7] Y. Liu and H. Antes, Application of visco-elastic boundary element method to creep problems in chemical engineering structures, *International Journal of Pressure Vessels and Piping*, 70, pp. 27–31, 1997.

[8] H. Taylor, Y. C. Lam and D. Boning, A computationally simple method for simulating the micro-embossing of thermoplastic layers, *Journal of Micromechanics Microengineering*, 19, 2009.

[9] C. Hung, R. H. Chen and C. R. Lin, The characterisation and finite-element analysis of a polymer under hot pressing, *International Journal of Advanced Manufacturing Technology*, 20, pp. 230–235, 2002.

[10] D. Drozdov, A model for the viscoelastic and viscoplastic responses of glassy polymers, *International Journal of Solids and Structures*, 38, pp. 8285–8304, 2001.

[11] F. Zaïri, M. Naït-Abdelaziz, K. Woznica and J. M. Gloaguen, Constitutive equations for the viscoplastic-damage behaviour of a rubber-modified polymer, *European Journal of Mechanics - A/Solids*, 24, pp. 169–182, 2005.

[12] D. W. Holmes and J. G. Loughran, Theoretical aspects of the testing of elasto-viscoelastic–viscoplastic materials, *Polymer Testing*, 27, pp. 189–203, 2008.

[13] G. Z. Voyiadjis, A. Shojaei and G. Li, A generalized coupled viscoplastic–viscodamage–viscohealing theory for glassy polymers, *International Journal of Plasticity*, 28, pp. 21–45, 2012.

[14] W. Zhao, M. Ries, P. Steinmann and S. Pfaller, A viscoelastic-viscoplastic constitutive model for glassy polymers informed by molecular dynamics simulations, *International Journal of Solids and Structures*, 226–227, 111071, 2021.

[15] C. Mareau, M. F. V. Berveiller, Micromechanical modeling coupling time-independent and time-dependent behaviors for heterogeneous materials, *International Journal of Solids and Structures*, 46, 223–237, 2009.

[16] J. S. Kim and A. H. Muliana, A combined viscoelastic–viscoplastic behavior of particle reinforced composites, *International Journal of Solids and Structures*, 47, pp. 580–594, 2010.

[17] D. Patil, A. Sharma, S. Aravindan and P.V. Rao, Development of hot embossing setup and fabrication of ordered nanostructures on large area of polymer surface for antibiofouling application, *Micro & Nano Letters*, 14(2), pp. 191–195, 2019.

[18] M. Hu, J. Xie, W. Li, Y. Niu, Theoretical and experimental study on hot-embossing of glass-microprism array without online cooling process, *Micromachines*, 11, 2020.

[19] D. Patil, S. Aravindan, A. Pal, W. H. Khan, P. Pragya, S. Pundir, I. Xess, S. Mohapatra, V. Perumal and P. V. Rao, Microtopographic superhydrophobic polymer surface to prevent urinary tract infections causing nosocomial drug-resistant bacterial adhesion, *Surfaces and Interfaces*, 41, 103239, 2023.

2 ABS Polymer Extrusion Joining

A Case Study on a Y-shaped Channel

Pintu Kumar

2.1 INTRODUCTION

An increase in pollution and rapidly depleting oil supplies have resulted in energy savings and environmental effect reduction policies over the years. These goals are achievable by utilising a range of materials. A diverse set of features enables designers to choose the best decision for achieving the specified structural performance. Advances in manufacturing technology led to the widespread adoption of lightweight alloy materials in a variety of sectors. Alloys such as magnesium, titanium, and aluminium have high strength-to-weight ratios [1, 2]. A set of components is used to create a simple structure at low expense [3]. There are numerous such features since the transportation industry values them. They provide advantages such as lower fuel use and emissions.

The moulding industry believes that customers are going to demand the latest advances in construction that are lightweight. Companies must address new production-related opportunities and difficulties, as well as quicken the creation and uptake of hybrid products, in order to be successful [4]. This evolution necessitates a fundamental shift in product design as well as a focus on the creation of alternative splicing technologies for the assembly and joining of components composed of various materials with wildly dissimilar physical and chemical properties.

Modern society depends on a wide range of parts and technologies, including automobiles, cell phones, computers, and medical equipment that can only be made using modern engineering polymers and composites with polymer matrices. Due to the complicated design of such items, welding becomes a crucial component of manufacturing. The value of mixing polymeric materials has been heavily highlighted. It has taken a lot of work to develop bonding techniques that improve flexibility in design, the mechanical characteristics of polymeric components, and durability, hence lowering failure rates and lengthening service life [5]. Today, structural and non-structural applications in thermal and coating applications frequently use polymer bonding processes. A variety of polymer encapsulating technologies are also available for applications where chemicals are limited, such as in the case of biocompatible components produced from biocompatible materials [6].

DOI: 10.1201/9781032665375-2

Since their initial synthesis at the start of the 20th century, technical or designed polymers have made significant advancements and, after many years of development, currently exhibit much better qualities. A lot of new mechanical, thermal, and photochemical qualities also occur as a result of this advancement, which also significantly lowers production costs. Additionally, more recent efforts have been made to enhance the performance of these materials in comparison to metals and ceramics [7].

Researchers are currently receiving serious attention. Effective new technology development tackles problems by integrating using a fusion-based approach. The process of "joining by creating" is one of the most remarkable mating strategies. In all these procedures, the linking of the components occurs as a result of plastic deformation caused by a strong connection, whether mating is done at warm or cold temperatures, with or without the addition of "filler" material.

In this study, the two basic types of techniques are polymer extrusion and hybrid joining methods. A brief description of each category is presented and discussed.

2.2 ABOUT THE PLASTICS

Polymer-based thermoplastics may be employed as innovative materials in a range of industries due to their lower density compared materials manufactured of metal. Three main types of thermoplastic materials are categorized as commercial plastics, engineering plastics, and high-performance plastics [8]. Metals and non-metallic materials have been displaced in applications needing great mechanical strength across a wide temperature range by engineering plastics and high-performance plastics [9]. Commercial plastics are frequently employed in production settings that need little in the way of mechanical sturdiness and normally operate at temperatures below 100°C [10]. Engineering plastics are less readily produced than commercial plastics, yet they are more valuable economically. Engineering plastics that fall within the polyamide category offer excellent mechanical strength as well as corrosion and slip resistance [11].

In the recent past, thermoplastic polymers have replaced metals in the aerospace and automotive sectors [7]. Due to this, manufacturing companies have been able to make significant financial savings over the past few decades. Other causes exist besides the lowering of production costs; weight reduction, thermal insulation, and flexibility are further crucial variables. New design possibilities are made possible by polymers. The aerospace industry's research and development on polymer-based metallurgy replacement, as well as related efforts, are heavily influenced by aircraft fuel efficiency. The use of polymers in computers and other consumer items has been profitable. Techniques for electromagnetic signal transfer and electrical isolation are also covered.

Polymeric materials have been used in the medical industry. In the early 20th century, polymer science advanced quickly, particularly that which was created with the outbreak of World War II, which served as a catalyst for the revolution [12]. When polymers were developed during the so-called polymer revolution, World War II was over. Those prohibited from military application first saw usage during peacetime.

Since metal was scarce during the war, scientists were obliged to consider substitute synthetic materials; as a result, nylon and polytetrafluoroethylene (PTFE) are the most often utilized polymers. Ropes, umbrellas, and parachutes are made of nylon.

Metal components are coated with PTFE due to its exceptional ability to resist corrosion. Poly(methyl methacrylate) (PMMA) was one of the first polymers to be explored as a potential implantable biomaterial [13, 14]. PMMA has now been extensively used for many other things, including prostheses, dental implants, bone fillers, and joint replacements [15]. No references have been found by the authors about the Y-shape ABS (acrylonitrile–butadiene–styrene) extrusion joint; ABS joining efforts have been made.

2.3 EXTRUSION PROCESS

In the process of extrusion, a kind of plastic deformation, the billet is passed through a separate hole using a die with a larger cross-sectional area than the original billet. The three types of condensation are hydrostatic, indirect, and direct. The interaction between the holder and the die wall, as well as the resulting interaction between the billet, results in high compressive stress. Effectively preventing the billet material from cracking at the first rupture from the billet is this compressive stress.

2.3.1 ABOUT DIRECT EXTRUSION PROCESS

The principle of direct extrusion is that the force applied to the billet and the movements of the billet follow the same direction, as indicated in Figure 2.1a, but it is the reverse in the case of indirect extrusion. The Y-shape die in this instance is created using the direct extrusion method. The following observations predict that direct extrusion is superior to indirect extrusion:

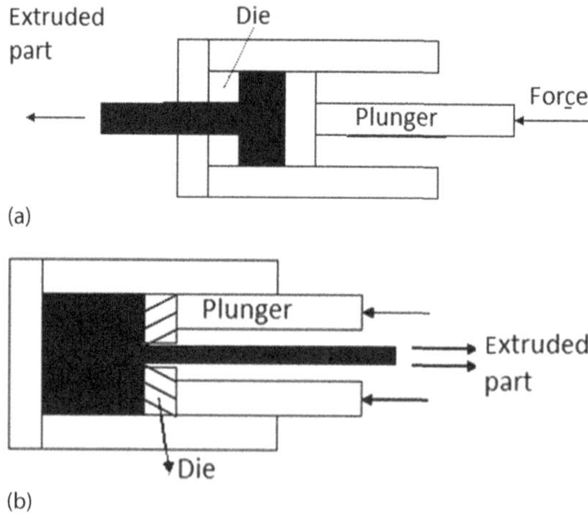

FIGURE 2.1   (a) Schematic of forward extrusion. (b) Schematic of backward extrusion.

- An internal plunger in the press that presses the billet applies compression force to this component.
- The mould consists of a tiny hole with the necessary cross-section. Due to its high compressive strength, the component may flow through the aperture of the mould and adapt its shape as necessary.
- The extruded component is taken out of the die and heated for improved mechanical characteristics.
- Direct extrusion is used in the production of solid rods, bars, hollow tubes, and hollow and solid sections depending on the layout and form of the die.

2.3.2 About Indirect Extrusion

In indirect extrusion, also known as backward and reverse extrusion, the die is mounted to the ram rather than at the opposite end of the container, as shown in Figure 2.1b. As the ram moves, the metal is forced to flow through the clearance in the opposite direction of the ram's motion. Because the billet is not forced to move in relation to the container, there is no friction at the container walls, and hence, the ram force is smaller than the direct extrusion. The decreased stiffness of the hollow ram and the difficulty in sustaining the extruded product as it exits the die are the limitations of indirect extrusion. Hollow (tubular) cross sections can be produced through indirect extrusion. There are practical restrictions to the length of the extruded part that can be manufactured using this procedure. As the job length increases, ram support becomes an issue. The friction between the container walls and the billet raises the ram pressure over that of indirect extrusion.

2.4 JOINING OF THERMOPLASTICS

Designers frequently challenge the choice of materials, procedures, and joining techniques in industrial applications. Utilising a variety of manufacturing and assembly processes, complex items are turned into their final forms.

Adhesive bonding, mechanical bonding, and direct tie-up are the three methods used to unite thermoplastics. Mechanical fastening, cutting, crimping, screwing, or crimping are all not supported on embedded devices when complete closure of the container is necessary. With some polymers, adhesive bonding of thermoplastics may be successful for non-embedding applications, but it is less successful for implants that need a lengthy service life. The aquatic ecosystem in adhesive has a propensity to degrade over time. The seal's durability and reliability are compromised. In order to weld thermoplastics, connecting surfaces made of homogeneous and diverse materials must be melted. The loading of pieces for a while on the joint surface causes non-static hardening, which leads to welding. Welding is often composed of interior and external procedures. Since the 1960s, hot plate welding, an external heating method, has been one of the most widely used welding techniques for thermoplastics [15–17]. Recent years have seen a rise in the use of friction stir welding for polymer welding.

2.5   A CASE STUDY: ABS POLYMER JOINING EMPLOYING Y-SHAPED EXTRUSION CHANNEL

2.5.1   IDENTIFICATION OF ABS MATERIALS

In 1954, Borg-Warner Corporation commercialised ABS, which had been invented in 1948. Its glass transition temperature is 105°C. Acrylonitrile, butadiene, and styrene are all present in it. Butadiene is a petroleum hydrocarbon derived from the C4 portion of steam cracking, tyrene monomers are created by the dehydrogenation of ethylbenzene, and acryliconitrile is a synthetic monomer comprising propylene and ammonia. It is anticipated that 13.49 million metric tonnes of ABS would be produced globally by 2024. Here, a billet with a diameter of 8 mm and a length of 100 mm is chosen for the experiment as shown in Figure 2.2.

2.5.2   DESIGN OF Y-SHAPE EXTRUSION CHANNEL

The Y-shape die channel design outperforms the die schematic for the extrusion channel when it comes to connecting polymers. How dies are created and utilised is described next.

The generated Y-shape die channel, seen in Figures 2.3a and 2.3b, is an embodiment of two comparable circular two-turn equal channel angular pressing channels [18]. Two input channels, two taper channels, and one common exit channel were all referred to as "Y" in the die design, as illustrated in Figures 2.3a and 2.3b). Because of the offset between the two intake channels, the tapered channel intersects with the common exit channel. Tri-junction (T) denotes the axial point of intersection of a common exit channel and a taper channel. Channel angle, corner angle, inner die radii, and outside die radii are constant at each die turn in symmetric channels. The Y-shape die channel is shown in Figure 2.3b to have a circular cross section with a diameter of 12 mm, a channel angle of 120°, a corner angle of 20°, and an internal die radii of 2 mm. Material from two independent input channels is combined into one at the tri-junction, where the first die turn is more effective than corner angles and outer die radii. Punch holders direct the distribution of load in both inflow channels.

2.5.3   JOINING PROCESS

Two ABS polymer rods must travel through a heated die channel in order to be joined together. It is made sure that the preheated die doesn't get any hotter than two thirds

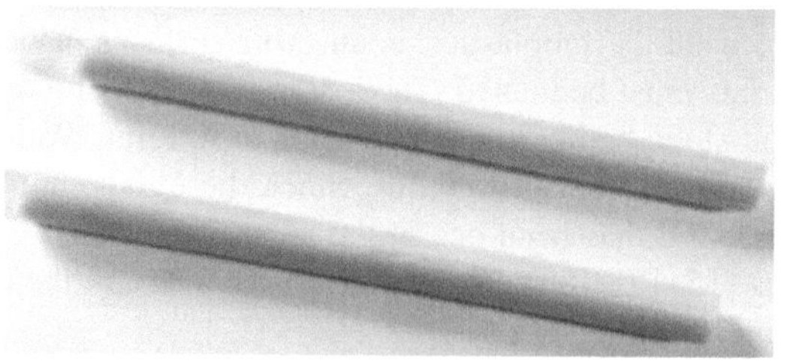

FIGURE 2.2   Billet for the extrusion and joining.

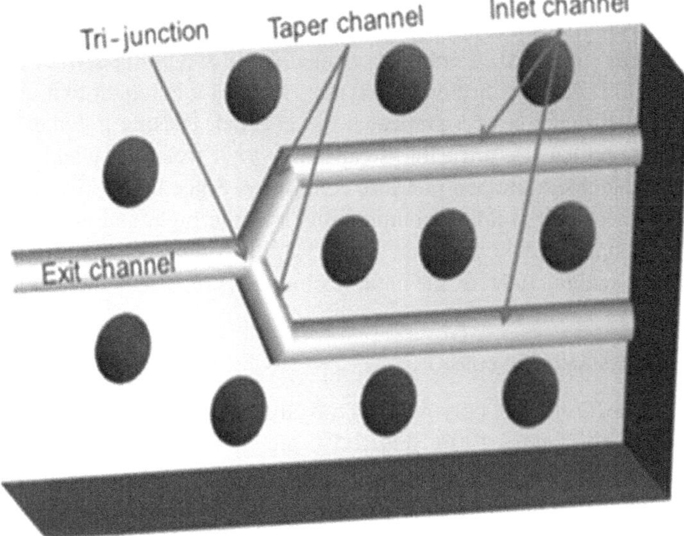

(a)

(b)

FIGURE 2.3 (a) Y-shape die channel computer-aided design model. (b) Y-shape die channel manufacture die.

of the polymer's melting point (245°C) when extrusion is being done. The muffle furnace's rope heater is used to pre-heat the die. For pressing both billets, two equivalent die punches are employed. A cold die channel is utilised with polymers to discover higher extrusion pressure. Softened polymers entered a common exit channel after passing via a tri-junction of a preheated die channel, causing polymer surfaces to disperse. Polymer surface diffusion is caused by their weldability and when treated under specific temperature and pressure conditions. Since both rods are constructed of ABS polymer, their welding compatibility is quite high, and as a consequence, their surfaces fuse quickly when they travel through a continuous die channel. To produce joints without defects, the entire technique is repeated.

2.5.4   RESULTS AND DISCUSSION

According to Section 2.5.3, polymer rods travelling through a constant die opening at two different temperatures 190°C and 165°C are shown in Figures 2.4a–c. Polymer exerts combined pressure when entering a common exit channel from two separate intake channels, softening the substance and causing it to fuse at the tri-junction. The typically produced joints were extruded using preheated die temperatures of 190°C, as shown in Figure 2.4a. Porous extruded rods with the defective joint are visible at higher temperatures. Incorrectly forcing molten polymer through the die at a higher die temperature results in porosity at the surfaces and the joint contact. At a preheated die temperature of 165°C, a defect-free junction is produced by several experiments as shown in Figure 2.4b. Figure 2.4c shows a typical macrograph of the joint along the transverse direction. Since there are no border regions in joint structures, fusion is more uniform and homogenous.

A point contact at the tri-junction serves as the initial connection of the billet's tip, and it remains unbroken during the whole extrusion process. The remaining piece of the billet connects at the surface while the extrusion load is continually applied. The continued passage of the billet down the exit channel until the end of the channel controls the uniform surface of the heat-affected zone (HAZ) on both sides of the joint. As a result, the joint thickness thins out along the escape route. Changes in friction coefficient, relative velocity, and heat loss from HAZ were the causes of this.

2.6   SUMMARY

Current research investigates the metal forming operation in the cutting-edge joining process using a Y-shape extrusion channel. When polymer rods travel through symmetrical die channels rotated at a 120° die angle and keeping the die temperature constant at 165°C, void-free, high-quality joints are produced. Concave joint section construction created within the exit channel ensured improved interlocking. When the die temperature exceeds 150°C, thick joint interface and cavities occur across the joint section.

2.7   FUTURE SCOPE

Over the last half-century, significant research has been conducted to improve process performance and control systems for diverse welding processes. However, there

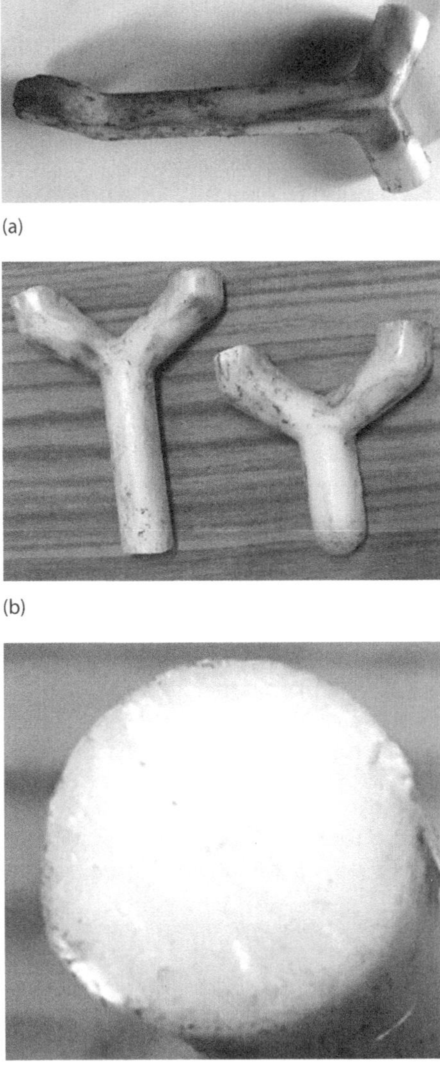

(a)

(b)

(c)

FIGURE 2.4 (a) Joint polymer at temperature 190°C. (b) Joint polymer at temperature 165°C. (c) Cross section of joint polymer in the transverse direction.

are still numerous issues to address. The following topics should receive priority attention in terms of research and development:

- Develop new welding processes and investigate their use
- Severe plastic deformation–based welding of polymer and inspection of the welded structures
- A substantial additive-free joint
- Research into potential methods for similar/dissimilar materials

REFERENCES

[1] D. P. Edwards and I. G. Crouch, Light alloys. In *The Science of Armour Materials* (pp. 117–166). Woodhead Publishing. 2017.

[2] K. Zheng, D. J. Politis, L. Wang and J. Lin, A review on forming techniques for manufacturing lightweight complex—shaped aluminium panel components. *International Journal of Lightweight Materials and Manufacture*, 1(2), pp. 55–80. 2018.

[3] A. A. Luo, A. K. Sachdev and D. Apelian, Alloy development and process innovations for light metals casting. *Journal of Materials Processing Technology*, 306, p. 117606. 2022.

[4] L. Josefson, K. Anyfantis, B.D.C. Pinheiro, B.Q. Chen, P. Dong, N. Ferrari, K. Gotoh, J. Huang, M. Krause, K. Liu, and S. Paboeuf, September. Committee V. 3: Materials and Fabrication Technology. *In International Ship and Offshore Structures Congress (p. D011S001R004). SNAME. 2022.*

[5] F. Ahmad, N. Yuvaraj, and P.K. Bajpai, Effect of reinforcement architecture on the macroscopic mechanical properties of fiberous polymer composites: A review. *Polymer Composites*, 41(6), pp. 2518–2534. 2020.

[6] D. Puppi and F. Chiellini, Biodegradable polymers for biomedical additive manufacturing. *Applied Materials Today*, 20, p. 100700. 2020.

[7] I.O. Oladele, T.F. Omotosho and A.A. Adediran, Polymer-based composites: An indispensable material for present and future applications. *International Journal of Polymer Science*, 2020, pp. 1–12. 2020.

[8] C. Al Christopher, Í.G. da Silva, K.D. Pangilinan, Q. Chen, E.B. Caldona and R.C. Advincula, High performance polymers for oil and gas applications. *Reactive and Functional Polymers, 162,* p. 104878. 2021.

[9] Q. Wang and Y. Li, Research advances in preparation, mechanism and application of thermally conductive and electrically insulating polymer composites in thermal management materials: A review. *High Performance Polymers*, 35(7), p. 09540083231164342. 2023.

[10] A. Lee, and M.S. Liew, Tertiary recycling of plastics waste: an analysis of feedstock, chemical and biological degradation methods. *Journal of Material Cycles and Waste Management*, 23(1), pp. 32–43. 2021.

[11] J.H. Mallinson, *Corrosion-resistant plastic composites in chemical plant design.* CRC Press. 2020.

[12] D.C. Mowery and N. Rosenberg, *Paths of innovation: Technological change in 20th-century America.* Cambridge University Press. 1999.

[13] Zhang, S., Liang, X., Gadd, G.M. and Zhao, Q., Advanced titanium dioxide-polytetrafluorethylene (TiO2-PTFE) nanocomposite coatings on stainless steel surfaces with antibacterial and anti-corrosion properties. *Applied Surface Science*, 490, pp. 231–241. 2019.

[14] S.J. Kim, B. Choi, K.S. Kim, W.J. Bae, S.H. Hong, J.Y. Lee, T.K. Hwang, and S.W. Kim, 2015. The potential role of polymethyl methacrylate as a new packaging material for the implantable medical device in the bladder. *BioMed Research International*, 2015(1), p. 852456. 2015.

[15] Ramakrishna, S., Mayer, J., Wintermantel, E. and Leong, K.W., 2001. Biomedical applications of polymer-composite materials: A review. *Composites Science and Technology*, 61(9), pp. 1189–1224.

[16] Kocatüfek, U., Yeni, Ç., Ülker, A. and Sayer, S., Optimization of hot plate welding parameters of glass fibered reinforced Polyamide 6 (PA6 GF15) composite material by Taguchi method. *Usak University Journal of Material Sciences*, 3(1), pp. 69–85. 2014.

[17] C. B. Bucknall, I. C. Drinkwater, and G. R. Smith, Hot plate welding of plastics: factors affecting weld strength. *Polymer Engineering & Science*, 20(6), pp. 432–440. 1980.

[18] P. Kumar, and S.S. Panda, An innovative method to join two polymer rods through Y-shape extrusion channel. *Measurement*, 119, pp. 270–282. 2018

3 Mechanical and Forming Behaviour of Sandwich Blanks

Nakka Anil Kumar and V. Satheeshkumar

3.1 INTRODUCTION

Sandwich blanks have an advantage over monolithic sheets in engineering applications due to their uniform stress distribution, unique mechanical behaviour, thermal and electrical insulation as well and superior acoustics properties. The selection of fabrication method is crucial since it affects the strength of interfacial adhesion and mode of fracture. Despite having such advantages, the complexity and time-consuming fabrication processes limited their usage in the automotive industry. So, instead of performing the sheet forming process first and following by joining them together, it will be advantageous to join them in the initial step and then follow the sheet metal–forming process. The forming of sandwich sheets is often more confined compared to monolithic sheets owing to the load-transferring mechanism through the layers. But incremental sheet forming (ISF) can extend the forming potential of sandwich sheets by employing simple tooling features. Even though sandwich composites have excellent mechanical qualities, delamination as well as wrinkling are commonly occurring defects due to geometrical imperfections and mechanical heterogeneity. This chapter presents the importance of sandwich sheets, fabrication methods, and mechanical behaviour of sandwich sheets from the literature. In particular, forming behaviour of sandwich blanks, incremental deformation of sandwich sheets, and remedies for possible defects have been extensively discussed.

3.2 SANDWICH STRUCTURES

Sandwich structures are a type of tailor-made blanks (TMBs) and have vast applications compared to other composite structures. Figure 3.1 shows the schematic of the sandwich structure comprising the lightweight core layer and two face sheets. Sandwich panels are a combination of two or more individual components with various properties, which when integrally joined together result in improved performance. By getting little weight from the core, the face sheets enhance the moment of inertia of the whole composite and resist bending and buckling. Sandwich structures are lightweight and can be employed in a variety of applications. The face sheets of sandwich sheets can be made of any metal, wood, plastic, fibres, or some sort of

DOI: 10.1201/9781032665375-3

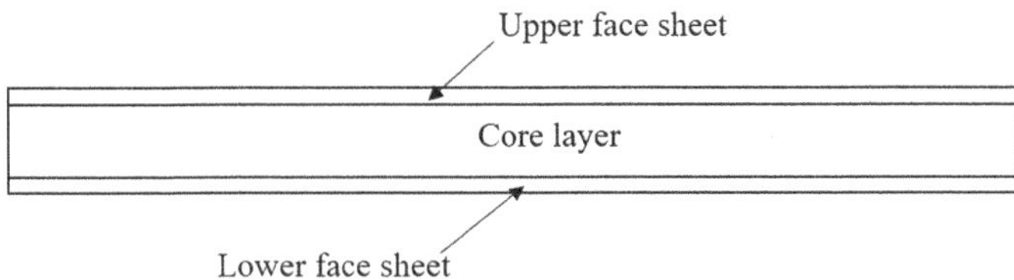

FIGURE 3.1 Schematic of sandwich panels.

fibre composite, while the core material can be made of wood, plastic, or metal foam with different structures like corrugated or truss, and honeycomb with various cell structures.

It is desirable to employ a lightweight material as a core with increased thickness to separate the face sheets since the increase in distance between the face sheets will increase the bending resistance of the whole structure. By referring to the analogy of an I-beam, this notion can be clearly understood. The idea is that while bending, the moment of inertia must be increased by pushing the material as far away from the centroidal axis of the beam which implies spacing the flanges widely from the middle. Sandwich beams effectively perform in a similar manner; however, the web will be replaced by the lightweight core. As a result, the idea of separating the faces with a lightweight core tends to improve the moment of inertia. Since the lightweight core is in the middle, the structure will have assured mechanical stiffness and bending strength by reduced weight for the structural requirements. The human skull is a good example of a sandwich panel in which there are two dense layers of compact bone and the little thin layer of trabecular bone in the middle. The wing of a bird is another example to demonstrate the importance of a sandwich structure having an almost truss-like structure in the middle. The same has been implemented in aircraft to create easy lift with reduced fuel cost.

3.3 STRATEGIES FOR FABRICATION OF SANDWICH BLANKS

Sandwich structures often have strong facings, including metal alloys, and fibre-reinforced polymers. The core can be processed of low-density wood, metal foams, and polymers to provide better mechanical performance. However, the weakest segment of the sandwich structure is the face–core interface. The bonding strength of the surfaces is controlled through the materials and structures involved, and it can vary for different processes fabrication processes. The classification of the fabrication processes for the sandwich structures can be based on the usage of mouldings. Traditionally, moulds are common for fabricating sandwich sheets. However, modern manufacturing processes extended to fabricate complex structures with additive manufacturing techniques. The selection of the technique depends on raw materials and required resolutions. The traditional techniques are time-consuming processes but the initial costs are low whereas additive manufacturing techniques can reduce the time and be economical for large production in the long run.

3.3.1 Different Categories of Joining Methods for Sandwich Blanks

The combining process is a crucial step in the fabrication of sandwich panels. As a result, numerous fabrication methods have been studied which are predominantly divided into two categories: adhesive bonding and fusion bonding (Grunewald et al., 2017).

Adhesive joining is applied to join similar, dissimilar, and different thicknesses of materials. Adhesive bonding has the advantage of a large area of bonding with uniform stress distribution and correspondingly leads to the elimination of distortion or residual stresses. Among various adhesives, Epoxies are widely used adhesives because of their ability to cure without producing volatile by-products and lower shrinkage during the curing (Ghosh et al., 2016). The adhesive-bonded TMBs are mainly utilizes in the construction of aircraft and automotive. Adhesive bonding involves different theories of adhesion, namely mechanical interlocking, diffusion, adsorption, chemical bonding, and wetting (Adams, 2021). The effectiveness of adhesive joining depends on geometrical parameters and surface preparation. Wingfield (1993) discussed the frequently used surface treatment methods, such as plasma treatment and grit blasting. To acquire a satisfactory bond of adhesives, the impurities on the sample surface, including releasing agents, machinery oils, or palm prints, must be removed instantly. Surface properties increase adhesion by three phenomena such as soaking the low-energy surfaces, chemical modification of surfaces, and an increase in the surface roughness. The adhesive bonding is limited due to the chemical and physical interaction between the glue and the skin or core is often not optimum, and the adhesive has a minor impact on the flexural strength of the sandwich panel, but it does have a substantial impact on the ductility of the whole structure (Ageorges et al., 2001, Grunewald et al., 2017).

The process works through the molecular interaction along the contact region between the layers. By interaction, the interface attains the high-volume characteristics of adherence. This process is well suited to thermoplastic materials. It is distinguished under the fact that the joint is formed by intermolecular diffusion of the components in the polymers. This is distinct from adhesive bonding where mechanical interlocking is the main constraint for adhesion. The initiation of fusion bonding happens while the molecules at the surface are becoming movable. Later, the bond is formed by a combination of surface rearrangement, wetting, diffusion, and polymer randomness. During fusion bonding, heat is used to weaken the polymers while pressure is used to bind the core layer and face sheets. For instance, vacuum moulding, compression moulding, and double-belt lamination are processes classified under fusion bonding. The specification of a procedural window for the fabrication of sandwich structure is prominent for fusion joining because it is restricted on one side by such a weak interfacial interaction during reduced temperatures while, on the other side, it may break the polymer core and converge to delamination at elevated temperatures. This method is very suitable when the skin and core materials have comparable thermal qualities since the molecule exchange process can degrade the counterpart materials when they have dissimilar thermal properties. Therefore, fusion bonding is considered to be ineffective when joining two different polymeric materials. The parameters including working temperature, chemical composition, miscibility, the distribution of molecular weight, the orientation of the chain, and the structure of polymer molecule affect molecule inter-diffusion. Joining the sheets through the use of vacuum moulding, compression moulding, or in situ foaming provides good interfacial strength out of available fusion bonding processes. (Young and Wool, 1983; Jabbari and Peppas, 1995; Trende et al., 1999; Brockholm and Bach, 2013; Grunewald et al., 2017). Roll bonding (Mousa and Kim, 2017; Carrado et al., 2010), heat pressing, direct injection molding (Ramani and Moriarty, 1998), over moulding (Gähde et al., 1992) are the other fusion bonding methods to fabricate the sandwiched TMBs.

3.3.2 Design Considerations for Fabrication of Sandwich Blanks

The face sheets of sandwich structures are typically designed to withstand normal loads, flexural forces, and bending moments while the core material is developed to withstand transverse shear forces acting on the structure. To meet the previously mentioned requirements, steel or aluminium alloys are often preferred for skin layers. In shear loading, the core must be strong enough to resist the face sheets from sliding over each other, and in compression, it should be rigid enough to hold them at with fixed distance away (Noor et al., 2009). For design purposes, Johnson and Sims (1986) divided the sandwich structures into two different types based on core flexibility. Shear distortion in bending is insignificant when the core is stiff in shear; therefore, classical bending theory can be applied with an adequate value for the flexural capacity. Shear distortion can be considered when the core is flexible in shear, contributing to panel deflections during transverse loading. In this instance, for beams and panels, classical bending calculations must be modified to account for thickness shearing phenomena.

Through a review of recent studies on a design concept for sandwich panels, Feng et al. (2020) recognized a few areas for further research, including structure design, material selection methods, performance evaluation technology, and the application of sandwich-structured TMBs with various core configurations. The design flexibility of sandwich structures provides the applicability of new core structures to improve mechanical qualities and attain reliability in final products. To address current demands, new forms of sandwich structures that are relatively low-density and used for multifunctionality should be developed. The design incorporates new materials such as shape memory alloys (SMAs), shape memory polymers, and piezoelectric since versatility and multifunctionality are the primary goals of sandwich structure design. To achieve this, adaptive or hybrid materials should be explored as core materials or even as face sheets in real-time applications.

Barre et al. (1995) investigated the static and fatigue behaviour of glass fibre sandwich composites by using balsa wood, polymethacrylimide foam, and phenolic corrugation as core materials. If the core thickness exceeds 90% of the nominal thickness, the mechanical characteristics of the core materials are the essential aspects for determining the static behaviour and fatigue breakdown while the adhesion strength between the face sheets and the adhesive core is essential for the endurance limit of sandwich structures. Naik et al. (2020) introduced a new technique for fabricating metal/polymer/metal (MPM) sandwich slabs for higher interface strength by using wire mesh in metal–polymer contacts during the fabrication phase. In addition to wire meshing, resistance welding is used to bond wire mesh to metal skin, and a vacuum hot-pressing method has been used to fill gaps in wire mesh with polymer to create a void-free mesh polymeric core. Unlike adhesive bonding, the bond strength is not weakened by moisture or water. Stretch forming as well as V-bending tests were used to evaluate the formability of fabricated sandwich composites. The results revealed that the integrated solidification method establishes higher bonding strength and higher inter-laminar shear strength.

3.4 MECHANICAL BEHAVIOUR OF SANDWICH BLANKS

The mechanical behaviour of sandwich sheets is different from the monolithic sheets majorly due to dissimilar structural composition and configurations. In this section, the tensile behaviour, flexural behaviour, and energy absorption capacity of the different sandwich structures are discussed.

3.4.1 TENSILE BEHAVIOUR

He et al. (2020) evaluated the influence of a number of layers and the layer direction on the tensile behaviour of multi-hole fibre metal laminates (FMLs). The results revealed that fracture strain increased with the increasing number of holes since the perforation enlarged the deformation region before fracture. Still, the layer direction effect is insignificant toward ultimate strength, but it changed the failure pattern from brittle fracture to delamination failure. Du et al. (2016) studied the effect of tensile loading on notch damage and failure mechanisms of thermoplastic-based FMLs and observed the fracture in the foremost fibre after yielding followed by the failure of the matrix. Skin characteristics and altering the portion of core volume increased the tensile strength and work-hardening rate of sandwich structures. The effect of core volume and surface properties of face sheets for metal/polypropylene/metal sandwich sheets was evaluated by Shin et al. (1999). Structures including a hard skin and a low-volume fraction of polypropylene core had a higher yield, ultimate tensile strengths, and lower work hardening rate. Even though work hardening is not dependent on core volume fraction, those with soft skin and a high-volume fraction of polypropylene core were shown the least strength. The amount of serration in the flow curve lowered due to an increase in the portion of polypropylene core volume in the sandwich structure and the tensile strength was become more sensitive to temperature. It is observed that the skin sheet failed first during tensile deformation, followed by the core and interface bonding.

Satheeshkumar and Narayanan (2014) evaluated the tensile properties of adhesive-bonded steel sheets by using two-part epoxy and acrylic-based adhesives as the core layers. It was asserted that the elongation and strain hardening exponent (n) was improved in the hardener-rich formulation of TMBs, which converted the adhesive to more ductile. The adhesive core delays the necking and it showed improved elongation when compared to the single and double sheets. Apart from the chemical properties, the geometrical parameters have an impact on the mechanical performance and deformation of sandwich composites. Zhang et al. (2022), experimentally investigated the mechanical behaviour of sandwich cylinders with square honeycomb cores, including metal liners along with theoretical and numerical simulations. With the rise of the relative density of square honeycomb, the conversion from local buckling of the shell to material yielding was pointed out. Meanwhile, the local buckling mode was increased with the initial failure load of the sandwich cylinder.

The weight-efficient, strong, and stiffer composites with metal liners can withstand axial, as well as transverse, loads. Gu et al. (2022) predicted damage and failure modes of a corrugated core sandwich cylinder with a thin wall metal liner reinforcement by quasi-static uniaxial compression tests. During, the testing local buckling was the initial failure that occurred in the metal liner, and even continued to spare the loads under circumferential support. Hou et al. (2021), studied the mechanical behaviour of corrugated sandwich panels under both quasi-static and dynamic loading conditions. The normal strength of corresponding sandwich composites was weakened and the shear strength had strengthened with an increase in shear compressive loading angle. It was drawn that the rate of loading had less influence on the deformation mode where a positive loading rate was a cause for the initial failure of the interface bond of panels. To strengthen the interface bonding, Shanshan et al. (2022) employed vacuum brazing and cementing during fabrication to determine the impact on the mechanical properties. It was demonstrated that the energy dissipation capacity improved after adding the reinforcement, and during the tensile–shear test, the peak load was improved due to an increase in the density of core material.

3.4.2 Flexural Behaviour

The flexural behaviour of sandwich structures depends on the material properties of face sheets and core, the orientation of the structure, and the loading distribution. The structural orientation of the core is significant for shear responses since flexural loading can cause core crushing. Apart from the core, the geometrical properties of face sheets also have a major influence on flexural strength. To meet the previously mentioned requirements, a few advanced materials were discussed for use as face sheets and core layers. Li et al. (2015) checked the flexural rigidity of E-glass 3D woven spacer fabric composites and observed that thicker face sheets provide high flexural strengths as well as local deformations since they continued to bear loads in spite of minute damages and fractures. Latour et al. (2021) explored the flexural behaviour of double-skin sandwich panels with steel face sheets and aluminium foam core for civil engineering applications such as primary decks in multi-storage buildings. Having the analogy of the previously mentioned thinner aluminium foam panels are susceptible to defective zones, it is suggested that thicker (>13 mm) panels should be used in steel and aluminium foam sandwich panels for improved performance. It was revealed that, if the density of the core is lowered, the flexural stiffness of metal foam sandwich structures drops, but the stiffness and ductility remain unchanged.

During bending, delamination is the dominant failure mode in sandwiched TMBs. Hassan et al. (2015) estimated the flexural strength of novel composite laminates consists of a different number of layers through the bending tests. The flexural strength was increased with the number of layers (tested up to 8 layers) used in the structure due to the higher levels of stiffness in the fibre layers and the face metals. The observed flexural strength is greater than the tensile strength since the total specimen has undergone uniform stress during the tensile tests. Lascoup et al. (2006) added stitching to fabricate the sandwich composite and concluded that the stitching improved the flexural resistance in out-of-plane loading conditions but that it decreased the in-plane properties. The core provided shear stiffness, and its geometry was shown a considerable effect on the behaviour the flexural rigidity. Sugiyama et al. (2018) fabricated continuous carbon fibre–reinforced thermoplastic sandwich composites using a three-dimensional (3D) printer, whereby the core structures were in the shape of hexagonal, rhombus, rectangular, and circular. From the observations, it was suggested that the rhombus core structure exhibits better flexural rigidity during bending.

3.4.3 Energy Absorption

If the number of layers in the core is raised for the same weight, the specific energy absorption of the sandwich sheet will improve significantly. The energy absorption of sandwich structures is impacted by the mechanical (density) and geometrical properties (thickness and structure of the foam core) of the structure. Once the deflection during axial stretching is greater than the thickness of the sandwich structure, it exhibits a significant role in the post-yield regime (Kazemi, 2021; Zhang et al., 2016). The inclusion of SMAs in the form of thin wires has a notable influence on the improvement of energy absorption capacity. Masoudi et al. (2021) used super-elastic SMA wires in hybrid composite face sheets and performed the quasi-static loading test. It was noticed that the face sheet which experiences the initial loading has effective resistance to damage. Even though the wires have increased the total weight of the composite, there was a substantial increase in levels of energy absorption before failure. Taghipoor et al. (2020) checked the influence of lattice core direction with polyurethane foam that was both filled and unfilled. It was pointed out that in both cases, the longitudinal direction of the lattice increased the energy absorption, mean force, and efficiency of crushing force while comparing similar parameters in the lateral direction.

Eyvazian et al. (2021) evaluated the mechanical behaviour of resin pin–reinforced composite sandwich structures under the indentation test and three-point bending conditions. From the results, it was concluded that the reinforcement increases the energy absorption capacity of sandwich composites. The energy absorption capacity increased to 31% under indentation, and it increased to 68% under three-point bending conditions compared to non-reinforcement sandwich structures. However, the resin pin diameter and arrangement of reinforcement affect the percentage change in the energy absorption. Ship- and offshore-applied sandwich structures are usually subjected to repeated dynamic loads. The energy absorption performances of sandwich layers under fatigue conditions can be efficiently modulated by the geometrical parameters. Guo et al. (2019) studied the mechanical behaviour of aluminium foam sandwich plates under impact loading, and it was observed that the peak impact force and energy absorption of the upper sheet were improved with thickness. But these were decreased in the aluminium foam core and the lower sheet with the increased thickness since the portion of the energy absorbed by the upper sheet is transformed into plastic deformation of the lower sheet. Zhang et al. (2021a), explored the dynamic behaviour of aluminium honeycomb sandwich panels and concluded that the levels of energy absorption can be controlled by wall thickness and the length of honeycomb cell cores.

Farrokhabadi et al. (2020) studied the parameters including contact force and energy absorption behaviour of multilayer corrugated core laminated sandwich sheets through quasi-static three-point bending. The prevailing failure mechanisms of sandwich structures during the load-bearing process are breakage of constituents, delamination, local indentation, and core de-bonding. Different geometries can exist as corrugated cores, and out of all, rectangular geometry will display better energy absorption results compared to others, such as trapezoidal and triangular geometries, due to the symmetrical configuration.

3.5 FORMABILITY OF SANDWICH BLANKS

Selecting a joining method and forming tools is important to achieve an accurate final product. The accuracy of the forming process is affected by a combination of elements, including

1. cross section of sandwich layers (homogeneous or non-homogeneous),
2. qualities and proportion of the constituents,
3. joining method,
4. geometry of the final product, and
5. loading conditions.

Mennecart et al. (2017) studied the forming performance of in situ drawn sandwich designs to check how the interlayer between skins influences the forming behaviour and developed a new technique that combines deep drawing and thermoplastic-resin transfer moulding. However, the viscosity of the resin and irregularity in component polymerization made the process difficult, despite the potential benefits. Due to the non-uniformity of the structure during the formation of circular cups, the inserted polymer did not stay in the initial region and produced accumulated bulging zones.

TMBs using dissimilar alloy materials would have better forming limits compared to TMBs with similar alloys because it improves the in-plane straining limits of TMBs. If, the metal fracture is the mode of failure, rather than the delamination, thus considerable improvement in forming is possible. So it is expected to employ more ductile adhesive as a core to achieve better formability in higher thickness ratios of TMBs. In addition, increasing the preheat temperature reduces the severity of the extreme fracture in the core but it raises the severity of the wrinkling defect. To ease understanding, the thicknesses in TMBs are separated by an imaginary boundary line named a transition line. If the force exerts orthogonal to the transition line, cohesion failure occurs prior to the failure of the face sheet, and in-plane plastic strain is limited by loading direction. If the in-plane strain is perpendicular to the transition line, it implies that the formability is strongly limited by the fracture energy of the core layer. Hence, it is recommended to choose a high-fracture-energy adhesive to achieve a successful final product. If the in-plane straining is parallel to the transition line, then it should assume the adhesive core is resisting the metal flow in forming, so a ductile adhesive is advisable to use as the core layer (Gresham et al., 2006; Zadpoor et al., 2009; Zadpoor et al., 2010).

3.5.1 Formability of MPM Sandwich Blanks

MPM sandwich composites have a high deformation potential than other comparable TMBs, which makes them excellent for complicated shape fabrication. Kim et al. (2009) determined the forming limit diagram (FLD) of AA5182/polypropylene/AA5182 sheets by using the revised M–K model to estimate the FLD. It was demonstrated that AA5182/polypropylene/AA5182 sandwich panels, which do have better ductility and deformation characteristics than monolithic aluminium sheets, would be used to create complex shapes for automotive components. Kim et al. (2009) achieved 17% of strain within the plane strain condition and successfully formed a model of a lightweight hood with aluminium alloy–polymer sandwich sheets. To establish the forming limit diagrams, Liu et al. (2012) conducted rigid punch dome tests and NAKAZIMA forming tests. The results show that the AA5052/polyethylene/AA5052 sandwich sheet has better formability than a monolithic AA5052 sheet and that the formability of the AA5052/polyethylene/AA5052 sandwich sheet increases with the thickness of the polyethylene core layer.

Liu and Zhuang (2018) examined the limiting draw ratio (LDR) and cylindrical cup attributes of an aluminium/polymer/aluminium sandwich sheet using experimental tests and numerical simulations. In the drawing test, shear forces occurred opposite to the drawing direction. It was observed that the biaxial tension is dominant in the deformation of the outer face sheet while the upper sheet was under compression. The formability of the sandwich sheet decreased with the increase in thickness and strength of the polymer core and demonstrated that the drawability of face sheets has a major impact on the LDR of sandwich sheets. Opposite of aluminium alloy sandwich sheets, steel alloy sandwich sheets showed lower forming limits compared to monolithic steel sheets. Harhash et al. (2014a) showed that the strain hardening exponent and strength coefficient of the 316L/polypropylene-polyethylene/316L steel sandwich sheets is lower than the steel monolithic sheet due to the influence of softcore on the strengthening characteristic in the sandwich sheet. The ductility of the steel/polymer/steel sandwich composite was evaluated by Forcellese and Simoncini (2020), who found that both upper and lower face sheets deformed at the same rate, despite the fact that the lower layer is significantly thicker than the upper layer. The polymer core is distinguished by the large reduction in thickness. It was noted that the thinning of the sandwich sheet is dependent on stroke length, and it was increased with each stroke, particularly at the top of the dome. In balanced biaxial stretching conditions, thinning in the polymer core is independent of angular orientation. But when the angle of orientation is 0° and 90°, the highest limits of the forming limit curve are attained, which is compatible with usual anisotropy behaviour. To determine the limit strains, it is suggested to use the fracture forming limit (FFL) graph through small-angle intervals since the multistage forming can increase the forming limitations of sandwiched sheets (Jackson et al., 2008; Harhash and Palkowski, 2021).

The formability of sandwich panels can be improved by identifying the material composition and operating temperatures. To get a successful final product, a few advanced materials were incorporated into sandwich structures followed by empirical suggestions. The structures fabricated with metallic face sheets and a short carbon fibre–reinforced thermoplastic (SCFRTP) core are easy to form in cold forming. If cores are made with ultrathin chopped carbon fibre–tape-reinforced thermoplastic (UTCTT) with 40% carbon fibre will easily form in warm forming by multi-stroke stamping at 180° without any defects. It is happening because replacing the SCFRTP with a UTCTT core gives improved specific peak load and shear modulus. If the core is altered with a low-ductile adhesive layer, then the structure should be supplemented with a high-ductile metal foam substance (Zhang et al., 2021b).

3.5.2 Factors Affecting the Formability of Sandwich Blanks

The formability of sandwich sheets is constrained due to several factors. Yet the formability can be improved by modifying the geometrical parameters of individual layers, such as the rolling direction of the face sheet, adhesion strength, core thickness, and others.

3.5.2.1 Rolling Direction of the Skin Sheets

Forcellese and Simoncini (2020) investigated the mechanical characteristics and formability of a three-layer MPM sandwich composite, in terms of the sample orientation axis with reference to the rolling direction. Tensile properties and anisotropic behaviour were determined by uniaxial tests and it was assessed the role of angular direction on formability with respect to the rolling direction (RD). Additionally, hemispherical punch tests were conducted to analyze the ductility and thinning behaviour in terms of both limiting dome height and forming limit curves (FLCs). The largest elongation was observed at 45° to RD, with a considerable post-necking displacement. Samples with a 90° to RD have a higher yield and ultimate tensile strengths whereas samples with a 45° angular orientation had a greater work hardening exponent value leading to higher limits of FLCs. The delamination happened on the surface when the axis of the sample axis was 0° and 90° to RD.

3.5.2.2 Adhesion Strength among the Layers

The role of interface adhesion strength on formability was examined by Liu et al. (2012), who built a numerical simulation model that includes the three different interface conditions, namely separation, adhesion, and sticking between the face sheet and the core. The samples were subjected to rigid punch tests and NAKAZIMA forming tests. The interface stresses eliminated the growth of voids at the interface, thus delaying the delamination. Sandwich sheets with the interface joint had higher forming limits than those with separation and the strong surface adhesion increases the levels of forming limits. Naik et al. (2022) used the vacuum hot-pressing method to fabricate the MPM sheets and reinforced them with steel wire meshes using resistance welding and observed good mechanical properties during peel and lap shear tests. When analysed, the initial fracture of the core rather than a failure of the interface improved the forming limits of TMBs.

3.5.2.3 Thickness of the Core Layer

Liu et al. (2012) used aluminium alloy sandwich structures to explore the effect of core thickness on the formability of sandwich sheets. The hot-pressing adhesive method was used to make three types of AA5052/polyethylene/AA5052 sandwich specimens with different core thicknesses as shown in Figure 3.2. The uniaxial experiments were conducted to look into the mechanical properties and stretching tests were performed to check the effect of polymer core thickness. The deformed AA5052 single sheet is shown in Figure 3.2a, where the ratio of width to diameter is varied to determine the forming limits. Similarly, Figures 3.2b–d present the tested sandwich sheets with core thicknesses of 1.5 mm, 2 mm, and 3 mm, respectively. It was noticed that the three sandwich sheets were shown three different FLCs while the forming limits of the sandwich sheet were greater than those of the monolithic AA5052 sheet, and it improved with the increase in the core thickness.

Since the core depth enhances the chances of crushing, changing the thickness of the skin/core layer does have a significant impact on the thinning tendency during the deep drawing process. However, in the stretching zone of the FLC, the thicker core sandwich sheets are likely to fail at lower strains (Harhash et al., 2014b). Due to mechanical characteristics and low ductility, the use of 3D (extended in the thickness direction) core–sandwiched sheets is limited. Yet with a few changes to the composition or methodology, they can produce good forming limits and lead to more applications. (Zhang et al., 2021b).

3.6 ISF OF SANDWICH BLANKS

For many applications, sandwich sheets are required as 3D shells rather than 2D sheets. The usage of three-dimensional structures is limited by the high costs and tooling required for the traditional forming process. The above limitations can be overcome by using the ISF process. ISF is a die-less forming process where the shape and accuracy of the final product purely depend on the movement of the tool along the defined path. ISF is generally a sort of forming process in which the deformation happens locally, progressively, and incrementally. Jackson et al. (2008) checked the feasibility of ISF of sandwich structures and concluded that ISF of sandwich sheets is quite possible. Figure 3.3 depicts the incremental deformation of the sandwich

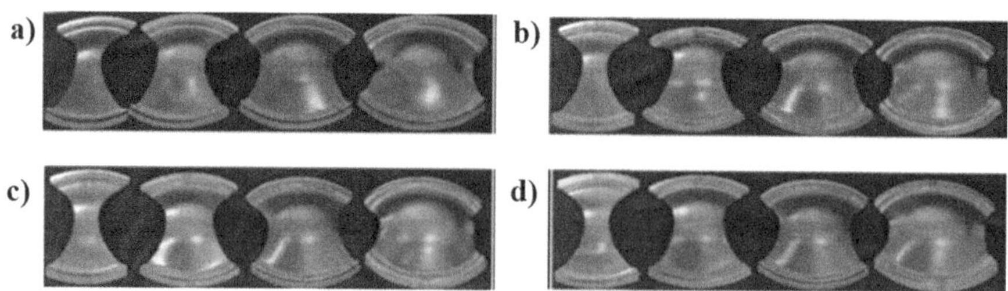

FIGURE 3.2 Deformed shapes of tested specimens with different thicknesses: (a) AA5052 single sheet (b) sandwich sheet with a thickness of 1.5 mm (c) sandwich sheet with a thickness of 2 mm (d) sandwich sheet with a thickness of 3 mm (Liu and Wei, 2013).

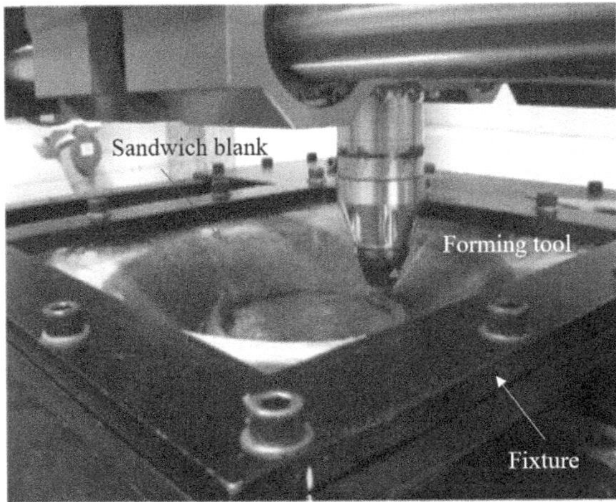

FIGURE 3.3 Experimental arrangement for single-point incremental sheet forming process (Jackson et al., 2008).

sheet in the Cambridge ISF machine at which the forming tool is pressing the sheet in the defined tool path. To achieve a successful forming process, the core should be ductile and largely incompressible, and to predict the thinning of sandwich sheets, the sine law gives similar results to a monolithic sheet. The vertical tool forces and through-thickness deformation of sandwich sheets will follow the same trends as monolithic sheets.

3.6.1 SINGLE-POINT INCREMENTAL FORMING

Due to customized mechanical properties and application-related properties, formable three or multilayer metal-polymer-metal sandwich sheets are replacing pure metallic sheets in several applications. In this section, the incremental forming of MPM sandwich structures and fibre metal laminates is discussed through the influencing parameters during deformation.

3.6.1.1 MPM Sandwich Blanks

Girjob et al. (2015) used circular and spiral tool paths to generate the truncated cone shape on aluminium–polypropylene–aluminium (Al-PP-Al) of 1.2 mm thickness. From the comparisons, it was suggested that the tool moving with the helicoidal path with varying radii produces accuracy in the final product. The distribution of maximum strains was uniform on the circumference with thinning of less than 5%. The formability as well as the mode of failure depend on different process parameters, such as incremental depth and thickness of the TMBs. Davarpanah and Malhotra (2018) explained the effects of the ISF process parameters for metal–polymer laminates. A higher polymer core thickness tends to have better formability since a higher polymer thickness can decrease the chances of delamination. The thinner metal sheets have a chance to fail in two modes of failure, including metal tearing and

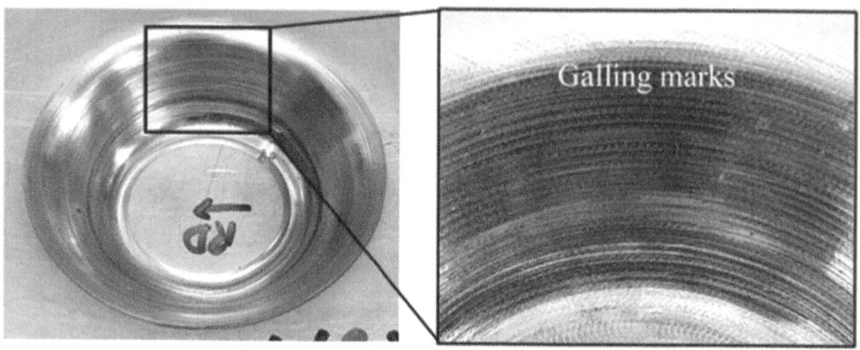

FIGURE 3.4   Galling marks on an upper sheet of metal/polymer/metal sandwich composite (Harhash et al., 2020).

delamination, whereas for thicker polymer sheets, galling failure is common instead of delamination. The formability of thinner core sandwich sheets is limited by the mode of failure. Harhash and Palkowski (2021) evaluated the forming limits of steel/polymer/steel sandwich sheets in terms of limiting wall angle, sample height, and strains on the surface. It was revealed that the sandwich sheets could deform into the required product but early failure would occur compared to the monolithic sheet. The preceding leads to a lower forming limit of the sandwich composite than single sheets. To produce high wall angles, a multistage process can extend the forming limits since small-angle intervals are less vulnerable to damaging the face sheets and delaying the failure. Galling of the upper sheet is the common failure mode of MPM sandwich sheets. Figure 3.4 represents the galling marks produced on the upper metal face sheet during the incremental forming process. Apart from that, it was noted cracking of metal sheets, delamination, and thinning of individual layers are the most occurred failure modes in the sandwich structures.

Stress triaxiality will decrease the formability since it induces hydrostatic pressure and restricts the sheet displacement. The formability of two-layer metal/polymer will have higher forming limits than metal/polymer/metal sheets. A higher triaxiality ratio resulted in lower formability for the MPM sandwich sheet. Formability is even affected by the tool path during the deformation. It is recommended to give a continuous tool path rather than a discontinuous tool path since a discontinuous tool path can delaminate the sheets and produce bump structures. The ISF of tailor-welded blanks can be affected by severe plastic deformation and dynamic recrystallization, and it will cause variable grain sizes in the deformed weld zone and reduce the limits of fracture strains (Kumar and Maji, 2022a, 2022b).

3.6.1.2   Fibre Metal Laminates

Fiorotto et al. (2010) investigated the formability of fibre metal laminates (FMLs) in ISF. To conform to the composite materials, it was used an aluminium sheet as a diaphragm along with a vacuum bag. Significant improvements in specific stiffness and strength were gained by using FMLs formed by the single-point incremental forming

(SPIF) process. During the incremental nature of deformation, wrinkles have been reduced due to the resin that tends to accumulate at the bottom of the vacuum bag. From the conclusions, it was realized that the SPIF can be used to produce intricate shapes on laminate composites quickly and economically.

3.6.2 Two-point Incremental Forming

Figure 3.5(a) shows the SPIL process which does not require the die and the shape of the final product purely depends on the tool path. Compared to the SPIF process, two-point incremental forming (TPIF) comprises a partial die or secondary tool. Figures 3.5b and 3.5c show the typical variations in the TPIF process, which involves the use of a partial die and a full die. Esmailian and Khalili (2021) investigated the influence of process parameters of ISF on Aluminium 5083–polypropylene–Aluminium 5083 sheets through TPIF. It was concluded that, for the same bending moment, the vertical force required to form the three layers is approximately 50% of the force required to form the single-layer aluminium sheet. It was due to the deformation of the sheet was perpendicular to tool motion correspondingly leading to stretching apart from bending. In TPIF, the process parameters were shown almost a similar effect on formability in SPIF. It was perceived that the geometrical accuracy and forming limits of the final product in TPIF were improved than in SPIF due to the lower elastic recovery during the loading.

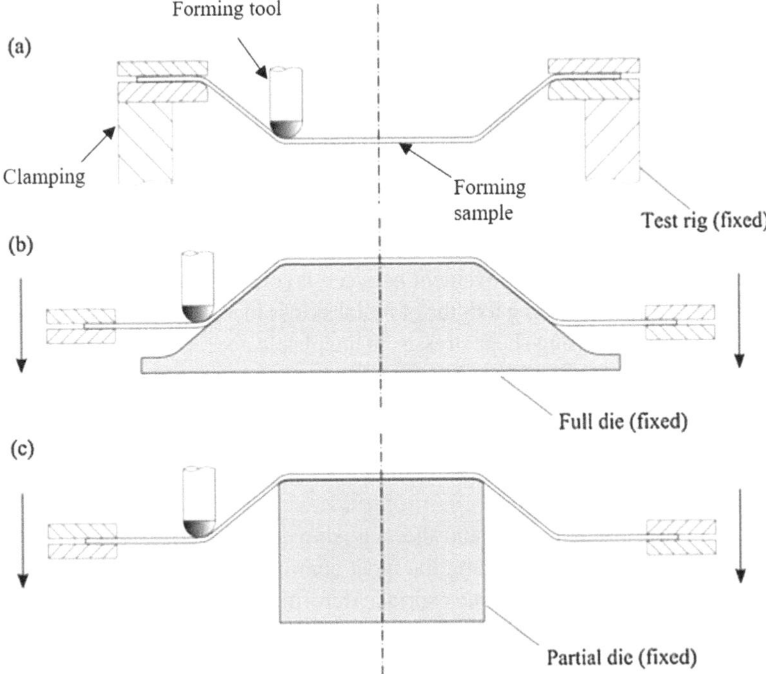

FIGURE 3.5 Variations of the two-point incremental forming process (a) No die, (b) full die, and (c) partial die (Lu et al., 2017).

3.6.3 Limitations in ISF of Sandwich Blanks

Even though it offers good formable limits, ISF has considerable limitations (Mahna, 2013; Afonso et al., 2019):

1. ISF is considered a time-consuming process, but forming multiple sheets at a time decreases the overall time.
2. It requires multistep processes to reach high forming inclinations nearer to 90°. However, multistage ISF produces lower spring-back angles which can provide accurate final products.

3.7 DEFORMATION MECHANISMS IN THE FORMING OF SANDWICH BLANKS

The formability of the sandwich sheet is typically disrupted by sliding and shearing that occurs between the skin sheet and the core material. The core has a significant impact on the mode of fracture and the moisture in the surroundings causes to weakening of the strength of the metal-polymer interface bond. (Ruokolainen and Sigler, 2008). For sandwich composites, adhesive strength at interfaces is meant to be critical, since surface contact establishes a system that can enable the sheets to function in different ways. Due to the inhomogeneous geometrical and mechanical properties of sandwich sheets, various failure mechanisms are expected.

3.7.1 Interlaminar Shearing (Delamination)

Considerable dissimilarity in mechanical characteristics among the layers creates non-uniform displacements in the forming regions, which would lead to delamination. For instance, a weak polymer core can act as a lubricant over metals, causing debonding at sheet–to–plastic core contacts, causing metal faces to float and early fracture during forming processes. By comparison, an extremely strong adhesive force between the polymer core and the metal sheet can cause a serious effect on formability since it prevents continuous smooth movement between layers. Likewise, delamination may happen because of the varying lengths of metal skins since they are twisted around all over the die, causing strong shear stresses in the plastic core. Unless adequate designing is being executed, delamination happens over the entire stretch of the composite sheets, though at small strains (Huang and Leu, 1995; Zadpoor et al., 2010).

3.7.2 Springback

Springback is the distortion of metal sheets arising due to internal stresses after loads have been relieved, and it is among the most common reasons for topological inefficiency in metal forming. It creates surface deformations and unpredictable out-of-shape constraints, which generate errors from predicted shapes, and, moreover, in assembling lines. The springback is influenced by several factors, such as fluctuation in mechanical behaviour, the stress–strain rate, the thickness of the sheet, initial anisotropy, work hardening, the shape of the tool, lubrication, and feed rate. But in forming sandwich sheets, sheet thickness and bending radius are the two significant elements that affect the springback behaviour. Since the sandwich panel resembles a comparable single sheet with the equivalent thickness, the forming accuracy is controllable during its minimal springback (Gardiner, 1957; Queener and De Angelis, 1968; Parsa et al., 2010; Liang et al., 2018). Parsa and Ettehad (2010) investigated the outcomes of the springback of AA3105/polypropylene/AA3105 sandwich metal sheets. It was demonstrated that the non-uniformity in the distribution of shear strain of the polypropylene layer could be the reason for the reported non-uniformity of sandwich sheet curvature after relieving the loads. From the results, it was confirmed the sandwich sheet could withstand more strain than that of the monolithic aluminium sheet. The springback behaviour of the monolithic sheet was compared to the sandwich sheet. For the same flexural rigidity, the springback of the monolithic aluminium sheet was higher than the aluminium/polypropylene/aluminium sandwich sheet.

3.7.3 Cracking of Core/Face Sheet

Pores inside the polymeric core which were created by entrapped air pockets during the fabrication process are the reason for the initiation of the fracture. An increase in the thickness of the polymer core reduces the propagation of fracture and yet the crack possibility further increases with thicker polymer films due to higher tensile forces on the top face metal. The cracking of the core portion can be delayed by the substantial plasticity of the polymer. Typically, tearing or cracking of metal sheets is a possible failure in sandwich sheets rather than core crushing. In comparison with monolithic sheets, the crack will develop soon in terms of fracturing heights, maximum wall angles, and changes in thickness (Wang et al., 2016; Carradò et al., 2011; Harhash et al., 2020; Harhash and Palkowski, 2021).

3.7.4 Wrinkling

Wrinkling occurs when the compressive stress in the face sheet equals local buckling stress. Harhash (2017) highlighted wrinkling as one of the failure modes of steel/polymer/steel sandwich sheets during the deep drawing process. It was revealed, that inhomogeneities could be generated when dissimilar materials were bonded together, causing a discontinuity in stress distribution in the thickness direction. The skin sheets were operated independently because of the weak interfacial adhesion, making them large wrinkling than a single thicker sheet.

The mode of failure is dependent on the blank holder force (BHF) and the preheat temperature of the blank. Wrinkling is a dominant failure in bonded sheets due to lower BHF. The number of wrinkling defects in the outer flange as well as in the sidewall increases with the preheat temperature of the sandwich blank. (Gresham et al., 2006). Morovvati et al. (2010) suggested the minimal BHF required to suppress wrinkling in two-layer sheet metal, which depends on blank dimensions, mechanical characteristics, lay-up, and punch diameter like a monolithic sheet. If the punch diameter is enlarged, the BHF for a given blank diameter tends to decrease, and vice versa. Fiorotto et al. (2010) asserted the importance of incremental sheet-forming techniques on composite materials. It was noticed that wrinkle-free components were formed after employing an aluminium sheet as the diaphragm and a vacuum bag. Commonly, face wrinkling occurs in sandwich TMBs, but the core works as a support to the face sheet. As the face sheet tries to buckle into the core, the core pushes back. So the core is referred to as an elastic foundation because it behaves like a spring that pushes to its original position.

3.8 SUMMARY

In this chapter, recent developments in the forming process of TMSBs are explained. It was perceived that face sheets and the core structure of sandwich sheets are designed to withstand the various loads during the service. But the efficiency of the TMSBs depends on the interfacial adhesion strength. So picking up the proper fabrication method and adding reinforcement is often required. For example, adhesive bonding with thin wire reinforcement is the preferred joining method for joining metal–polymer composites rather than fusion bonding. The mechanical benefits of sandwich sheets are mostly concerned with tensile behaviour, flexural behaviour, and energy absorption. It was pointed out that the formability of sandwich sheets is affected by the geometrical parameters as well as the mechanical properties of the core and metal face sheets. In general, incremental forming of sandwich sheets is possible when the core is a thermoplastic material such as polypropylene and polyvinyl chloride. This is because, while using a thermoset core, the initial fracture occurs in the core, and the desired strength of the sandwich sheet will be lost. So, the formability of metal/foam–metal/metal sandwich sheets has not been discussed in this study since they cannot be formed into intricate shapes due to the high incompressibility of the core layer. Still, there is a scope to study the formability of sandwich blanks, since most real-time applications require formed parts. From the literature, it was realized that ISF has the potential to form multilayer TMBs with high forming limits since it involves localized plastic deformation. The forming limits of sandwich blanks are limited due to interlaminar shearing and early failure, so it is suggested to use a non-porous polymer since it can delay the possible earlier failure and provide substantial plasticity for the total sandwich blank.

REFERENCES

Adams, Robert D. (2021). *Adhesive bonding: science, technology, and applications*. Woodhead Publishing.

Afonso, D., Alves de Sousa, R., Torcato, R., Pires, L. Incremental sheet forming. In: Afonso, D., Alves de Sousa, R., Torcato, R., Pires, L., editors. *Incremental forming as a rapid tooling process*. Switzerland: Springer Nature; 2019. p. 23–43.

Ageorges, C., Ye, L., & Hou, M. (2001). Advances in fusion bonding techniques for joining thermoplastic matrix composites: a review. *Composites Part A: Applied Science and Manufacturing, 32*(6), 839–857.

Barre, S., Benezeggah, M. L., & Matthieh, P. *Proceedings of the ICCM10*, Vol. I (1995) pp. 633–640.

Brockholm, Juhl T., & Bach, D. (2013). Predicting the laser weldability of dissimilar polymers. *Polymer, 54*, 3891–3897.

Carradò, A., Faerber, J., Niemeyer, S., Ziegmann, G., & Palkowski, H. (2011). Metal/polymer/metal hybrid systems: Towards potential formability applications. *Composite Structures*, *93*(2), 715–721.

Carrado, A., Sokolova, O., Ziegmann, G., & Plalkowski, H. (2010). Press joining rolling process for hybrid systems. *Key Engineering Materials*, *425*, 271–281.

Davarpanah, M. A., & Malhotra, R. (2018). Formability and failure modes in single point incremental forming of metal-polymer laminates. *Procedia Manufacturing*, *26*, 343–348.

Du, D., Hu, Y., Li, H., Liu, C., & Tao, J. (2016). Open-hole tensile progressive damage and failure prediction of carbon fiber-reinforced PEEK–titanium laminates. *Composites Part B: Engineering*, *91*, 65–74.

Esmailian, M., & Khalili, K. (2021). Two-Point Incremental Forming of Metal–Polymer Three-Layer Sheets. *Iranian Journal of Science and Technology, Transactions of Mechanical Engineering*, *45*(1), 181–196.

Eyvazian, A., Moeinifard, M., Musharavati, F., Taghizadeh, S. A., Mahdi, E., Hamouda, A. M., & Tran, T. N. (2021). Mechanical behavior of resin pin-reinforced composite sandwich panels under quasi-static indentation and three-point bending loading conditions. *Journal of Sandwich Structures & Materials*, *23*(6), 2127–2145.

Farrokhabadi, A., Taghizadeh, S. A., Madadi, H., Norouzi, H., & Ataei, A. (2020). Experimental and numerical analysis of novel multi-layer sandwich panels under three-point bending load. *Composite Structures*, *250*, 112631.

Feng, Y., Qiu, H., Gao, Y., Zheng, H., & Tan, J. (2020). Creative design for sandwich structures: A review. *International Journal of Advanced Robotic Systems*, *17*(3), 1729881420921327.

Fiorotto, M., Sorgente, M., & Lucchetta, G. (2010). Preliminary studies on single point incremental forming for composite materials. *International Journal of Material Forming*, *3*(1), 951–954.

Forcellese, A., & Simoncini, M. (2020). Mechanical properties and formability of metal–polymer-metal sandwich composites. *The International Journal of Advanced Manufacturing Technology*, *107*(7), 3333–3349.

Gardiner, F. J. (1957). The spring back of metals. *Transactions of the ASME*, *79*, 33–42.

Ghosh, P. K., Patel, A., & Kumar, K. (2016). Adhesive joining of copper using nano-filler composite adhesive. *Polymer*, *87*, 159–169.

Girjob, C., Racz, G., Bologa, O., & Biris, C. (2015). FEM Simulation of Laminated Lightweight Materials Processed through Single Point Incremental Forming. In *Applied Mechanics and Materials* (Vol. 772, pp. 38–43). Trans Tech Publications Ltd.

Gresham, J., Cantwell, W., Cardew-Hall, M. J., Compston, P., & Kalyanasundaram, S. (2006). Drawing behavior of metal–composite sandwich structures. *Composite Structures*, *75*(1–4), 305–312.

Grunewald, J, Parlevliet, P, and Altstadt, V. (2017). Manufacturing of thermoplastic composite sandwich structures: a review of the literature. *Journal of Thermoplastic Composite Materials*, *30*(4), 437–464.

Gu, F., Chen, L., Zhu, X., Lu, X., & Fang, D. (2022). Fabrication and uniaxial compression mechanical behavior of composite corrugated-core sandwich cylinder with thin-wall metal liner. *Mechanics of Advanced Materials and Structures*, *29*(7), 949–962.

Guo, K., Zhu, L., Li, Y., & Yu, T. X. (2019). Numerical study on the mechanical behavior of foam core sandwich plates under repeated impact loadings. *Composite Structures*, *224*, 111030.

Harhash, M. Forming behaviour of multilayer metal/polymer/metal systems [Dissertation]. Clausthal-Zellerfeld. Deutschland: TU Clausthal; 2017.

Harhash, M., Carrado, A., & Palkowski, H. (2014b). Forming limit diagram of steel/polymer/steel sandwich systems for the automotive industry. In *Advanced composites for aerospace, marine, and land applications* (pp. 243–254). Springer, Cham.

Harhash, M., Gilbert, R. R., Hartmann, S., & Palkowski, H. (2020) Experimental characterization, analytical and numerical investigations of metal/polymer/metal sandwich composites – part 2: free bending. *Composite Structures, 232*, 111421.

Harhash, M., & Palkowski, H. (2021). Incremental sheet forming of steel/polymer/steel sandwich composites. *Journal of Materials Research and Technology, 13*, 417–430.

Harhash, M., Sokolova, O., Carradó, A., & Palkowski, H. (2014a) Mechanical properties and forming behavior of laminated steel/polymer sandwich systems with local inlays – part 1. *Composite Structures, 118*, 112–120.

Hassan, M. K., Abdellah, M. Y., Azabi, S. K., & Marzouk, W. (2015). Investigation of the mechanical behavior of novel fiber metal laminates. *International Journal of Mechanical & Mechatronics Engineering IJMME-IJENS, 15*(3), 112–118.

He, W., Wang, C., Wang, S., Yao, L., Wu, J., & Xie, D. (2020). Tensile mechanical behavior and failure mechanisms of multihole fiber metal laminates—Experimental characterization and numerical prediction. *Journal of Reinforced Plastics and Composites, 39*(13–14), 499–519.

Hou, B., Wang, Y., Tang, Z. B., Zhao, H., Xi, X. L., & Li, Y. L. (2021). The mechanical behaviors of the corrugated sandwich panel under quasi-static and dynamic shear-compressive loadings. *International Journal of Impact Engineering, 156*, 103956.

Huang, Y. M., & Leu, D. K. (1995). Finite-element simulation of the bending process of steel/polymer/steel laminate sheet. *Journal of Adhesion Science and Technology, 52*(3), 319–337.

Jabbari, E., & Peppas, N. A. (1995). A model for interdiffusion at interfaces of polymers with dissimilar physical properties. *Polymer, 36*(3), 575–586.

Jackson, K. P., Allwood, J. M., & Landert, M. (2008). Incremental forming of sandwich panels. *Journal of Materials Processing Technology, 204*(1–3), 290–303.

Joachim Gähde, J. F. F., Gehrke, Rolf, Loeschcke, Ingrid, Sachse, Jörg, 1992. Adhesionof polyurethane to surface-modified steel. *Journal of Adhesion Science and Technology, 6*, 569–586.

Johnson, A. F., & Sims, G. D. (1986). Mechanical properties and design of sandwich materials. *Composites, 17*(4), 321–328.

Kazemi, M. (2021, February). Experimental analysis of sandwich composite beams under three-point bending with an emphasis on the layering effects of foam core. In *Structures* (Vol. 29, pp. 383–391).

Kim, K. J., Rhee, M. H., Choi, B. I., Kim, C. W., Sung, C. W., Han, C. P., … Won, S. T. (2009). Development of application technique of aluminum sandwich sheets for automotive hood. *International Journal of Precision Engineering and Manufacturing, 10*(4), 71–75.

Kumar, G., & Maji, K. (2022a). Forming limit analysis of friction stir tailor welded AA5083 and AA7075 sheets in single point incremental forming. *International Journal of Material Forming, 15*(3), 1–23.

Kumar, G., & Maji, K. (2022b). Investigations on formability of tailor laminated sheets in single point incremental forming. *Proceedings of the Institution of Mechanical Engineers, Part B: Journal of Engineering Manufacture*, 09544054221076244.

Lascoup, B., Aboura, Z., Khellil, K., & Benzeggagh, M. (2006). On the mechanical effect of stitch addition in sandwich panel. *Composites Science and Technology, 66*(10), 1385–1398.

Latour, M., D'Aniello, M., Landolfo, R., & Rizzano, G. (2021). Experimental and numerical study of double-skin aluminium foam sandwich panels in bending. *Thin-Walled Structures, 164*, 107894.

Li, D. S., Zhao, C. Q., Jiang, N., & Jiang, L. (2015). Fabrication, properties and failure of 3D integrated woven spacer composites with thickened face sheets. *Materials Letters, 148*, 103–105.

Liang, X. B., Cai, Z. Y., & Zhang, X. (2018). Forming characteristics analysis and springback prediction of bi-directional trapezoidal sandwich panels in the multi-point bend-forming. *The International Journal of Advanced Manufacturing Technology, 98*(5), 1709–1720.

Liu, J., & Zhuang, L. (2018). Cylindrical cup-drawing characteristics of aluminum-polymer sandwich sheet. *The International Journal of Advanced Manufacturing Technology, 97*(5–8), 1885–1896.

Liu, J. G., Wei, L., & Wang, J. X. (2012). Influence of interfacial adhesion strength on formability of AA5052/polyethylene/AA5052 sandwich sheet. *Transactions of Nonferrous Metals Society of China, 22*, s395–s401.

Liu, J. G., & Wei, X. U. E. (2013). Formability of AA5052/polyethylene/AA5052 sandwich sheets. *Transactions of Nonferrous Metals Society of China, 23*(4), 964–969.

Lu, H., Kearney, M., Wang, C., Liu, S., & Meehan, P. A. (2017). Part accuracy improvement in two-point incremental forming with a partial die using a model predictive control algorithm. *Precision Engineering, 49*, 179–188.

Mahna, A. Formability analysis of polymers in incremental sheet forming process [Master of Science]. Gazima_gusa. North Cyprus: Eastern Mediterranean University; 2013.

Masoudi Moghaddam, S. A., Yarmohammd Tooski, M., Jabbari, M., & Khorshidvand, A. R. (2021). Mechanical behavior of sandwich panels with hybrid face sheets and embedded super-elastic SMA wires under quasi-static loading: An experimental investigation. *International Journal of Damage Mechanics, 30*(2), 196–213.

Mennecart, T., Hiegemann, L., & Khalifa, N. B. (2017). Analysis of the forming behavior of in-situ drawn sandwich sheets. *Procedia Engineering, 207*, 890–895.

Morovvati, M. R., Mollaei-Dariani, B., & Asadian-Ardakani, M. H. (2010). A theoretical, numerical, and experimental investigation of plastic wrinkling of circular two-layer sheet metal in the deep drawing. *Journal of Materials Processing Technology, 210*(13), 1738–1747.

Mousa, S., & Kim, G. Y. (2017). A direct adhesion of metal-polymer-metal sandwich composites by warm roll bonding. *Journal of Materials Processing Technology, 239*, 133–139.

Naik, R. K., Panda, S. K., & Racherla, V. (2020). A new method for joining metal and polymer sheets in sandwich panels for highly improved interface strength. *Composite Structures, 251*, 112661.

Naik, R. K., Panda, S. K., & Racherla, V. (2022). Failure analysis of metal-polymer-metal sandwich panels with wire mesh interlayers: Finite element modeling and experimental validation. *Composite Structures, 280*, 114813.

Noor, AK, Burton, WS, Bert, CW. (2009). Computational models for sandwich panels and shells. *Applied Mechanics Reviews, 49*(3), 155.

Parsa, M. H., & Ettehad, M. (2010). Experimental and finite element study on the springback of double-curved aluminum/polypropylene/aluminum sandwich sheet. *Materials & Design, 31*(9), 4174–4183.

Parsa, M. H., Ettehad, M., Matin, P. H., Ahkami, A., & Nasher, S. (2010). Experimental and numerical determination of limiting drawing ratio of Al3105-polypropylene-Al3105 sandwich sheets. *Journal of Engineering Materials and Technology, 132*(3), 031004.

Queener, C. A., & De Angelis, R. J. (1968). Elastic spring back and residual stresses in sheet metal formed by bending. *Transactions of the ASME, 61*, 757–768.

Ramani, K., Moriarty, B., 1998. Thermoplastic bonding to metals via injection molding for macro-composite manufacture. *Polymer Engineering & Science, 38*, 870–877.

Ruokolainen, RB, Sigler, DR (2008) The effect of adhesion and tensile properties on the formability of laminated steels. *Journal of Materials Engineering and Performance, 17*, 330–339.

Satheeshkumar, V., & Narayanan, R. G. (2014). Investigation on the influence of adhesive properties on the formability of adhesive-bonded steel sheets. *Proceedings of the Institution of Mechanical Engineers, Part C: Journal of Mechanical Engineering Science, 228*(3), 405–425.

Shanshan, W. A. N. G., Yuhan, W. E. I., Xin, X. U. E., Zhangbin, W. U., & Hongbai, B. A. I. (2022). Connection technology and mechanical properties of sandwich structure with the core of elastic damping metal spiral wire mesh. *Journal of Composite Materials. (Acta Materiae Compositae Sinica), 39*(3), 1308–1321.

Shin, K. S., Kim, K. J., Choi, S. W., & Rhee, M. H. (1999). Mechanical properties of aluminum/polypropylene/aluminum sandwich sheets. *Metals and Materials, 5*(6), 613–618.

Sugiyama, K., Matsuzaki, R., Ueda, M., Todoroki, A., & Hirano, Y. (2018). 3D printing of composite sandwich structures using continuous carbon fiber and fiber tension. *Composites Part A: Applied Science and Manufacturing, 113*, 114–121.

Taghipoor, H., Eyvazian, A., Musharavati, F., Sebaey, T. A., & Ghiaskar, A. (2020, December). Experimental investigation of the three-point bending properties of sandwich beams with polyurethane foam-filled lattice cores. *Structures, 28*, 424–432.

Trende, A., Astrom, B. T., & Woginger, A., et al. (1999). Modelling of heat transfer in thermoplastic composite manufacturing: double-belt press lamination. *Compos A, 30*, 935–943.

Wang, W., Dulieu-Barton, J. M., & Thomsen, O. T. (2016). A methodology for characterizing the interfacial fracture toughness of sandwich structures using high-speed infrared thermography. *Experimental Mechanics, 56*(1), 121–132.

Wingfield, J. (1993). Treatment of composite surfaces for adhesive bonding. *The International Journal of Adhesion and Adhesives, 13*(3), 151–156.

Young, H. K., & Wool, R. (1983). A theory of healing at a polymer-polymer interface. *Macromolecules, 16*, 1115–1120.

Zadpoor, A. A., Sinke, J., & Benedictus, R. (2009). The mechanical behavior of adhesively bonded tailor-made blanks. *International Journal of Adhesion and Adhesives, 29*(5), 558–571.

Zadpoor, A. A., Sinke, J., & Benedictus, R. (2010). Elastoplastic deformation of dissimilar-alloy adhesively-bonded tailor-made blanks. *Materials & Design, 31*(10), 4611–4620.

Zhang, J., Qin, Q., Xiang, C., & Wang, T. J. (2016). Plastic analysis of multilayer sandwich beams with metal foam cores. *Acta Mechanica, 227*(9), 2477–2491.

Zhang, J., Taylor, T., & Yanagimoto, J. (2021b). Mechanical properties and cold and warm forming characteristics of sandwich sheets with a three-dimensional CFRTP core. *Composite Structures, 269*, 114048.

Zhang, Y., Li, Y., Guo, K., & Zhu, L. (2021a). Dynamic mechanical behavior and energy absorption of aluminum honeycomb sandwich panels under repeated impact loads. *Ocean Engineering, 219*, 108344.

Zhang, Z. J., Wang, Y. J., Huang, L., Fu, Y., Zhang, Z. Q., Wei, X., & Jin, F. (2022). Mechanical behaviors and failure modes of sandwich cylinders with square honeycomb cores under axial compression. *Thin-Walled Structures, 172*, 108868.

4 Hot Isostatic Pressing of Transparent Ceramics
An Overview

V. Murugabalaji, B. N. Sreeharan, and A. Kannan

4.1 INTRODUCTION

Ceramics are generally brittle in nature, and introducing even small cracks will cause catastrophic failure [1]. So researchers concentrated on developing dense ceramics free of defects to achieve superior thermo-mechanical and optical characteristics. The applications of transparent ceramics include armours [2, 3], spacecraft windows [4], solid-state lasers [5], healthcare [6], electrical devices [7] and so on primarily because of their exceptional thermo-mechanical properties, superior optical transmittance, affordability of raw materials and high flexibility in shape complexity [8]. High transparency of the ceramic materials can be achieved only when high-purity material is densified through the appropriate densification method [4]. Usually, air pre-sintering followed by the hot isostatic pressing (HIP) of the ceramics is carried out to achieve the required properties [9]. Various ceramics are employed in this process depending on the applications as listed in Table 4.1. This chapter provides an introduction to the HIP process and its significance in achieving the required properties. The microstructural characteristics achieved for the various ceramics after HIP are discussed in relation to the densification rate. The optical and mechanical properties achieved in this process are also discussed briefly in the subsequent sections.

4.2 HIP

HIP is a method that employs the isostatic (triaxial) nature of pressure to assist in the sintering of the pre-sintered compacts [19]. HIP is a near-net shape manufacturing process in which there is no necessity for further machining processes thereby reducing the overall manufacturing cost. In the late 1980s, HIP was introduced for the economical manufacturing of alumina inserts for cutting tools. Since then, its application for the manufacturing of oxide ceramics has been developed for Mn-Zn ferrite, partially stabilized zirconia (PSZ) and lead zirconate titanate (PZT) [20]. Later, the process gained significance for producing a wide variety of ceramics including transparent ceramics.

DOI: 10.1201/9781032665375-4

TABLE 4.1

Ceramic Materials Processed through HIP and Their Applications

S.No	Material	Applications	References
1.	Yttrium aluminium garnet (YAG), Yb-doped YAG, Yb:SC$_2$O$_3$ ceramics, Yb-doped Lu$_2$O$_3$, Yb-doped CaF$_2$	Solid-state laser devices	Liu et al. [9], Toci et al. [10], Maksimov et al. [11]
2.	Y$_2$O$_3$ (Ceramics)	Shedding light on their prospective use as thermal barriers and high-temperature optical windows	Huang et al. [12]
3.	Magnesium aluminium oxynitride (MgAlON) ceramics	Optical windows, light-emitting diodes and armour materials	Zeng et al. [13]
4.	KNbTeO$_6$ transparent ceramics	Infrared lenses in imaging systems of cameras in thermal and medical applications	Zhai et al. [14]
5.	Gadolinium gallium garnet (Gd3Ga5O12) ceramics	Laser systems and magneto-optical devices	Wang et al. [4]
6.	Aluminium oxynitride (AlON) ceramics	Optoelectronic and safety window technology	Li et al. [6], Shan et al. [15], Chen et al. [16]
7.	Magnesium aluminate spinel (MgAl$_2$O$_4$), Magnesium oxide (MgO) ceramics	Transparent armour and infrared windows	Itatani et al. [1], Liu et al. [17, 18]

The schematic diagram of the HIP unit showing the application of pressure and temperature is shown in Figure 4.1. The arrowheads in Figure 4.1 indicate the direction of application of isostatic pressure on the workpiece held on the support. The HIP unit generally looks like an electric furnace maintained in a pressure vessel. The striking feature is the gas (Argon commonly) employed in the vessel for maintaining isostatic pressure should possess high density and low viscosity to favour the natural convection of heat [21]. The furnace in which HIP is carried out can be categorized into three types as follows: (1) Kanthal heating elements that can withstand oxygen and operate at temperatures up to 1200°C enable to load and unload workpieces safely while they are hot. (2) Heating elements made of molybdenum, suitably used for temperatures up to 1450°C, are primarily utilized to densify materials prone to surface contamination. (3) Heating elements made of graphite specially designed for temperatures ranging from 2000°C or higher allow the processing of materials in an atmosphere of gases like argon or nitrogen [21, 22]. The HIP method without a container is used for processing ceramics such as cemented carbides, alumina, ferrites, porous ceria-doped tetragonal zirconia and piezoelectric ceramics [21].

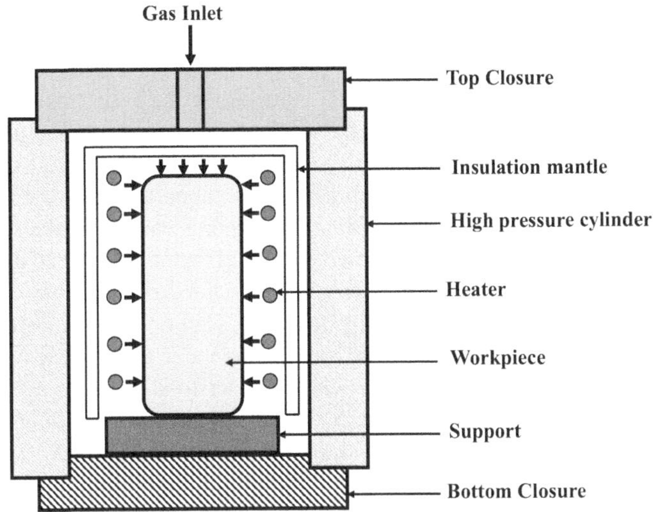

FIGURE 4.1 Schematic drawing of an HIP unit.

4.3 CHARACTERIZATIONS

Most of the researchers are interested in examining the relative density and microstructural aspects of the HIP-processed samples. The impact of HIP depends on achieving better properties of the processed samples.

4.3.1 RELATIVE DENSITY

Relative density measurements are usually carried out by Archimedes principle. Relative density is obtained by calculating the ratio of bulk density to the theoretical density of the ceramic [1]. The change in relative density with respect to the temperature range of pre-sintering gives a vivid picture of the densification rate. Liu et al. found that there was an increase in the relative density of $MgAl_2O_4$ when the microwave pre-sintering temperature increases. The slope was steep in the beginning (1330°C–1400°C), and then it was gentle at higher temperatures (1400°C–1550°C). From their results, it can be concluded that densification occurs rapidly at low temperatures less than 1400°C. The air pre-sintering of $MgAl_2O_4$ by Liu et al. in the temperature range of 1450°C–1650°C also exhibited a similar trend in densification. The densification was rapid below 1550°C. The relative density studies by Maksimov et al. [11] on Yb-doped Lu_2O_3 ceramics also indicated the same trend of variation in relative density with respect to sintering temperature.

4.3.2 MICROSTRUCTURE

Microstructural characterizations of air pre-sintered $MgAl_2O_4$ by Liu et al. [17] indicated that the samples exhibit residual pores along the grain boundaries. As the pre-sintering temperature was raised, the quantity of pores along the grain boundary decreased due to the densification process, while the grain size experienced a gradual growth. At temperatures exceeding 1550°C, abnormal growth of grains was noted, resulting in the formation of intragranular pores caused by the rapid rate of grain growth. They also observed the microstructural features of HIPed ceramics and found that pre-sintering at 1500°C and 1525°C yielded better densifications after the HIP process. Another work by Maksimov et al. [11] on Yb-doped Lu_2O_3 ceramics also exhibited the same trend of increase in grain size at higher temperatures of pressure-less sintering and HIP. Thus, it can be understood that in general, the same trend occurs for all transparent ceramics during pre-sintering and HIP treatments. Optimal temperatures should be selected to obtain superior characteristics.

From the microstructural analysis of the ceramics, the porosity can be determined as follows: The total volume fraction of the pores (P) can be calculated from Equation 4.1 by Boulesteix et al. [5]:

$$P = C_n \frac{\pi}{6} \int_0^\infty N(\varnothing)\varnothing^3 d\varnothing, \tag{4.1}$$

where C_n represents the volume density of pores, $\varnothing$ denotes the equivalent disc diameter of the pores and $N(\varnothing)$ is the pore size distribution function given by Equation 4.2. Scanning electron microscopy (SEM) and confocal laser scanning microscopy (CLSM) are viable techniques for acquiring the necessary parameters utilized in the calculation of P.

$$N(\varnothing) = \frac{1}{\sqrt{2\pi}\,\sigma \varnothing_{max} \exp(0.5\sigma^2)} \exp\left[-\frac{\left(\ln(\varnothing) - \ln(\varnothing_{max})\right)^2}{2\sigma^2} \right] \tag{4.2}$$

Here, σ corresponds to the width of the pore size distribution and denotes pore diameter at the maximum value of $N(\varnothing)$.

From the preceding equations, the mean diameter of the pores "$\varnothing_{mean}$" will be calculated as given in Equation 4.3:

$$\varnothing_{mean} = \varnothing_{max} \exp(1.5\sigma^2). \tag{4.3}$$

Thus, the mean diameter of pores is evaluated by the researchers in analysing the porosity of transparent ceramics through microstructural methods.

4.3.3 THERMAL PROPERTIES

The HIP processing of ceramics significantly affects the thermal conductivity (κ) of the ceramics also due to the increase in grain size. The calculation of thermal conductivity is based on the following Equation 4.4 by Itatani et al. [1]:

$$\kappa = a\rho C_p, \tag{4.4}$$

where a represents the thermal diffusivity, ρ denotes the density of the ceramic and C_p corresponds to the specific heat capacity at constant pressure. The thermal conductivity of MgO ceramics exhibits a linear increase as the grain size expands. This can be attributed to the reduction in grain boundary area, which enables unrestricted transmission of phonons and enhances thermal conductivity [1]. The thermal conductivity studies of doped and undoped Y_2O_3 ceramics by Li et al. [23] were marked by the reduction in thermal conductivity (κ) with the increment in temperature. This decrease was less in doped ceramics than in the pure form. The thermal conductivity of doped ceramics is less compared to undoped forms due to the reduction in the mean free path of phonon caused by the dopant in ceramic lattice [24]. In another similar study by Liu et al. [25] on Tm-doped Lu_2O_3, the thermal conductivity of doped ceramic also was reduced due to the increased phonon scattering rate due to dopant action in atomic level and increase in temperature. The lattice vibration concept of thermal conductivity can be related by parameter a given by Equation 4.5:

$$a = \frac{1}{3} v_s l_{tot}, \tag{4.5}$$

where v_s is the mean velocity of the phonon and l_{tot} is the total mean free path of the phonon.

4.4 OPTICAL PROPERTIES

4.4.1 OPTICAL APPLICATIONS IN DIVERSE FIELDS

Numerous applications require ceramics' optical characteristics, including the following:

Ceramics are used to create optical fibres because they have a high refractive index and are transparent to visible light. As a result, there is little loss in the long-distance transmission of light. Ceramics are utilized to create laser components like lenses and mirrors because they can survive the intense heat and pressure that lasers produce. To ensure that the laser beam is properly focused, they also have good optical qualities [4, 5]. Because ceramics are biocompatible and have strong optical qualities, they are utilized to produce medical equipment like lenses and implants. They are therefore perfect for uses that call for physical contact with the body [4]. As ceramics are transparent to a wide variety of wavelengths, including visible, infrared and ultraviolet lights, they are utilized in aeronautical applications, such as windows and domes. This makes it possible to see clearly and shield yourself from the sun's rays. Since ceramics have good optical characteristics and can resist the high temperatures and pressures produced by these devices, they are utilized in displays like LCDs and OLEDs [4, 7].

Ceramics' optical characteristics can be modified to match the demands of a given application. Because of this, ceramics are a desirable and adaptable material for a variety of uses. The following applications of ceramics' optical characteristics are significant in addition to those already mentioned. Because ceramics can be manufactured to

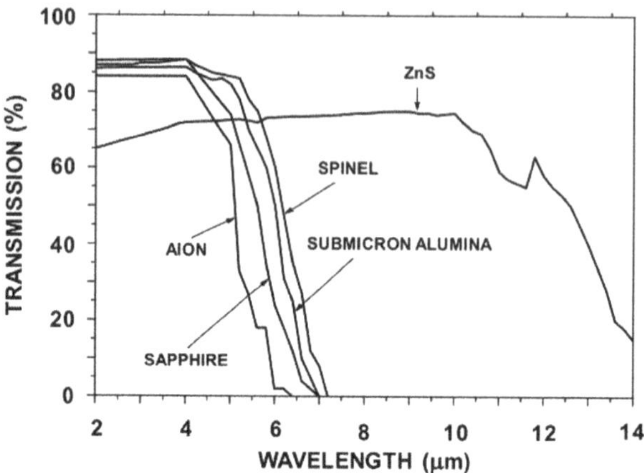

FIGURE 4.2   Properties of transmission of transparent ceramics and their wavelength ranges. **Adapted from Johnson et al. [26].**

radiate light in a variety of colours, they are used to make light bulbs, fluorescent tubes and other lighting fixtures. Ceramics are utilized to create solar cells because of their capacity to absorb sunlight and transform it into electrical energy. Ceramics are employed in security applications because they may be used to identify and authenticate things, such as optical filters and lenses.

Transparent ceramics must be manufactured in a way that matches the requirements, as was previously indicated. HIP, a procedure that produces products with a close shape, can do this. Transparent ceramic processing is difficult, however, there have been substantial improvements recently. The improved performance of transparent ceramics can be attributed to advancements in materials and production techniques [4]. The transmission properties of various ceramics and their wavelength ranges are depicted in Figure 4.2.

4.4.2 Transmittance and Loss Coefficients

Transmittance and loss coefficients are crucial optical characteristics to consider while producing transparent ceramics [6]. The ability of ceramics to permit the passage of light through their structure is referred to as their transmittance optical characteristics [15]. Ceramics have a wide variety of transmittance behaviours that are impacted by elements like composition, crystal structure and manufacturing processes [27]. Particularly transparent ceramics have high transmittance, which allows light to flow through with little absorption or scattering. These translucent ceramics are used in optical equipment including windows, lenses and lasers because of their outstanding optical clarity and thermal stability. By comparison, many ceramics are naturally opaque because of their light-scattering and light-absorbing microstructure. In applications like coatings, thermal insulation and structural components, where the ceramics' exceptional mechanical strength and tolerance to extreme temperatures are of utmost importance, their opacity is useful [28, 29].

The behaviour of light in ceramics is mostly determined by the optical parameters of transmittance and loss coefficients. The term *transmittance* describes a material's capacity to let light pass through with little absorption or scattering [15]. Conversely, loss coefficients show how much light is absorbed or scattered inside the ceramic substance [15]. Reduced transmittance results from larger loss coefficients, which indicate greater light absorption or dispersion. Numerous elements, such as composition, crystal structure and manufacturing processes, have an impact on these qualities [30]. Ceramic materials can be improved for certain applications, such as optical coatings, sensors and waveguides, by comprehending and managing transmittance and loss coefficients. Ceramics can be created to satisfy the requirements of various optical systems and help advance science by adjusting their qualities. Li et al. [6] employed the technique of lowering the quantity of gas inclusions, by the addition of SiO_2 to the AlON powder in order to enhance the transparency of the ceramics. The gas inclusions that remained in the ceramics after pre-sintering were removed using the HIP procedure. HIP ceramics can be annealed using this technique at standard pressure without losing optical transmittance. Second, because the pre-sintering process does not need high temperatures or pressures to be used, the method is more effective. Ma et al. [13] utilized fine and low agglomerated MgAlON powder to achieve high transparency in MgAlON ceramics. Pressureless sintering at elevated temperatures for suitable hours is sufficient to achieve high density in MgAlON ceramics. It further improves the transparency of MgAlON ceramics. In the case of aluminium oxynitride (AlON) ceramics, density and grain size both will rise and decrease as the HIP temperature and pressure are raised. Consequently, the incorporation of SiO_2 into the AlON powder will lead to an increase in the optical transmittance of the resulting ceramics. The microstructure and optical characteristics of the AlON ceramics are barely impacted by the HIP time. For making translucent AlON ceramics with high optical transmittance, HIP will be a promising technique [31]. The addition of Y_2O_3/La_2O_3 additives to AlON powder promoted sintering and resulted in AlON ceramics with finer grain size and higher density. This led to an improvement in the optical transmittance of the AlON ceramics. The type and amount of additive had a significant influence on the microstructural and optical characteristics of the AlON ceramics. The addition of Y_2O_3/La_2O_3 additives is a promising approach for enhancing the optical transmittance of transparent AlON ceramics [16]. AlON ceramics' leftover pores were removed via the HIP process, increasing optical transmittance. The AlON ceramics' mechanical qualities were not diminished by the HIP process. The process holds promise for producing very transparent AlON ceramics for a range of uses [32].

With respect to Figure 4.3, in comparison to samples obtained by traditional solid-state reactive sintering (SSRS) under vacuum at a temperature of 1750°C for 3 hours, post-HIP treatment at a lower temperature (1650°C for 6 hours) enables the creation of Nd:YAG ceramics with noticeably greater transmittance. Based on the findings, it can be concluded that employing post-HIP treatment shows great promise as a method to produce defect-free Nd:YAG transparent ceramics with excellent optical quality [33].

Polycrystalline ceramics typically contain opaque material that scatters or absorbs visible light. Ceramics can, nevertheless, acquire a transparency comparable to glass

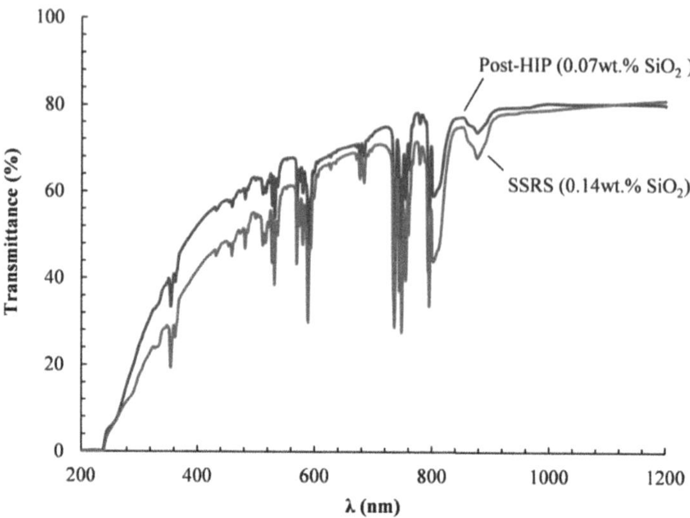

FIGURE 4.3 Optical transmittance of transparent Nd: YAG ceramics in the visible range for SSRS and post-HIP conditions.

Adapted from Chretien et al. [33].

or single-crystal sapphire through extremely regulated methods. In the visible to medium infrared band, transparent ceramics have been found to transmit light at a rate of about 84% for YAG and 81% for Y_2O_3. These ratios correspond to the transparency levels determined using the refractive index of the material [34].

4.5 MECHANICAL PROPERTIES

Flexural strength and hardness are the key parameters of interest among various mechanical properties of transparent ceramics. The influences of pre-sintering temperature on the mechanical properties like flexural strength and hardness of $MgAl_2O_4$ were studied by Liu et al. [17]. The maximum value of the flexural strength obtained was 281 ± 16 MPa and the hardness value was 13.23 ± 0.07 GPa. These values were achieved when the $MgAl_2O_4$ ceramic sample was pre-sintered for 3 hours at 1525°C and thereafter HIP as a post-treatment for 3 hours at 1600°C. Under these conditions, the sample exhibited an average grain size of 1.5 μm. Also, it was found that a further increment in the pre-sintering temperature of the ceramic sample up to 1550°C shows a small decrement in the value of flexural strength and hardness of the prepared $MgAl_2O_4$ ceramic sample. This phenomenon could be understood by the name Hall–Petch effect given in Equation 4.6 [17]:

$$\sigma = \sigma_0 + kd^{-0.5}. \tag{4.6}$$

Here, σ represents the yield strength, σ_0 denotes the lattice friction due to single dislocation k is the constant related to the material and d represents the grain size.

The Hall–Petch expression proves the fact that with the increment in the value of grain size from 1.4 µm to 1.8 µm, the mechanical properties were decreased when the sample pre-sintering temperature increased up to 1550°C and an HIP temperature at 1600°C [17].

Xu et al. prepared the novel $ZnO_3Al_2O_3$ transparent ceramic material using reactive HIP for their study. A 100-g load was applied to find the Vickers hardness of the prepared ceramic sample with the help of a micro-hardness tester. The obtained result of Vickers hardness was 13.0 ± 0.1 GPa and the Young's modulus was 287.4 GPa. Following the results obtained in the study, Young's modulus and hardness were having good agreement with $MgAl_2O_4$ [27]. In another study, Zhao et al. [35] prepared $KNbTeO_6$ transparent ceramic material using pseudo-HIP along with a pressure-less sintering method. The elastic modulus tester was used to find the Young's modulus of the prepared specimen. Micro-hardness tester was used to measure the Vickers hardness of the material. A 50-g load was applied for 15 seconds while measuring the hardness value. The tests were conducted at room temperature. The obtained results of the Young's and shear moduli values were 196 GPa and 80.6 GPa, respectively. The hardness value achieved was 6.0 ± 0.03 GPa. The obtained results exhibited a significant agreement with the findings reported in the study carried out by Dolhen et al. [36]. The impact of the pressure less post sintering in H_2 on the mechanical and structural properties of the HIPed Al_2O_3 produced by oxidized AlN powder was investigated by Balazsi et al. [37]. The samples of the post-sintered material have delivered an increase in hardness value to 17–18 GPa with very good densification. The hardness test was conducted using a micro-hardness tester. The applied load was 19.61 N during the hardness test. They characterized hardness values as a function of oxidation time as depicted in Figure 4.4.

Ning et al. [38] fabricated laser-grade Y_2O_3 ceramic material with ZrO_2 additive using HIP. They explored that, by reducing the grain size of transparent ceramic material, the hardness and toughness values could be improved up to 5–7 times higher than a single crystal ceramic. So, an improvement in mechanical properties

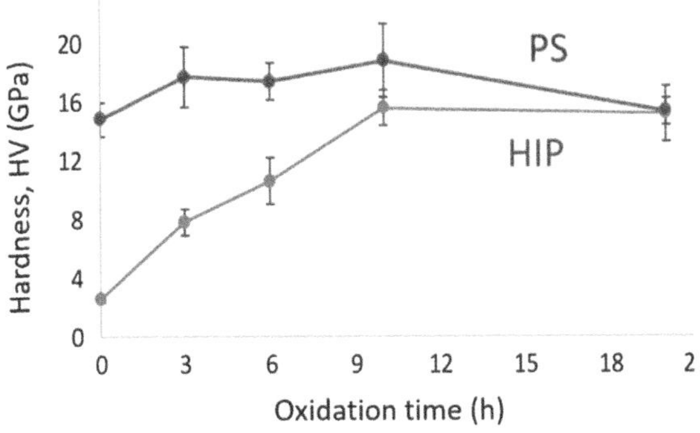

FIGURE 4.4 Hardness values of HIP and pressureless pre-sintered Al_2O_3 as a function of oxidation time of AlN base powder.

Adapted from Balazsi et al. [37].

aids in getting the betterment in resistance to thermal shock where the material can be used in laser applications of high power range.

An et al. [39] fabricated AlON transparent ceramics with the inclusion of Si_3N_4 additive using pressure-less pre-sintering with HIP and investigated the prepared sample for flexural strength and optical properties. The temperature was maintained at 1800°C. The flexural strength increased with the increase in the wt% of the Si_3N_4 between 0.02 and 0.20 wt%. There was an increase in flexural strength from 322 ± 67 MPa to 417 ± 59 MPa in their study. The preceding studies clearly indicate the significance of the HIP process in obtaining excellent mechanical properties in transparent ceramics.

4.6 CONCLUDING REMARKS

This chapter presented an overview of the significance of HIP in the production of transparent ceramics and the various properties of the materials achieved that are vital for the appropriate applications in various sectors. The following key ideas are evident from the discussions:

- The transparent ceramics undergo rapid densification at intermediate temperatures which is important to be understood to achieve pore-free products with superior optical properties. At high temperatures, grain growth is more and leads to intra-granular porosity in almost all transparent ceramic materials, which is highly deleterious.
- The thermal conductivity of the ceramics demonstrates a positive correlation with the increase in grain size and a negative correlation with the rise in test temperatures. Also, dopant addition significantly reduces thermal conductivity due to the reduced mean free path of the phonon and increased phonon-scattering rate.
- The transmittance property of the ceramics can be enhanced by the addition of sintering aids and dopants in the case of ceramics used for laser applications. These additions promote fine grain size and high density during the HIP process, thereby helping in achieving better optical properties.
- The transparent ceramics that are HIPed were found to possess superior flexural strength, hardness, toughness, Youngs and shear moduli due to grain refinement. Attaining superior mechanical properties heavily relies on the choice of pre-sintering temperatures.

REFERENCES

[1] Itatani, K., Tsujimoto, T., Kishimoto, A. (2006), Thermal and optical properties of transparent magnesium oxide ceramics fabricated by post hot-isostatic pressing. *Journal of the European Ceramic Society*, 26, 639–645. doi: 10.1016/j.jeurceramsoc.2005.06.011

[2] McCauley, J. W., Patel, P., Chen, M., Gilde, G., Strassburger, E., Paliwal, B., Ramesh, K. T., Dandekar, D. P. (2009), AlON: A brief history of its emergence and evolution. *Journal of the European Ceramic Society*, 29, 223–236. doi: 10.1016/j.jeurceramsoc.2008.03.046

[3] Krell, A., Klimke, J., Hutzler, T. (2009), Advanced spinel and sub-μm Al_2O_3 for transparent armour applications. *Journal of the European Ceramic Society*, 29, 275–281. doi: 10.1016/j.jeurceramsoc.2008.03.024

[4] Wang, S. F., Zhang, J., Luo, D. W., Gu, F., Tang, D. Y., Dong, Z. L., Tan, G. E. B., Que, W. X., Zhang, T. S., Li, S., Kong, L. B. (2013), Transparent ceramics: Processing, materials and applications. *Progress in Solid State Chemistry*, 41, 20–54. doi: 10.1016/j. progsolidstchem.2012.12.002

[5] Boulesteix, R., Maître, A., Baumard, J.-F., Rabinovitch, Y., Reynaud, F. (2010), Light scattering by pores in transparent Nd:YAG ceramics for lasers: correlations between microstructure and optical properties. *Optics Express*, 18, 14992. doi: 10.1364/oe.18.014992

[6] Li, J., Zhang, B., Tian, R., Mao, X., Zhang, J., Wang, S. (2021), Hot isostatic pressing of transparent AlON ceramics assisted by dissolution of gas inclusions. *Journal of the European Ceramic Society*, 41, 4327–4336. doi: 10.1016/j.jeurceramsoc.2021.02.035

[7] Messing, G. L., Stevenson, A. J. (2008), Toward pore-free ceramics. *Science*, 322, 383–384. doi: 10.1126/science.1160903

[8] Dericioglu, A. F., Boccaccini, A. R., Dlouhy, I., Kagawa, Y. (2005), Effect of chemical composition on the optical properties and fracture toughness of transparent magnesium aluminate spinel ceramics. *Materials Transactions*, 46, 996–1003. doi: 10.2320/matertrans.46.996

[9] Liu, Z., Wei, J., Toci, G., Pirri, A., Patrizi, B., Feng, Y., Xie, T., Hreniak, D., Vannini, M., Li, J. (2022), Microstructure and laser emission of Yb:CaF$_2$ transparent ceramics fabricated by air pre-sintering and hot isostatic pressing. *Optical Materials*, 129, 112540. doi: 10.1016/j.optmat.2022.112540

[10] Toci, G., Hostaša, J., Patrizi, B., Biasini, V., Pirri, A., Piancastelli, A., Vannini, M. (2020), Fabrication and laser performances of Yb:Sc$_2$O$_3$ transparent ceramics from different combination of vacuum sintering and hot isostatic pressing conditions. *Journal of the European Ceramic Society*, 40, 881–886. doi: 10.1016/j.jeurceramsoc.2019.10.059

[11] Maksimov, R. N., Shitov, V. A., Platonov, V. V., Yurovskikh, A. S., Toci, G., Patrizi, B., Vannini, M., Pirri, A. (2021), Hot isostatic pressing of transparent Yb^{3+}-doped Lu$_2$O$_3$ ceramics for laser applications. *Ceramics International*, 47, 5168–5176. doi: 10.1016/j.ceramint.2020.10.094

[12] Huang, X., Zhang, X., Hu, Z., Feng, Y., Wei, J., Liu, X., Li, X., Chen, H., Wu, L., Pan, H., Li, J. (2019), Fabrication of Y$_2$O$_3$ transparent ceramics by hot isostatic pressing from precipitated nanopowders. *Optical Materials*, 92, 359–365. doi: 10.1016/j. optmat.2019.04.047

[13] Ma, B., Zhang, W., Wang, Y., Xie, X., Song, H., Yao, C., Zhang, Z., Xu, Q. (2018), Hot isostatic pressing of MgAlON transparent ceramic from carbothermal powder. *Ceramics International*, 44, 4512–4515. doi: 10.1016/j.ceramint.2017.12.023

[14] Zhai, C., Xu, J., Hou, Y., Sun, G., Zhao, B., Yu, H. (2022), Effect of fiber orientation on surface characteristics of C/SiC composites by laser-assisted machining. *Ceramics International*, 48, 6402–6413. doi: 10.1016/j.ceramint.2021.11.183

[15] Shan, Y., Ma, L., Sun, X., Han, X., Wang, X., Li, J., Xu, J. (2022), Fast fabrication of AlON ceramics with ultra-high transmittance by the assistance of hot isostatic pressing treatment. *Journal of the European Ceramic Society*, 42, 5538–5544. doi: 10.1016/j. jeurceramsoc.2022.06.065

[16] Chen, F., Zhang, F., Wang, J., Zhang, H., Tian, R., Zhang, Z., Wang, S. (2015), Hot isostatic pressing of transparent AlON ceramics with Y$_2$O$_3$/La$_2$O$_3$ additives. *Journal of Alloys and Compounds*, 650, 753–757. doi: 10.1016/j.jallcom.2015.08.028

[17] Liu, Q., Jing, Y., Su, S., Li, X., Liu, X., Feng, Y., Chen, X., Li, J. (2020), Microstructure and properties of MgAl$_2$O$_4$ transparent ceramics fabricated by hot isostatic pressing. *Optical Materials*, 104, 109938. doi: 10.1016/j.optmat.2020.109938

[18] Liu, Y., Zhu, J., Dai, B. (2020), Transparent MgAl$_2$O$_4$ ceramics prepared by microwave sintering and hot isostatic pressing. *Ceramics International*, 46, 25738–25740. doi: 10.1016/j.ceramint.2020.07.051

[19] Riley, F. L. (1983), *Progress in Nitrogen Ceramics*. Martinus Nijhoff Publishers, Leeds, UK.

[20] Yoshinaka, M., Hirota, K., Ito, M., Takano, H., Yamaguchi, O. (1999), Hot isostatic pressing of reactive SnO_2 powder. *Journal of the American Ceramic Society*, 82, 216–218. doi: 10.1111/j.1151-2916.1999.tb01746.x

[21] Bocanegra-Bernal, M. H. (2004), Hot Isostatic Pressing (HIP) technology and its. *Journal of Materials Science*, 9, 6399–6420.

[22] Barinov, S. M. (1997), *Effect of Hot Isostatic Pressing on the Mechanical Properties of Aluminum Oxide Ceramics*. 9–12.

[23] Li, X., Zhu, Q., Xu, Y., Mao, X., Feng, M., Jiang, B., Zhang, L. (2019), Optical and thermal properties of TiO_2-doped Y_2O_3 transparent ceramics synthesized by hot isostatic pressing. *Journal of the American Ceramic Society*, 102, 2021–2028. doi: 10.1111/jace.16032

[24] Li, X., Xu, Y., Mao, X., Zhu, Q., Xie, J., Feng, M., Jiang, B., Zhang, L. (2018), Investigation of optical, mechanical, and thermal properties of ZrO_2-doped Y_2O_3 transparent ceramics fabricated by HIP. *Ceramics International*, 44, 1362–1369. doi: 10.1016/j.ceramint.2017.08.204

[25] Liu, Z., Toci, G., Pirri, A., Patrizi, B., Feng, Y., Hu, D., Chen, H., Hreniak, D., Vannini, M., Li, J. (2022), Fabrication and characterizations of $Tm:Lu_2O_3$ transparent ceramics for 2 μm laser applications. *Optical Materials*, 131, 112705. doi: 10.1016/j.optmat.2022.112705

[26] Johnson, R., Biswas, P., Ramavath, P., Kumar, R. S., Padmanabham, G. (2012), Transparent polycrystalline ceramics: An overview. *Transactions of the Indian Ceramic Society*, 71, 73–85. doi: 10.1080/0371750X.2012.716230

[27] Xu, P., Wang, H., Zheng, K., Chen, B., Yang, M., Chen, Q., Wang, B., Tu, B., Wang, W., Fu, Z. (2022), Novel transparent $ZnO·3Al_2O_3$ ceramics prepared by reactive hot isostatic pressing. *Journal of the European Ceramic Society*, 42, 724–728. doi: 10.1016/j.jeurceramsoc.2021.10.060

[28] Li, D., Jia, D., Yang, Z., Zhou, Y. (2022), Principles, design, structure and properties of ceramics for microwave absorption or transmission at high-temperatures. *International Materials Reviews*, 67, 266–297. doi: 10.1080/09506608.2021.1941716

[29] Cao, X. Q., Vassen, R., Stoever, D. (2004), Ceramic materials for thermal barrier coatings. *Journal of the European Ceramic Society*, 24, 1–10. doi: 10.1016/S0955-2219(03)00129-8

[30] Wang, M., Wang, R., Dai, H., Li, T., Sun, Y., Liu, D., Yan, F., Xing, X. (2022), The effects of hot isostatic pressing temperature on the structure and properties in GdMnO3 ceramics. *Ceramics International*, 48, 3685–3694. doi: 10.1016/j.ceramint.2021.10.150

[31] Chen, F., Zhang, F., Wang, J., Zhang, H., Tian, R., Zhang, J., Zhang, Z., Sun, F., Wang, S. (2014), Microstructure and optical properties of transparent aluminum oxynitride ceramics by hot isostatic pressing. *Scripta Materialia*, 81, 20–23. doi: 10.1016/j.scriptamat.2014.02.009

[32] Jiang, N., Liu, Q., Xie, T., Ma, P., Kou, H., Pan, Y., Li, J. (2017), Fabrication of highly transparent AlON ceramics by hot isostatic pressing post-treatment. *Journal of the European Ceramic Society*, 37, 4213–4216. doi: 10.1016/j.jeurceramsoc.2017.04.028

[33] Chrétien, L., Boulesteix, R., Maître, A., Sallé, C., Reignoux, Y. (2014), Post-sintering treatment of neodymium-doped yttrium aluminum garnet (Nd:YAG) transparent ceramics. *Optical Materials Express*, 4, 2166. doi: 10.1364/ome.4.002166

[34] https://www.coorstek.co.jpengrddetail_04.html

[35] Zhao, Q., Wang, H., Tu, B., Xu, P., Jing, Z., Wang, W., Fu, Z. (2023), $KNbTeO_6$ transparent ceramics prepared by the combination of pressure-less sintering and pseudo hot isostatic pressing. *Journal of the European Ceramic Society*, 43, 4226–4231. doi: 10.1016/j.jeurceramsoc.2023.03.025

[36] Dolhen, M., Carreaud, J., Delaizir, G., Duclère, J. R., Vandenhende, M., Tessier-Doyen, N., Tantot, O., Passerieux, D., Coulon, P. E., Thomas, P., Allix, M., Chenu, S. (2020), New $KNbTeO_6$ transparent tellurate ceramics. *Journal of the European Ceramic Society*, 40, 4164–4170. doi: 10.1016/j.jeurceramsoc.2020.04.005

[37] Balázsi, K., Varanasi, D., Horváth, Z. E., Furkó, M., Cinar, F. S., Balázsi, C. (2022), Effect of the pressureless post-sintering on the hot isostatic pressed Al_2O_3 prepared from the oxidized AlN powder. *Scientific Reports*, 12, 1–9. doi: 10.1038/s41598-022-12456-2

[38] Ning, K., Wang, J., Ma, J., Dong, Z., Kong, L. B., Tang, D. (2020), Fabrication of laser grade Yb: Y_2O_3 transparent ceramics with ZrO_2 additive through hot isostatic pressing. *Materials Today Communications*, 24, 101185. doi: 10.1016/j.mtcomm.2020.101185

[39] An, L., Shi, R., Mao, X., Zhang, B., Li, J., Zhang, J., Wang, S. (2023), Fabrication of AlON transparent ceramics with Si_3N_4 sintering additive. *Journal of Advanced Ceramics*, 12, 1361–1370. doi: 10.26599/JAC.2023.9220760

5 Machining of Polymer-based Composites
The Outlook, Challenges and Opportunities

Sandhyarani Biswas and Jasti Anurag

5.1 INTRODUCTION

Recently, polymers and their composites have been used widely in many applications due to their high strength-to-weight ratio, light weight, resistance to corrosion, design flexibility, ease of fabrication, better coefficient of friction, wear resistance and self-lubricating properties. Polymer matrix composites (PMCs), which consist of a polymer reinforced with either fibres or particles, provide superior mechanical performance and enhanced functionalities in comparison to conventional materials. However, the composition and structure of PMCs pose unique challenges when it comes to machining and shaping these materials to precise specifications (Hsissou et al., 2021). PMCs are usually manufactured to near net shapes; however, few postprocessing operations like trimming, drilling or turning are often needed to achieve dimensional accuracy, surface quality and other requirements for assembly. Machining composites varies significantly compared to machining conventional metals in many aspects. Since the polymers and reinforcements have different properties, PMCs often exhibit anisotropy and inhomogeneity. The interaction of the cutting tool is mainly alternatively with the matrix and reinforcement and therefore the response to the PMC machining is entirely different. The mechanism in the chip formation process in conventional metals is mainly due to the shearing action and plastic deformation. However, in PMCs, the mechanism deals with the fibre fracture, matrix failure and the debonding of matrix and fibres (Wan et al., 2019). Also, the chips produced mainly are continuous and curled in metal cutting, whereas they crumble and fragment in PMC cutting. In addition, very rapid wear in the cutting tool also occurred. These problems mainly limit the application of composite materials, and therefore, research work is going on by many researchers to analyze the various aspects of PMC machining. Significant damage to the workpiece like delamination, burr, fibre pull out, fuzzing, burning and fibre breakage may be introduced. In PMC machining, delamination is considered serious damage, which adversely affects the

DOI: 10.1201/9781032665375-5

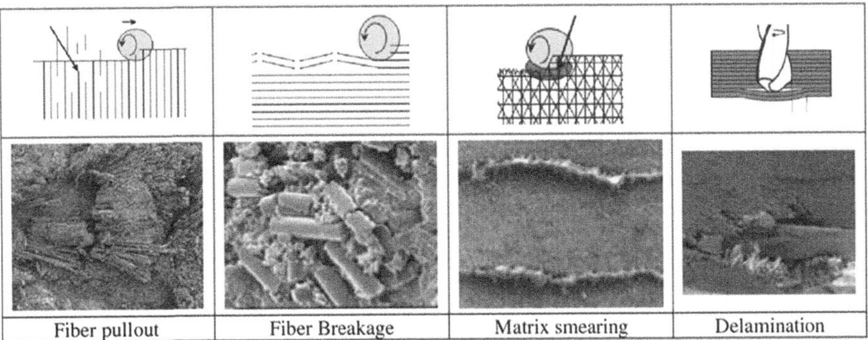

| Fiber pullout | Fiber Breakage | Matrix smearing | Delamination |

FIGURE 5.1 Surface failures during the machining process of polymeric composites (Altin Karataş & Gökkaya, 2018).

structural integrity and surface quality. Figure 5.1 illustrates different failure mechanisms, including matrix smearing, delamination, fibre pull-out and fibre breaks.

Conventional machining like milling, drilling, turning and the like can be used for machining PMCs. However, difficulties arise when machining PMCs, as conventional knowledge and experience are not sufficient for these materials as their machinability is completely different from that of conventional metals (Palanikumar, Karunamoorthy, & Karthikeyan, 2006). The ability to machine PMCs depends mainly on factors such as the properties of the matrix and fibre, the amount of fibre present and the orientation of the fibres. Therefore, proper parameter selection, tool design and enhanced interfacial bonding between matrix and fibre are required to achieve reduced delamination and increased machinability. This chapter reviews the recent developments in the machining of PMCs. The influence of different factors on the performance of machined parts is discussed. The various approaches, as well as recent optimization methods, statistical techniques and numerical modelling used for machining polymer composites, are demonstrated. An overview of the different issues related to PMC machining is presented in this chapter. Finally, the recent challenges and future scope in the machining of PMCs are also highlighted.

5.2 CONVENTIONAL MACHINING OF PMCs

Conventional machining is a subtractive manufacturing technique that removes material from a workpiece in order to get the desired shape and dimensions. This category encompasses a variety of operations applicable to PMCs, including cutting, drilling, milling and turning. Composites are typically composed of fibres oriented in particular directions, with the composite's properties varying accordingly. As a result, controlling cutting forces and managing chip formation during the machining process becomes complex, which may result in damage to both the tool and workpiece. Because of the anisotropic and nonhomogeneous properties, machining PMC poses many difficulties. The machinability of composites is affected by several factors, including fibre type, fibre content, matrix material and fibre orientation.

By selecting suitable cutting tools and optimizing operating conditions, it is possible to achieve favourable results when machining PMCs. Despite the increasing popularity of nonconventional machining techniques like laser cutting and water jet machining, conventional machining techniques continue to be widely employed due to their cost-effectiveness, adaptability and ease of implementation. A few commonly used conventional machining processes for PMCs are described next.

5.2.1 Turning of PMCs

The turning of PMCs is an essential machining process that is used to convert composites to the desired size. Numerous PMC parts undergo finishing through the process of turning. These components include steering columns, drag links, rolls, bearings, columns and axles. Furthermore, many axisymmetric parts need turning operation as a vital procedure for the precise finishing of high-accuracy components and exceptionally accurate joint regions. The effect of different factors on the surface quality, tool wear, cutting forces and cutting temperatures of PMCs have been studied by many researchers. The most significant factors that influence the machinability of PMCs are fibre volume fraction, orientation and type (Sheikh-Ahmad, 2009). Turning of fibre composites results in various flaws appearing on the composite surface, including matrix degradation, poor surface finish, delamination and severe fibre pull-out (Vigneshwaran et al., 2021). Kechagias et al. (2009) and Ntziantzias et al. (2011) analysed the impact of turning parameters like feed rate and speed on the surface quality of PMCs. It was seen that the effect of feed rate on the surface roughness is more significant than the cutting speed. The surface roughness increased with the feed rate. Conversely, the decrease in the surface roughness is seen with the increase in the cutting speed. Similar results are also reported by Zajac et al. (2014), Palanikumar et al. (2006) and Hutyrová et al. (2014, 2015). Sivakiran et al. (2018) analysed the influence of machining parameters on the material removal rate (MRR) and surface quality of banana fibre-based PMCs. It was seen that the feed rate is influencing significantly the surface roughness and MRR in comparison to the depth of cut and cutting speed. Similar results are also reported by Rajasekaran et al. (2011, 2013) for a surface roughness study of PMCs reinforced with carbon fibre using fuzzy logic and the Taguchi method. Rajasekaran et al. (2022) and Mahesha et al. (2022) also analysed the impact of turning parameters on the surface quality of PMCs using artificial intelligence techniques and reported similar results. The effect of fillers on the surface quality of pineapple leaf fibre–based PMCs using response surface methodology was investigated by Dilli Babu et al. (2020), who concluded that a filler reduces surface roughness under a turning operation. Petropoulos et al. (2008) analysed the influence of fibre type on the surface quality of polyether ether ketone (PEEK) composites after turning. It was seen that fibres increased the surface roughness, while composites reinforced with glass fibres had higher surface roughness than composites reinforced with carbon fibres. The solidity surface ratio is illustrated in Figure 5.2. Hussain et al. (2014) investigated the influence of fibre orientation and process parameters on the surface roughness of PMCs reinforced with glass fibre. It was reported that the surface roughness is low when the feed force is parallel to the fibre orientation and concluded that fibre orientation has less of an

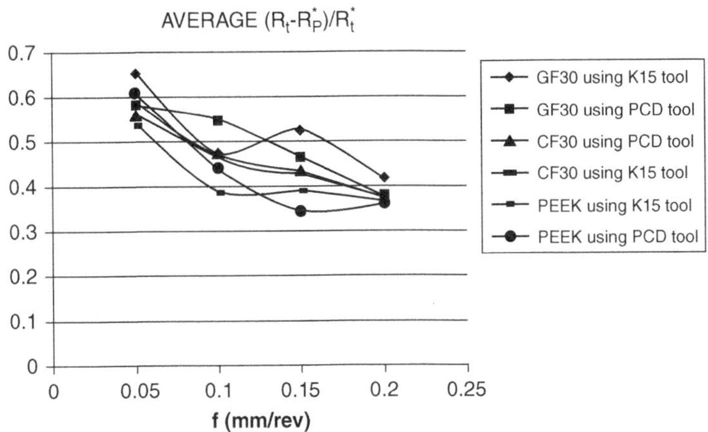

FIGURE 5.2 Solidity surface ratio of PEEK composites (Petropoulos et al., 2008).

effect compared to feed rate and cutting speed. Somsakova et al. (2012) investigated the influence of tool radius on the surface quality of wood plastic composite during turning operation and concluded that a turning tool with a larger nose radius shows a better surface finish. A similar result is also reported by Kini and Chincholkar (2010).

5.2.2 Milling of PMCs

Milling is a widely used technique for removing materials in the machining of PMC parts. However, unlike the high-speed MRRs observed in metal milling, polymer composite milling occurs on a much smaller scale. This is primarily because components fabricated with polymer composites are typically manufactured to their desired shape, and any subsequent milling is primarily focused on tasks such as deburring, trimming and achieving precise contour shapes. Unlike turning, the cutting tool rotates in milling, often with multiple cutting edges engaged in simultaneous cutting. This introduces complexities to the milling process, including variations in chip size, cutting forces and fibre orientation that fluctuate with the rotation of the tool. The machinability of PMCs in milling is usually characterized by factors such as surface roughness, delamination and tool wear (Ozkan et al., 2019; Sheikh-Ahmad, 2009). Azmi et al. (2013) studied the effect of machining parameters on the surface quality of glass fibre–based PMCs in end milling. It was reported that feed rate had a high negative effect while depth of cut had the least. It was also observed that the machining force is highly influenced by the feed rate and least affected by spindle speed. Wang et al., (2016a) analysed the effect of process parameters on the cutting force and cutting temperature of carbon fibre–based PMCs during milling operation. It was seen that the feed rate and cutting speed significantly affected the cutting temperature and cutting force, whereas the depth of cut had the least influence on the cutting temperature and cutting force. It was concluded that a high depth of cut, a low feed rate and speed is optimal to obtain better surface quality. Similar results are also reported by Wang et al., (2016b) and Jia et al. (2018). The impact of the

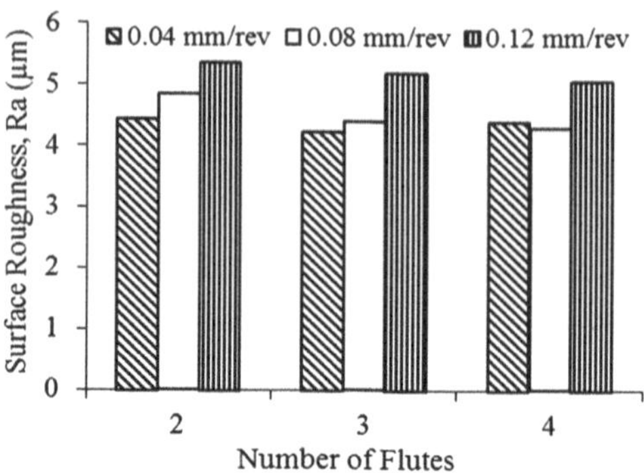

FIGURE 5.3 Effect of number of flutes on surface quality (Çelik et al., 2019).

orientation of the fibre on the surface quality in end milling of flax fibre–based PMCs was studied by Slamani et al. (2022). It was seen that a high surface finish is obtained when the orientation of fibres is perpendicular to the direction of milling, that is, 90°, whereas the worst surface roughness was observed when the fibre orientation is 45°. Çelik et al. (2019) analysed the effect of the number of flutes, speed and feed rate on the surface quality and delamination factor of the jute fibre–based PMCs in milling. It was seen that an increase in the feed rate and speed resulted in a higher delamination factor and surface roughness, whereas an increase in the number of flutes reduced the surface roughness (represented in Figure 5.3).

Kumar and Prakash (2021) studied the impact of machining parameters and fibre orientation on the PMCs in the milling operation. It was reported that the depth of cut is highly influencing the cutting force and the surface quality, while the machinability index is highly influenced by feed rate. The contribution of all these parameters is represented in Figure 5.4. Kishore et al. (2021) analysed the effect of graphene nano-filler on the surface roughness of milled basalt–jute-hybrid composites. It was observed that the composites reinforced with nanofillers (up to 0.6% w/w) had lower surface roughness when compared to composites with no nanofillers. However, composites with nanofillers at a weight percentage of 1% had a high surface roughness compared to composites with no nanofillers. It was concluded that nanofillers with a weight fraction of 0.2% had the lowest surface roughness for all feed rates and cutting speeds.

5.2.3 Drilling of PMCs

Drilling operation is one of the most important machining processes which is often required for making the holes in PMC structures to assemble different parts into a final desired component. The processes of countersinking, counterboring and drilling are frequently necessary to adequately prepare composite components for the purpose of joining and assembly. Despite its widespread utilization, drilling continues to be regarded as one of the most complex machining processes. Delamination, tool wear and thermal management are significant factors that need careful consideration (Khashaba, 2013; Sheikh-Ahmad, 2009). The low thermal conductivity value of both the polymer and the fibre material creates a favourable condition for accumulating

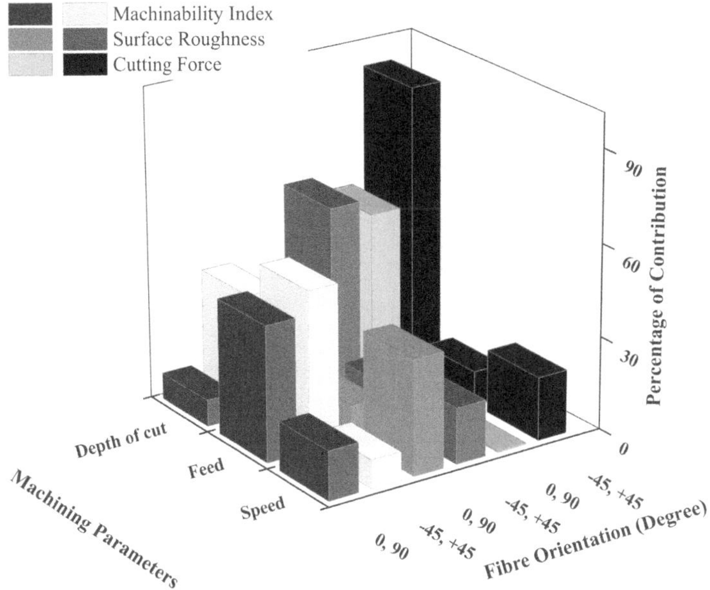

FIGURE 5.4 Effect of milling parameters on performance output (Kumar & Prakash, 2021).

heat in the cutting region. The heat produced during the drilling process is influenced by the feed rate and cutting speed, thus requiring the use of optimum process parameters when machining PMCs to prevent heat-induced damage (Krishnaraj et al., 2013). In certain instances, the utilization of authorized coolant may be employed to mitigate cutting temperatures and regulate the generation of machining dust. In addition, the presence of a distinct coefficient of thermal expansion values of fibre and the polymer poses challenges in achieving precise dimensional accuracy when drilling holes. The potential for the reduction in the size of the holes after drilling may result in suboptimal assembly tolerance. The presence of reinforcement fibres leads to significant wear through the process of abrasion on the cutting edges (Abrão et al., 2007). The wearing down of the cutting edge subsequently leads to an augmentation in the thrust force. The dominant factor influencing the initiation of delamination is the thrust force. Through the meticulous selection of cutting parameters and tool geometry, it is possible to attain a drilling process devoid of defects (Chen, 1997). Debnath et al. (2014) evaluated the impact of machining parameters and drill geometry on the performance of drilled PMCs. It was seen that as the speed and feed rate increase, the torque and thrust force increase. It was also reported that a parabolic drill produced the least thrust force and torque, while a step drill produced the highest. El-Sonbaty et al. (2004) evaluated the influence of fibre content and machining parameters on the surface quality of drilled glass fiber-reinforced polymer composites (FRPCs). It was observed that the surface roughness of epoxy resin is least influenced by the feed rate and cutting speed. Whereas the surface roughness of composites increased with an increase in the fibre volume fraction and cutting speed. The influence of drill size and feed rate on the torque and thrust force of

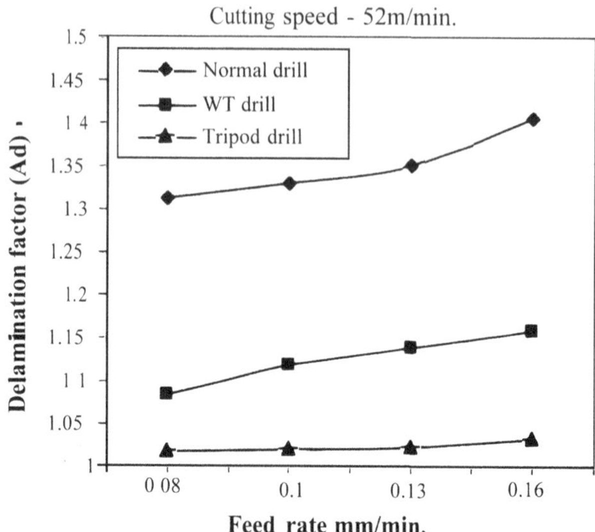

FIGURE 5.5   Effect of point geometry on delamination factor (Velayudham & Krishnamurthy, 2007).

glass fibre–based PMCs was investigated by Mohan et al. (2005). It was concluded that the increase in the feed rate increased the torque and thrust force, whereas an increase in the drill size decreased them. Wang et al. (2018) analysed the impact of the drilling area temperature on the delamination and surface roughness of carbon fibre–based PMCs. It was seen that when the drilling area temperature was more than the glass transition temperature of the matrix, the delamination and surface roughness increased. Velayudham and Krishnamurthy (2007) analysed the impact of point geometry on the delamination of PMCs in drilling and saw that the delamination was least for the tripod geometry drill (illustrated in Figure 5.5).

The influence of tool material, process parameters and content of nanofillers on the surface roughness of drilled polymer nanocomposites was studied by Verma et al. (2021). It was reported that the surface roughness is least when the weight percentage of nanofillers is 1%. The surface roughness increased when the feed rate increased, while it decreased with an increase in spindle speed. Drilling with drill bits made of titanium aluminium nitride (TiAlN) produced the least surface roughness, while drill bits made of high-speed steel (HSS) produced the highest surface roughness. Ramesh and Gopinath (2017) analysed the effect of drill diameter and the process parameters on the thrust force generated due to drilling sisal/glass fibre–based PMCs. It was seen that with an increase in drill diameter and feed rate, the thrust force increases, whereas with an increase in spindle speed, it decreases.

5.2.4 Abrasive Machining of PMCs

Abrasive machining processes involve material removal through the motion of small, rigidly attached hard particles. It helps eliminate defects, such as flash or burrs, which can compromise the functionality and aesthetics of the final product. The effect of abrasive types and grinding parameters on the grinding force of glass fibre–based PMCs was studied by Chockalingam and Kuang (2015). It was observed that an alumina wheel produced a lower grinding force than that of a cubic boron nitride (CBN) wheel. At a low depth of cut and speed, it was observed that with an increase in feed rate, grinding force ratios increased. Khoran et al. (2022) investigated the impact of fibre type and concluded that the highest grinding forces were obtained for glass fibre–based PMCs.

5.3 NON-CONVENTIONAL MACHINING OF PMCs

The increasing demand for high-performance composites necessitates the integration of reinforcement materials that possess enhanced hardness, stiffness and strength characteristics into advanced composite structures. Nevertheless, there are inherent difficulties associated with the secondary machining of these materials. Conventional machining techniques encounter challenges when applied to composites due to their high abrasiveness, heat sensitivity, low thermal conductivity, anisotropic properties and heterogeneous nature. Non-conventional machining processes refer to methodologies that predominantly utilize non-mechanical means of interaction energy to remove material. This contrasts with the shearing action commonly employed in tool-based systems such as grinding, turning and milling. In contrast, unconventional methods encompass the use of electrical and thermal as feasible substitutes for conventional tool-based techniques (Jackson & Toward, 2021). The advantages of non-conventional machining techniques are high surface quality for complex geometries, high precision and burr-free surfaces, among others.

5.3.1 ELECTRICAL DISCHARGE MACHINING OF PMCs

Electrical discharge machining (EDM) needs thermal energy wherein electrically conductive substances undergo thermal erosion through the utilization of spark discharges. The highly concentrated and intense heat generated by the spark discharge leads to the vaporization or melting of infinitesimal fragments of the workpiece material. The contours of the electrode establish the spatial domain wherein spark erosion takes place, thereby dictating the resultant cavity formation. As the tool and workpiece refrain from establishing direct physical contact, EDM permits the facile fabrication of precision components from challenging-to-machine and electrically conductive materials (Islam Shyha, 2021). Shlykov et al. (2020) studied the influence of a perforated masking sheet on the surface finish of PMCs after EDM. It was concluded that there are no torn layers of fibres due to masking tape. The effect of electrode material on the surface finish was investigated by Ablyaz et al. (2019). It was observed that composite tool electrodes produced a better surface finish, while graphite produced a rough surface. Dutta et al. (2021) determined the influence of process parameters on the machined surface morphology and cutting time of carbon fibre–based PMCs. The results showed that the peak current and pulse duration have a substantial influence on the cutting time and surface morphology, while the pulse interval and wire tension have less effect. The influence of pulse duration and gap current on the machining damage was studied by Sheikh-Ahmad (2016). The magnitude of the heat-affected zone (HAZ) was determined to be primarily influenced by the duration of the applied pulse, with the minimum HAZ observed at an intermediate pulse duration. Conversely, delamination predominantly transpired at the entry point of the hole and was chiefly impacted by the current gap. The tapering of the hole, by comparison, was primarily influenced by the pulse duration, followed by the current gap and their interplay. Singh et al. (2020) analysed the influence of LASER assistance on the precision of micro-hole creation in carbon FRPCs using EDM. The investigation demonstrated that the utilization of LASER technology results in a reduction of approximately 10–12% in both the overcut and taper of the hole. The micro-hole created through the process of LASER-assisted EDM demonstrated enhanced machining quality in relation to the surface damage, dimensional accuracy and circularity of the hole. The influence of pulse duration, pulse current and multiwalled carbon nanotube (MWCNT) content on machining damages was studied by Yazdanfar and Shahrajabian (2021). It was observed that introducing MWCNT into epoxy resin improves MRR and decreases the taper hole and delamination factor. Habib and Okada (2016) determined the influence of voltage, pulse duration, peak current and pulse-off time on the cutting parameters of carbon fibre–based PMCs. It was reported that the MRR increases with an increase in any parameter. It was also observed that the surface roughness increased with an increase in the open circuit voltage, peak current and pulse duration and decreased with an increase in the speed of the electrode and pulse-off time. It was concluded that the graphite electrode yielded an inferior surface finish and a higher MRR than the copper electrode. With a decrease in the rotation speed, the gap size decreased, while it increased with an increase in the rest of the parameters. Kumaran et al. (2020) evaluated the impact of carbon fillers on the machinability of carbon fibre–based PMCs. It was observed that the addition of fillers produced an increased MRR compared to the control. The evaluated fillers showed signs that might point to their ability to lessen electrode wear. Notably, compared to the control, the composites that included carbon black filler consistently showed less tool wear. Using a filler in carbon fibre–based PMCs also reduced the thermal damage at the exit hole. The influence of cutting direction on the machining of carbon fibre–based PMCs was studied by Habib et al. (2013). It was seen that the MRR was lower for a one-directional composite when compared to two-directional composites. It was also observed that the machinability of a two-directional composite is lower than a one-directional composite.

5.3.2 Laser Jet Machining of PMCs

The laser machining technique is classified as a non-conventional method of machining, wherein heat is generated using a laser beam for a thermal machining procedure. The laser machining system encompasses four primary elements, namely the focusing mechanism, laser delivery, laser discharge tube and excitation source, as well as auxiliary devices. In the absence of tool-to-workpiece contact, laser cutting eliminates tool wear, cutting forces and mechanical loading–induced component distortion. This procedure is not affected by the material's compressive or tensile strength but, rather, by its thermal energy. Therefore, laser cutting is ideally suitable for machining heterogeneous materials consisting of distinct phases with differing mechanical properties. It features high machining rates, a narrow kerf width and the ability to cut intricately contoured shapes. Nevertheless, laser cutting has a number of disadvantages, such as material alterations and decreased strength due to the HAZ formation, the presence of kerf taper and decreased cutting efficiency with increasing workpiece thickness. Salama et al. (2016, 2016) studied the impact of repetition rate, scanning speed and laser power on the ablation depth and heat-affected zones of laser-machined carbon fibre–based PMCs. It was seen that as the laser power increases, the ablation depth increases, and as the scanning speed increases, it decreases. It was also observed that the HAZ increases with an increase in laser power and repetition rate and decreases with an increase in scanning speed. Riveiro et al. (2017) investigated the effect of laser parameters on the machining of carbon fibre–based PMCs. It was seen that the material cannot be cut using low linear energies and low pulse energies together. However, at high linear and pulse energies, the cuts that are obtained are only partially or somewhat complete, and no acceptable outcomes are obtained as shown in Figure 5.6. It was also observed that laser processing parameters had no effect on taper angle. As the pulse energy increased, a decrease in the heat-affected zone was observed, while it increased with an increase in assist gas pressure.

The effect of various lasers on the HAZ extension of carbon fibre–based PMCs was studied by Herzog et al. (2008), who concluded that Nd:YAG laser achieved a low heat-affected zone while using a CO_2 laser resulted in a higher heat-affected zone. Shyha (2013) and I. S. Shyha et al. (2014) analysed the influence of the laser beam power, workpiece material, cutting speed and gas pressure on the MRR, surface roughness, kerf width and angle of laser-trimmed polymer composites. It was reported that the MRR rate increases with an increase in all parameters for both carbon and glass fibre–based composites, except that the beam power and gas pressure did not have much effect on the carbon fibre–based PMCs. It was also reported

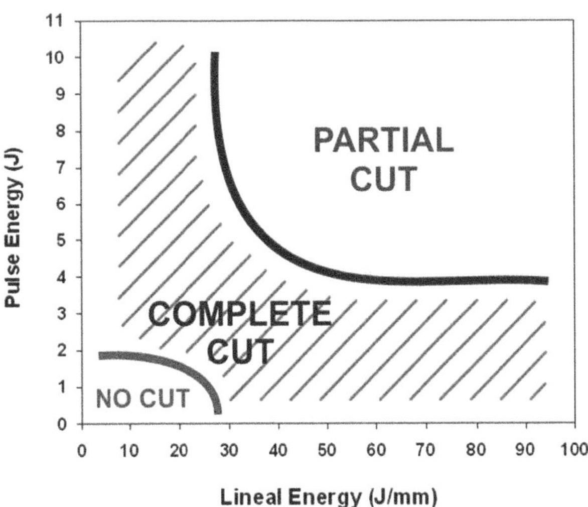

FIGURE 5.6 Processing map in relation to lineal and pulse energy (Riveiro et al., 2017).

that the surface roughness decreases with an increase in laser power and increases with an increase in cutting speed. The surface roughness of glass fibre–based composite was observed to be higher than carbon FRPCs. The kerf width and angle decrease with an increase in cutting speed. The kerf width was observed to be higher for glass FRPCs, while the kerf angle was higher for carbon FRPCs. Hejjaji et al. (2016) analysed the influence of the fibre type (unidirectional and multidirectional) on the laser drilling of carbon and glass fibre–based PMCs. It was reported that the surface roughness and damage factor of multidirectional composites were higher than unidirectional composites.

5.3.3 Water Jet and Abrasive Water Jet Machining of PMCs

Water jet (WJ) machining is a useful method of composite machining which involves pressurizing water up to extremely high levels and directing it through a small orifice at high velocity onto the composite surface. The abrasive water jet (AWJ) machining is an unconventional method that uses a highly pressurized stream of water infused with aluminium oxide or fine-grained garnet abrasive particles in a suitable ratio to remove material. This is achieved by directing the stream through a nozzle onto the specimen, resulting in surface erosion. Numerous advantages related to AWJ cutting make it preferable to conventional and non-conventional machining techniques. AWJ can cut virtually any material without significant heat damage or deformation (Islam Shyha, 2021). Due to the absence of cutting tools and minimal cutting forces, the setup time is reduced relative to conventional processes, and the requirements for fixture are either minimal or non-existent. As per the capabilities of the machine tool, the AWJ process can be used to cut both complex two-dimensional and three-dimensional (3D) contours. In addition, AWJ is recognized as an environmentally friendly method of material removal as the method does not require any cutting fluids and does not generate fumes or hazardous waste like conventional machining techniques. The utilized abrasives and water can also be recycled or eliminated naturally. However, the operating expenses are more due to replacing abrasive particles and other wear components, such as the mixing tube and orifice. In addition, the WJ, which travels at supersonic speeds, produces significantly more noise compared to non-traditional machining techniques, such as EDM and laser cutting (Sheikh-Ahmad, 2009). Numerous research studies have been done on the WJ and AWJ machining of PMCs. Madhu and Balasubramanian (2017) investigated the effect of nozzle diameter on the output parameters of WJ of glass fibre–based PMCs and concluded that nozzle diameter of 2.5 mm was optimal as it produced maximum MRR. In contrast, a nozzle diameter of 3.5 mm was found to produce the least MRR. Ramraji et al. (2020) analysed the influence of water pressure, traverse speed and cut-off distance on the surface quality and kerf taper of AWJ-machined PMCs. It was observed that the surface roughness and kerf taper decrease with an increase in water pressure and increase with an increase in traverse speed and cut-off distance. Wong (2018) analysed the influence of AWJ parameters on the delamination of FRP-hybrid composites. It was concluded that the delamination factor increases with an increase in stand-off distance and traverse rate and decreases with an increase in abrasive flow rate and hydraulic pressure. Jagadish and Rajakumaran (2018) analysed the impact

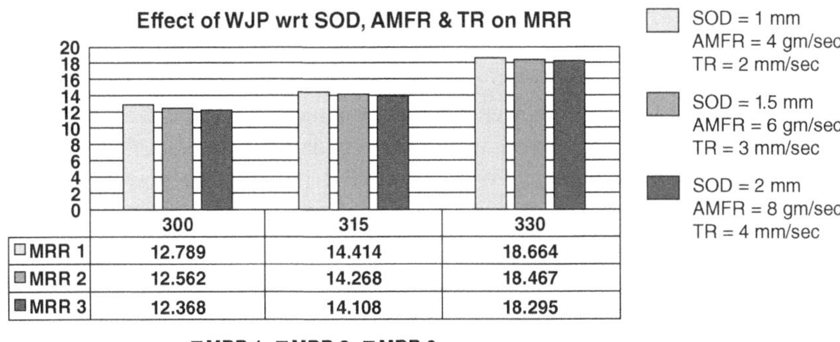

	300	315	330
□ MRR 1	12.789	14.414	18.664
▨ MRR 2	12.562	14.268	18.467
▣ MRR 3	12.368	14.108	18.295

FIGURE 5.7 Effect of abrasive material flow rate on MRR (Ashok Kumar et al., 2020).

of abrasive grain size on the surface quality and MRR of AWJ-machined pineapple fibre–based PMCs. It was seen that the MRR increases with an increase in abrasive grain size. It was also observed that a grain size of 90 mesh produced the least surface roughness. Ashok Kumar et al. (2020) analysed the influence of abrasive mass flow rate on the output parameters of WJ-machined glass fibre–based PMCs and concluded that MRR, kerf width and taper angle increase with an increase in the abrasive mass flow rate (Figure 5.7).

5.4 ADVANCED MACHINING OF PMCs

Advanced machining techniques surpass the limitations of conventional methods by utilizing cutting-edge technologies and procedures. These methods increase the machinability of polymer composites by employing cutting-edge tools, equipment and techniques. By combining advanced materials, precise control systems and intelligent algorithms, the difficulties associated with machining composite materials can be effectively conquered.

5.4.1 Ultrasonic Machining of PMCs

Ultrasonic machining is an advanced machining technique that removes material from workpieces using high-frequency vibrations. This machining offers unique advantages in terms of precision, efficiency and the capacity to machine hard and brittle materials (Feucht et al., 2014). Ultrasonic machining involves vibrating a tool at ultrasonic frequencies. Often in the presence of abrasive particles suspended in a slurry or liquid medium, the tool's vibrations are transmitted to the workpiece. As the tool presses against the workpiece, the abrasive particles carried by the medium act as a micro-hammer, causing micro-chipping or erosion to remove material. Due to the microscopic nature of the tool's vibrations, ultrasonic machining reduces the risk of thermal damage to the workpiece, making it suitable for heat-sensitive materials. In addition, the process permits the precise machining of intricate shapes and fine details, enabling the production of complex components that may be difficult to produce using other methods (J. Wang et al., 2017).

Wang, et al. (2018) and Hui Wang, Wang, et al. (2018) analysed the influence of machining parameters on the surface quality, torque and cutting force of ultrasonic-machined carbon fibre–based PMCs. It was observed that with an increase in tool rotation speed and ultrasonic power, the torque and cutting force decreased. It was also observed that with an increase in depth of cut and feed rate, surface roughness, torque and cutting force increase. With an increase in ultrasonic power, the surface roughness increased, and with an increase in total rotation speed, the surface roughness decreased. Debnath et al. (2015) studied the effect of power rating, abrasive grit number and slurry concentration on the tool wear rate, MRR and surface quality of ultrasonic drilling of glass fibre–based PMCs. It was seen that the output parameters increased with an increase in power rating and slurry concentration and a decrease in abrasive grit number. The variation of surface roughness is represented in Figure 5.8. Cong, et al. (2012) studied the effect of coolant on output parameters of ultrasonic-machined carbon FRPCs and concluded that using cutting fluid over cold air reduced the surface roughness, tool wear, torque and cutting force. The influence of grinding direction on the output parameters was analysed by Li et al. (2019). It was reported that the difference between the cutting force generated by up grinding is considerably higher than that obtained by down grinding. In contrast, the surface roughness was not affected by grinding direction. Hui Wang et al. (2019) analysed the influence of ultrasonic frequency on the machining of carbon fibre–based PMCs and observed that with an increase in frequency, the surface roughness and cutting forces decreased. Cong, Pei, et al. (2012) analysed the impact of process parameters on the power consumption of the ultrasonic machine. It was observed

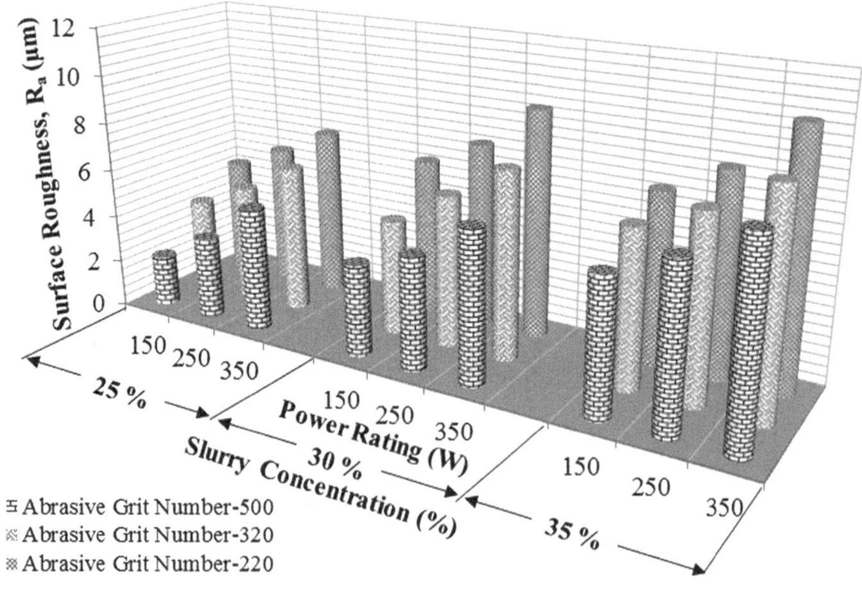

FIGURE 5.8 Effect of surface roughness with input parameters (Debnath et al., 2015).

that with an increase in feed rate, the total power consumption decreased. However, the variations in ultrasonic power and total rotation speed had less effect on the total power consumption. It was concluded that the coolant pump had the highest power consumption. The influence of the process parameters on the cutting temperature of ultrasonic machining of carbon fibre–based PMC was studied by Cong, Zou, et al. (2012). It was seen that with an increase in ultrasonic power and tool rotation speed, the cutting temperature increases. In contrast, the cutting temperature decreases with an increase in feed rate. Shi et al. (2019) studied the effect of the sampling array on the surface roughness of rotary ultrasonic-machined carbon fibre–based composites and concluded that the surface roughness decreases with an increase in the sampling array.

5.4.2 Cryogenic Machining of PMCs

Cryogenic machining, also known as cryo-machining, involves the application of extremely low temperatures during the machining process. By utilizing cryogenic fluids, such as liquid nitrogen, this innovative technique enhances cutting performance, improves surface quality and extends tool life. With its ability to address the thermal challenges associated with machining PMCs, cryogenic machining has gained significant attention and recognition. The primary goal of cryogenic machining of polymer composites is to mitigate the detrimental effects of heat, which can arise during traditional machining processes. Heat generation not only affects the structural integrity of the material but also damages the accuracy and surface finish of the machined parts. By employing cryogenic cooling, the excessive heat is efficiently dissipated, reducing the HAZ and minimizing thermal damage to the material. Moreover, cryogenic machining offers many advantages over conventional techniques. The extreme cooling effect not only prevents thermal degradation but also provides a lubrication effect, reducing the friction between the workpiece and the cutting tool (Yildiz & Sundaram, 2012). Morkavuk et al. (2018) analysed the influence of cryogenic cooling on the output parameters of machining. It was seen that using cryogenic cooling increases the cutting force and decreases the surface roughness and delamination factor. Similar results are also reported by some researchers (Basmaci et al., 2017; Cococcetta et al., 2021; Cunningham et al., 2018; Domingo et al., 2019; Kaynak et al., 2014; Khanna et al., 2020; Koklu & Morkavuk, 2019; Shokrani et al., 2019). The effect of cryogenic cooling on tool wear was also investigated by Cococcetta et al. (2021), who concluded that tool wear was reduced with the use of cryogenic cooling.

5.5 STATISTICAL APPROACH IN MACHINING OF PMCs

The statistical approach to machining polymer composites entails the collection and analysis of machining data, followed by the application of statistical methods to gain insights and make informed decisions. Engineers and researchers can comprehend the relationships between machining parameters and performance outcomes by utilizing statistical tools. This allows them to develop predictive models, optimize machining processes and improve the productivity and quality of machined components. In addition, using statistical methods in the machining of polymer composites permits the identification of critical factors that influence machining performance. Engineers can determine the optimum combination of machining parameters to achieve desired results, such as reduced tool wear, improved surface finish and increased dimensional accuracy by analysing the data. In addition, statistical analysis aids in detecting any potential flaws or anomalies in the machined components, allowing for timely adjustments and enhancements to the manufacturing process. This data-driven strategy not only improves productivity but also reduces the costs associated with trial-and-error procedures. In addition, statistical tools permit engineers to evaluate the variability of machining processes and establish control measures to ensure output of a consistent quality. By continuously monitoring and analysing data, manufacturers can implement preventive maintenance strategies and address any deviations or problems before they become severe. The optimization of surface roughness due to turning of glass fibre–based PMCs using RSM and genetic algorithm was studied by Hussain et al. (2014). Rajasekaran et al. (2011, 2013) analysed and predicted the surface roughness of carbon fibre–based PMCs using fuzzy logic and Taguchi analysis. Sakthivel et al. (2015) investigated the influence of drilling parameters using an analysis of variance (ANOVA) and a regression model and concluded that feed rate significantly influences the delamination and thrust force. It was also reported that the optimal feed rate, speed and drill bit diameter are 0.1 mm/rev, 300 RPM and 3 mm, respectively. Ramesh and Gopinath (2017) developed a regression model for the prediction of the thrust force based on the input parameters and analysed the input parameters using the ANOVA method. It was concluded that the thrust force was mostly influenced by feed rate. A similar methodology was used to predict the thrust force and torque for drilling natural fibre–based PMCs (Joe Patrick Gnanaraj et al., 2021). Malik et al. (2022) used RSM to predict the cutting temperature, thrust force and delamination factor based on thickness, feed rate and speed. Davim et al. (2004) analysed the influence of process parameters on the machining force and delamination factor of milled glass fibre–based PMCs using ANOVA. It was concluded that feed rate had a significant effect on the output parameters. Sreenivasulu (2013) studied the significance of end milling input parameters on the delamination factor and surface roughness using the Taguchi method and predicted these values using an artificial neural network (ANN). It was observed that cutting speed and depth of cut had a significant influence on the delamination factor and surface roughness. It was also observed that predicted values were close to experimental values. Chandrasekaran and Devarasiddappa (2014) analysed and predicted the surface roughness using ANOVA and an ANN. It was observed that grinding wheel velocity is the most significant factor and experimental results predicted results are in good agreement. Uysal et al. (2012) analysed the influence of process parameters on the tool wear in drilling PMCs using the Taguchi method. The optimal values for drilling were determined in their study. Hassan et al. (2022) analysed and predicted the machining parameters using ANOVA and the Taguchi method. It was concluded that the gap voltage and peak current significantly influence the radial overcut and electrode wear rate. The significance of machining parameters on the MRR was studied by Abdallah et al. (2018), who concluded that the ignition current had the highest significance.

5.6 NUMERICAL MODELLING AND SIMULATION IN MACHINING PMCs

Research on machining PMCs has mainly been conducted by using experimental means. An analysis of machining through experimentation with every possible combination in order to find the optimum parameter setting is time-consuming and not economical. Therefore, research has been started for the numerical modelling and simulation of the machinability of PMCs with advanced technology and computational abilities to save time and cost. Lasri et al. (2009) studied the progressive failure of unidirectional glass FRPCs using finite element analysis (FEA) in orthogonal cutting. A damage assessment was made using three failure criteria, such as Hashin, maximum stress and Hoffman. Using different failure criteria, the influence of the fibre orientation on sub-surface damage and cutting forces was studied. Zenia et al. (2015) did a numerical analysis to find out the most influential process parameters as well as their interactions on the damage and cutting force for carbon fibre–based PMCs. It was reported that the depth of cut and fibre orientation affect more than the tool-edge radius and tool-rake angle. Shinde et al. (2022) did a review on the numerical modelling using finite element analysis of drilling of FRPCs. The influence of machining parameters, material, tool geometry and temperature on the performance output was reviewed in their study. Isbilir and Ghassemieh (2014) developed a 3D numerical model for drilling carbon FRPCs. A progressive failure model was utilized in FEA modelling for drilling, and it was concluded that the numerical results were in good agreement with the experimental values. Santiuste et al. (2014) predicted the thermal effects of orthogonal machining using FEA. It was predicted that the fibre orientation strongly influenced the mechanical and thermal interlaminar damage. Cepero-Mejías et al. (2019) investigated the influence of cutting geometry on the surface damage of orthogonal cutting of PMCs and reported that the high relief angle reduced the surface damage, while tool edge radius did not show any significant influence on the surface damage. A comparison of analytical and numerical modelling of cryogenic drilling was done by Balan et al. (2021). It was reported that experimental results are in good agreement with the values of numerical and analytical models. Negarestani et al. (2010) simulated the laser machining of carbon FRPCs using FEA. It was observed that the fibre pull-out values were over-predicted by FEA, whereas ablation depth values were under-predicted. The machining behaviour of natural FRPCs was predicted using micromechanical modelling and observed that cutting and thrust force values obtained from micromechanical modelling are in good agreement with experimental results.

5.7 CHALLENGES AND OPPORTUNITIES IN MACHINING PMCs

Due to the heterogeneous structure, machining of PMCs, specifically conventional machining processes, causes many health- and structural-related issues like matrix smearing, fibre pull-out, fibre break, delamination, reduced tool life and unhealthy dust formation. Although much work has been done on the modelling and simulation of PMC machining, some aspects still exist that need to be addressed like more studies required to analyse the effect of fibre content, fibre orientation, fibre type, polymer type and other process parameters on the temperature distribution, tool wear, cutting forces and the quality of the machined surface. In the future, the influence of different factors in machining PMCs needs to be studied by considering friction in different cases, tool deformation and the like for accurate numerical simulation. To simulate the interface cracks and other defects that occur in the processing of PMCs, advanced approaches should be developed. Furthermore, the simulation results should be validated with the experimental results. Furthermore, modelling and analysis of more complex PMC machining other than orthogonal cutting, for example, oblique cutting needs to be done in order to optimize these processes. Damage-free machining of PMCs specifically with conventional machining processes is a highly challenging task even under proper operating conditions, because of issues such as thermal sensitivity and heterogeneity. A full understanding of the material removal mechanism and chip formation mechanism is required in order to design the tool and select the optimum parameters to achieve the desired output. Advanced optimization techniques need to be explored for machining PMCs for optimum machining response. The future outlook of machining of PMCs requires substantial research work on the development of cutting tools, machine tools and inspection techniques.

5.8   CONCLUSION

PMC is an important class of materials widely used for many applications. The problems encountered in PMC machining are still challenging in terms of dimensional accuracy and surface quality achievement. This chapter gave a detailed review of the machining of PMCs. The machining of PMC is more complicated than metal machining due to the inhomogeneous and anisotropic properties of PMCs. Due to the abrasive nature of the reinforcing fibres, harder tools are desired for machining PMCs. Carbide tools, diamond tools and coated carbide tools give good results in terms of tool life and tool wear during the machining of PMCs. Delamination should be avoided in the workpiece by reducing tool wear in order to keep low machining forces. Numerical models and their analysis are more suitable to simulate complicated PMC machining and optimize the machining parameters. Different parameters like feed, cutting speed, depth of cut, tool wear, tool geometry, fibre type, fibre content and fibre-stacking sequence are the effecting factors on the delamination and quality of the machined surface. The conventional machining of PMCs involves the problems of dimensional inaccuracies, delamination and higher surface roughness. The non-conventional machining process has been shown to be a more convenient method than other conventional methods. More research work is going on the use of non-conventional processes for PMC machining to avoid the typical issues involved in high tool wear and workpiece damage due to mechanical forces. However, the optimum parameter selection for machining PMCs is required in order to attain the desired machined part quality.

REFERENCES

Abdallah, R., Soo, S. L., & Hood, R. (2018). A feasibility study on wire electrical discharge machining of carbon fiber reinforced plastic composites. *Procedia CIRP*, *77*, 195–198. https://doi.org/10.1016/j.procir.2018.08.284

Ablyaz, T. R., Muratov, K. R., Shlykov, E. S., Shipunov, G. S., & Shakirzyanov, T. V. (2019). Electric-discharge machining of polymer composites. *Russian Engineering Research, 39*(10), 898–900. https://doi.org/10.3103/S1068798X19100058

Abrão, A. M., Faria, P. E., Rubio, J. C. C., Reis, P., & Davim, J. P. (2007). Drilling of fiber reinforced plastics: A review. *Journal of Materials Processing Technology, 186*(1), 1–7. https://doi.org/10.1016/j.jmatprotec.2006.11.146

Altin Karataş, M., & Gökkaya, H. (2018). A review on machinability of carbon fiber reinforced polymer (CFRP) and glass fiber reinforced polymer (GFRP) composite materials. *Defence Technology, 14*(4), 318–326. https://doi.org/10.1016/j.dt.2018.02.001

Ashok Kumar, U., Mehtab Alam, S., & Laxminarayana, P. (2020). Influence of abrasive water jet cutting on glass fiber reinforced polymer (GFRP) composites. *Materials Today: Proceedings, 27*, 1651–1654. https://doi.org/10.1016/j.matpr.2020.03.554

Azmi, A. I., Lin, R. J. T., & Bhattacharyya, D. (2013). Machinability study of glass fiber-reinforced polymer composites during end milling. *The International Journal of Advanced Manufacturing Technology, 64*(1), 247–261. https://doi.org/10.1007/s00170-012-4006-6

Balan, A. S. S., Kannan, C., Jain, K., Chakraborty, S., Joshi, S., Rawat, K., Alsanie, W. F., & Thakur, V. K. (2021). Numerical modelling and analytical comparison of delamination during cryogenic drilling of CFRP. *Polymers, 13*(22). https://doi.org/10.3390/polym13223995

Basmaci, G., Yoruk, A. S., Koklu, U., & Morkavuk, S. (2017). Impact of Cryogenic Condition and Drill Diameter on Drilling Performance of CFRP. *Applied Sciences, 7*(7). https://doi.org/10.3390/app7070667

Çelik, Y. H., Kilickap, E., & Kilickap, A. İ. (2019). An experimental study on milling of natural fiber (jute)- reinforced polymer composites. *Journal of Composite Materials, 53*(22), 3127–3137. https://doi.org/10.1177/0021998319826373

Cepero-Mejías, F., Curiel-Sosa, J. L., Zhang, C., & Phadnis, V. A. (2019). Effect of cutter geometry on machining induced damage in orthogonal cutting of UD polymer composites: FE study. *Composite Structures, 214*, 439–450. https://doi.org/10.1016/j.compstruct.2019.02.012

Chandrasekaran, M., & Devarasiddappa, D. (2014). Artificial neural network modeling for surface roughness prediction in cylindrical grinding of Al-SiCp metal matrix composites and ANOVA analysis. *Advances in Production Engineering & Management, 9*(2), 59–70. https://doi.org/10.14743/apem2014.2.176

Chen, W.-C. (1997). Some experimental investigations in the drilling of carbon fiber-reinforced plastic (CFRP) composite laminates. *International Journal of Machine Tools and Manufacture, 37*(8), 1097–1108. https://doi.org/10.1016/S0890-6955(96)00095-8

Chockalingam, P., & Kuang, K. C. (2015). Effects of abrasive types on grinding of chopped strand mat glass fiber-reinforced polymer composite laminates. *Machining Science and Technology, 19*(2), 313–324. https://doi.org/10.1080/10910344.2015.1018534

Cococcetta, N., Jahan, M. P., Schoop, J., Ma, J., Pearl, D., & Hassan, M. (2021). Post-processing of 3D printed thermoplastic CFRP composites using cryogenic machining. *Journal of Manufacturing Processes, 68*, 332–346. https://doi.org/10.1016/j.jmapro.2021.05.054

Cong, W. L., Feng, Q., Pei, Z. J., Deines, T. W., & Treadwell, C. (2012). Rotary ultrasonic machining of carbon fiber-reinforced plastic composites: Using cutting fluid vs. cold air as coolant. *Journal of Composite Materials, 46*(14), 1745–1753. https://doi.org/10.1177/0021998311424625

Cong, W. L., Pei, Z. J., Deines, T. W., Srivastava, A., Riley, L., & Treadwell, C. (2012). Rotary ultrasonic machining of CFRP composites: A study on power consumption. *Ultrasonics, 52*(8), 1030–1037. https://doi.org/10.1016/j.ultras.2012.08.007

Cong, W. L., Zou, X., Deines, T. W., Wu, N., Wang, X., & Pei, Z. J. (2012). Rotary ultrasonic machining of carbon fiber reinforced plastic composites: An experimental study on cutting temperature. *Journal of Reinforced Plastics and Composites, 31*(22), 1516–1525. https://doi.org/10.1177/0731684412464913

Cunningham, C. R., Shokrani, A., & Dhokia, V. (2018). Edge trimming of carbon fiber reinforced plastic. *Procedia CIRP, 77*, 199–202. https://doi.org/10.1016/j.procir.2018.08.285

Davim, J. P., Reis, P., & António, C. C. (2004). A study on milling of glass fiber reinforced plastics manufactured by hand-lay up using statistical analysis (ANOVA). *Composite Structures, 64*(3), 493–500. https://doi.org/10.1016/j.compstruct.2003.09.054

Debnath, K., Singh, I., & Dvivedi, A. (2014). Drilling characteristics of Sisal fiber-reinforced epoxy and polypropylene composites. *Materials and Manufacturing Processes, 29*(11–12), 1401–1409. https://doi.org/10.1080/10426914.2014.941870

Debnath, K., Singh, I., & Dvivedi, A. (2015). Rotary mode ultrasonic drilling of glass fiber-reinforced epoxy laminates. *Journal of Composite Materials, 49*(8), 949–963. https://doi.org/10.1177/0021998314527857

Dilli Babu, G., Jagadish Babu, B., Bintu Sumanth, K., & Sivaji Babu, K. (2020). Experimental investigation on surface roughness of turned Nano-Khorasan based pineapple leaf fiber-reinforced polymer composites using response surface methodology. *Materials Today: Proceedings, 27*, 2213–2217. https://doi.org/10.1016/j.matpr.2019.09.099

Domingo, R., de Agustina, B., & Marín, M. M. (2019). An analysis of the forces in the cryogenic peripheral milling of composites reinforced with carbon fiber. *Procedia Manufacturing, 41*, 423–429. https://doi.org/10.1016/j.promfg.2019.09.028

Dutta, H., Debnath, K., & Sarma, D. K. (2021). Investigation on cutting of thin carbon fiber-reinforced polymer composite plate using sandwich electrode-assisted wire electrical-discharge machining. *Proceedings of the Institution of Mechanical Engineers, Part E: Journal of Process Mechanical Engineering, 235*(5), 1628–1638. https://doi.org/10.1177/09544089211013318

El-Sonbaty, I., Khashaba, U. A., & Machaly, T. (2004). Factors affecting the machinability of GFR/epoxy composites. *Composite Structures, 63*(3), 329–338. https://doi.org/10.1016/S0263-8223(03)00181-8

Feucht, F., Ketelaer, J., Wolff, A., Mori, M., & Fujishima, M. (2014). Latest machining technologies of hard-to-cut materials by ultrasonic machine tool. *Procedia CIRP, 14*, 148–152. https://doi.org/10.1016/j.procir.2014.03.040

Habib, S., & Okada, A. (2016). Influence of electrical discharge machining parameters on cutting parameters of carbon fiber-reinforced plastic. *Machining Science and Technology, 20*(1), 99–114. https://doi.org/10.1080/10910344.2015.1133914

Habib, S., Okada, A., & Ichii, S. (2013). Effect of cutting direction on machining of carbon fiber reinforced plastic by electrical discharge machining process. *International Journal of Machining and Machinability of Materials, 13*(4), 414–427. https://doi.org/10.1504/IJMMM.2013.054272

Hassan, A., He, Y. L., Rehman, M., Ishfaq, K., Zahoor, S., Hussain, M. Z., Siddique, F., & Wang, D. C. (2022). Machinability investigation in electric discharge machining of carbon fiber reinforced composites for aerospace applications. *Polymer Composites, 43*(11), 7773–7788. https://doi.org/10.1002/pc.26878

Hejjaji, A., Singh, D., Kubher, S., Kalyanasundaram, D., & Gururaja, S. (2016). Machining damage in FRPs: Laser versus conventional drilling. *Composites Part A: Applied Science and Manufacturing, 82*, 42–52. https://doi.org/10.1016/j.compositesa.2015.11.036

Herzog, D., Jaeschke, P., Meier, O., & Haferkamp, H. (2008). Investigations on the thermal effect caused by laser cutting with respect to static strength of CFRP. *International Journal of Machine Tools and Manufacture, 48*(12), 1464–1473. https://doi.org/10.1016/j.ijmachtools.2008.04.007

Hsissou, R., Seghiri, R., Benzekri, Z., Hilali, M., Rafik, M., & Elharfi, A. (2021). Polymer composite materials: A comprehensive review. *Composite Structures, 262*, 113640. https://doi.org/10.1016/j.compstruct.2021.113640

Hussain, S. A., Pandurangadu, V., & Kumar, K. P. (2014). Optimization of surface roughness in turning of GFRP composites using genetic algorithm. *International Journal of Engineering, Science and Technology, 6*(1), 49–57. https://doi.org/10.4314/ijest.v6i1.6

Hutyrová, Z., Zajac, J., & Fečová, V. (2014). Machining of Non-Homogeneous Composite Material with Natural Fibers Reinforcement and HDPE Matrix. *Precision Machining VII, 581*, 95–99. https://doi.org/10.4028/www.scientific.net/KEM.581.95

Hutyrová, Z., Zajac, J., Mital, D., Harničárová, M., & Valíček, J. (2015). Evaluation of share material after turning of wood plastic composite. *2015 International Conference on Solar Energy and Building (ICSoEB)*, 1–5. https://doi.org/10.1109/ICSoEB.2015.7244946

Isbilir, O., & Ghassemieh, E. (2014). Three-dimensional numerical modelling of drilling of carbon fiber-reinforced plastic composites. *Journal of Composite Materials, 48*(10), 1209–1219. https://doi.org/10.1177/0021998313484947

Islam Shyha, D. H. (2021). *Advances in Machining of Composite Materials*. In I. Shyha & D. Huo (Eds.), Springer International Publishing. https://doi.org/10.1007/978-3-030-71438-3

Jackson, M. J., & Toward, M. J. (2021). Grinding and abrasive machining of composite materials. In I. Shyha & D. Huo (Eds.), *Advances in Machining of Composite Materials: Conventional and Non-conventional Processes* (pp. 459–483). Springer International Publishing. https://doi.org/10.1007/978-3-030-71438-3_17

Jagadish, Gupta K., & Rajakumaran, M. (2018). Evaluation of machining performance of pineapple filler based reinforced polymer composites using abrasive water jet machining process. *IOP Conference Series: Materials Science and Engineering, 430*(1), 12046. https://doi.org/10.1088/1757-899X/430/1/012046

Jia, Z., Fu, R., Wang, F., Qian, B., & He, C. (2018). Temperature effects in end milling carbon fiber reinforced polymer composites. *Polymer Composites, 39*(2), 437–447. https://doi.org/10.1002/pc.23954

Joe Patrick Gnanaraj, S., Balasubramanian, V., Abdul Munaf, A., Mago Stalany, V., Chandrasekar, K., & Saravana Kumar, M. (2021). Investigation on drilling parameters of jute and cotton fiber strengthened epoxy composites. *Materials Today: Proceedings, 45*, 7107–7112. https://doi.org/10.1016/j.matpr.2021.01.946

Kaynak, Y., Lu, T., & Jawahir, I. S. (2014). Cryogenic Machining-Induced Surface Integrity: A Review and Comparison with Dry, MQL, and Flood-Cooled Machining. *Machining Science and Technology, 18*(2), 149–198. https://doi.org/10.1080/10910344.2014.897836

Kechagias, J., Petropoulos, G., Iakovakis, V., & Maropoulos, S. (2009). An investigation of surface texture parameters during turning of a reinforced polymer composite using design of experiments and analysis. *International Journal of Experimental Design and Process Optimisation, 1*(2–3), 164–177. https://doi.org/10.1504/IJEDPO.2009.030317

Khanna, N., Pusavec, F., Agrawal, C., & Krolczyk, G. M. (2020). Measurement and evaluation of hole attributes for drilling CFRP composites using an indigenously developed cryogenic machining facility. *Measurement, 154*, 107504. https://doi.org/10.1016/j.measurement.2020.107504

Khashaba, U. A. (2013). Drilling of polymer matrix composites: A review. *Journal of Composite Materials, 47*(15), 1817–1832. https://doi.org/10.1177/0021998312451609

Khoran, M., Azarhoushang, B., & Amirabadi, H. (2022). Evaluating the influence of reinforcing fiber type on the grinding process of PEEK's composites. *The International Journal of Advanced Manufacturing Technology, 119*(3), 2187–2200. https://doi.org/10.1007/s00170-021-08384-6

Kini, M. V., & Chincholkar, A. M. (2010). Effect of machining parameters on surface roughness and material removal rate in finish turning of ±30° glass fiber reinforced polymer pipes. *Materials & Design, 31*(7), 3590–3598. https://doi.org/10.1016/j.matdes.2010.01.013

Kishore, M., Amrita, M., & Kamesh, B. (2021). Experimental investigation of milling on basalt-jute hybrid composites with graphene as nanofiller. *Materials Today: Proceedings, 43*, 726–730. https://doi.org/10.1016/j.matpr.2020.12.847

Koklu, U., & Morkavuk, S. (2019). Cryogenic drilling of carbon fiber-reinforced composite (CFRP). *Surface Review and Letters, 26*(09), 1950060. https://doi.org/10.1142/S0218625X19500604

Krishnaraj, V., Zitoune, R., & Davim, J. P. (2013). Drilling of Composites. In *Drilling of Polymer-Matrix Composites* (pp. 13–35). Springer Berlin Heidelberg. https://doi.org/10.1007/978-3-642-38345-8_2

Kumar, A. L., & Prakash, M. (2021). The effect of fiber orientation on mechanical properties and machinability of GFRP composites by end milling using cutting force analysis. *Polymers and Polymer Composites*, 29(9_suppl), S178–S187. https://doi.org/10.1177/0967391121991289

Kumaran, V. U., Kliuev, M., Billeter, R., & Wegener, K. (2020). Influence of carbon based fillers on EDM machinability of CFRP. *Procedia CIRP*, 95, 437–442. https://doi.org/10.1016/j.procir.2020.03.153

Lasri, L., Nouari, M., & El Mansori, M. (2009). Modelling of chip separation in machining unidirectional FRP composites by stiffness degradation concept. *Composites Science and Technology*, 69(5), 684–692. https://doi.org/10.1016/j.compscitech.2009.01.004

Li, Y., Ren, C., Wang, H., Hu, Y., Ning, F., Wang, X., & Cong, W. (2019). Edge surface grinding of CFRP composites using rotary ultrasonic machining: comparison of two machining methods. *The International Journal of Advanced Manufacturing Technology*, 100(9), 3237–3248. https://doi.org/10.1007/s00170-018-2901-1

Madhu, S., & Balasubramanian, M. (2017). Effect of abrasive jet process parameters on machining glass fiber reinforced polymer composite. *Materialwissenschaft Und Werkstofftechnik*, 48(11), 1146–1157. https://doi.org/10.1002/mawe.201600744

Mahesha, C. R., Suprabha, R., Puthilibai, G., Devatarika, V., Jeyashri, R., & Deepashri, R. (2022). Investigation of Surface Roughness in Machine using Artificial Intelligence Techniques. *2022 International Conference on Communication, Computing and Internet of Things (IC3IoT)*, 1–4. https://doi.org/10.1109/IC3IOT53935.2022.9767952

Malik, K., Ahmad, F., Keong, W. T., & Gunister, E. (2022). The effects of drilling parameters on thrust force, temperature and hole quality of glass fiber reinforced polymer composites. *Polymers and Polymer Composites*, 30, 09673911221131113. https://doi.org/10.1177/09673911221131113

Wong, M. M., Azmi, A. I., Lee, C. C., & Mansor, A. F. (2018). Kerf taper and delamination damage minimization of FRP hybrid composites under abrasive water-jet machining. *The International Journal of Advanced Manufacturing Technology*, 94(5), 1727–1744. https://doi.org/10.1007/s00170-016-9669-y

Mohan, N. S., Ramachandra, A., & Kulkarni, S. M. (2005). Machining of Fiber-reinforced Thermoplastics: Influence of Feed and Drill Size on Thrust Force and Torque during Drilling. *Journal of Reinforced Plastics and Composites*, 24(12), 1247–1257. https://doi.org/10.1177/0731684405049865

Morkavuk, S., Köklü, U., Bağcı, M., & Gemi, L. (2018). Cryogenic machining of carbon fiber reinforced plastic (CFRP) composites and the effects of cryogenic treatment on tensile properties: A comparative study. *Composites Part B: Engineering*, 147, 1–11. https://doi.org/10.1016/j.compositesb.2018.04.024

Negarestani, R., Sundar, M., Sheikh, M. A., Mativenga, P., Li, L., Li, Z. L., Chu, P. L., Khin, C. C., Zheng, H. Y., & Lim, G. C. (2010). Numerical simulation of laser machining of carbon-fiber-reinforced composites. *Proceedings of the Institution of Mechanical Engineers, Part B: Journal of Engineering Manufacture*, 224(7), 1017–1027. https://doi.org/10.1243/09544054JEM1662

Ntziantzias, J., Kechagias, J., Pappas, M., & Vaxevanidis, N. M. (2011). An experimental study of cutting force system during turning of a glass fiber reinforced polymer composite. *Proceedings of the 4th International Conference on Manufacturing Engineering--ICMEN*, 5–7.

Ozkan, D., Sabri Gok, M., Oge, M., & Cahit Karaoglanli, A. (2019). Milling Behavior Analysis of Carbon Fiber-Reinforced Polymer (CFRP) Composites. *Materials Today: Proceedings*, 11, 526–533. https://doi.org/10.1016/j.matpr.2019.01.024

Palanikumar, K., Karunamoorthy, L., & Karthikeyan, R. (2006). Assessment of factors influencing surface roughness on the machining of glass fiber-reinforced polymer composites. *Materials & Design, 27*(10), 862–871. https://doi.org/10.1016/j.matdes.2005.03.011

Palanikumar, K., Karunamoorthy, L., Karthikeyan, R., & Latha, B. (2006). Optimization of machining parameters in turning GFRP composites using a carbide (K10) tool based on the taguchi method with fuzzy logics. *Metals and Materials International, 12*(6), 483–491. https://doi.org/10.1007/BF03027748

Petropoulos, G., Mata, F., & Davim, J. P. (2008). Statistical study of surface roughness in turning of peek composites. *Materials & Design, 29*(1), 218–223. https://doi.org/10.1016/j.matdes.2006.11.005

Rajasekaran, T., Palanikumar, K., & Arunachalam, S. (2013). Investigation on the Turning Parameters for Surface Roughness using Taguchi Analysis. *Procedia Engineering, 51*, 781–790. https://doi.org/10.1016/j.proeng.2013.01.112

Rajasekaran, T., Palanikumar, K., & Latha, B. (2022). Investigation and analysis of surface roughness in machining carbon fiber reinforced polymer composites using artificial intelligence techniques. *Carbon Letters, 32*(2), 615–627. https://doi.org/10.1007/s42823-021-00298-3

Rajasekaran, T., Palanikumar, K., & Vinayagam, B. K. (2011). Application of fuzzy logic for modeling surface roughness in turning CFRP composites using CBN tool. *Production Engineering, 5*(2), 191–199. https://doi.org/10.1007/s11740-011-0297-y

Ramesh, M., & Gopinath, A. (2017). Measurement and analysis of thrust force in drilling sisal-glass fiber reinforced polymer composites. *IOP Conference Series: Materials Science and Engineering, 197*(1), 12056. https://doi.org/10.1088/1757-899X/197/1/012056

Ramraji, K., Rajkumar, K., Dhananchezian, M., & Sabarinathan, P. (2020). Key Experimental Investigations of cutting dimensionality by Abrasive Water Jet Machining on Basalt Fiber/Fly ash Reinforced Polymer Composite. *Materials Today: Proceedings, 22*, 1351–1359. https://doi.org/10.1016/j.matpr.2020.01.428

Riveiro, A., Quintero, F., Lusquiños, F., del Val, J., Comesaña, R., Boutinguiza, M., & Pou, J. (2017). Laser cutting of Carbon Fiber Composite materials. *Procedia Manufacturing, 13*, 388–395. https://doi.org/10.1016/j.promfg.2017.09.026

Sakthivel, M., Vijayakumar, S., & Prasad, N. K. (2015). Drilling analysis on basalt/sisal reinforced polymer composites using ANOVA and regression model. *Applied Mathematical Sciences, 9*(66), 3285–3290. https://doi.org/10.12988/ams.2015.54315

Salama, A., Li, L., Mativenga, P., & Sabli, A. (2016). High-power picosecond laser drilling/machining of carbon fiber-reinforced polymer (CFRP) composites. *Applied Physics A, 122*(2), 73. https://doi.org/10.1007/s00339-016-9607-8

Salama, A., Li, L., Mativenga, P., & Whitehead, D. (2016). TEA CO2 laser machining of CFRP composite. *Applied Physics A, 122*(5), 497. https://doi.org/10.1007/s00339-016-0025-8

Santiuste, C., Díaz-Álvarez, J., Soldani, X., & Miguélez, H. (2014). Modelling thermal effects in machining of carbon fiber reinforced polymer composites. *Journal of Reinforced Plastics and Composites, 33*(8), 758–766. https://doi.org/10.1177/0731684413515956

Sheikh-Ahmad, J. Y. (2009). *Machining of polymer composites* (Vol. 387355391). Springer.

Sheikh-Ahmad, J. Y. (2016). Hole Quality and Damage in Drilling Carbon/Epoxy Composites by Electrical Discharge Machining. *Materials and Manufacturing Processes, 31*(7), 941–950. https://doi.org/10.1080/10426914.2015.1048368

Shi, H., Yuan, S., Li, Z., Song, H., & Qian, J. (2019). Evaluation of surface roughness based on sampling array for rotary ultrasonic machining of carbon fiber reinforced polymer composites. *Measurement, 138*, 175–181. https://doi.org/10.1016/j.measurement.2019.02.002

Shinde, A. S., Siva, I., Munde, Y., Hameed Sultan, M. T., Hua, L. S., & Shahar, F. S. (2022). Numerical modelling of drilling of fiber reinforced polymer matrix composite: a review. *Journal of Materials Research and Technology, 20*, 3561–3578. https://doi.org/10.1016/j.jmrt.2022.08.063

Shlykov, E. S., Ablyaz, T. R., & Oglezneva, S. A. (2020). Electrical Discharge Machining of Polymer Composites. *Russian Engineering Research*, *40*(10), 878–879. https://doi.org/10.3103/S1068798X20100275

Shokrani, A., Leafe, H., & Newman, S. T. (2019). Cryogenic drilling of carbon fiber reinforced plastic with tool consideration. *Procedia CIRP*, *85*, 55–60. https://doi.org/10.1016/j.procir.2019.10.008

Shyha, I. (2013). An Investigation into CO2 Laser Trimming of CFRP and GFRP Composites. *Procedia Engineering*, *63*, 931–937. https://doi.org/10.1016/j.proeng.2013.08.200

Shyha, I. S., Kuo, C.-L., & Soo, S. L. (2014). Workpiece surface integrity and productivity when cutting CFRP and GFRP composites using a CO_2 laser. *International Journal of Mechatronics and Manufacturing Systems*, *7*(2–3), 93–107. https://doi.org/10.1504/IJMMS.2014.064746

Singh, M., Singh, S., & Kumar, S. (2020). Investigating the impact of LASER assistance on the accuracy of micro-holes generated in carbon fiber reinforced polymer composite by electrochemical discharge machining. *Journal of Manufacturing Processes*, *60*, 586–595. https://doi.org/10.1016/j.jmapro.2020.10.056

Sivakiran, G., Gangwal, Y., Venkatachalam, G., Pandivelan, C., & Ayyappan, S. (2018). Investigations on Machining of Banana Fiber Reinforced Hybrid Polymer Matrix Composite Materials. *Materials Today: Proceedings*, *5*(2, Part 2), 7908–7914. https://doi.org/10.1016/j.matpr.2017.11.472

Slamani, M., Chafai, H., & Chatelain, J. F. (2022). Effect of milling parameters on the surface quality of a flax fiber-reinforced polymer composite. *Proceedings of the Institution of Mechanical Engineers, Part E: Journal of Process Mechanical Engineering*, 09544089221126087. https://doi.org/10.1177/09544089221126087

Somsakova, Z., Zajac, J., Michalik, P., & Kasina, M. (2012). Machining of Wood Plastic Composite (Pilot Experiment). *Materiale Plastice*, *49*, 55–57.

Sreenivasulu, R. (2013). Optimization of Surface Roughness and Delamination Damage of GFRP Composite Material in End Milling Using Taguchi Design Method and Artificial Neural Network. *Procedia Engineering*, *64*, 785–794. https://doi.org/10.1016/j.proeng.2013.09.154

Uysal, A., Altan, M., & Altan, E. (2012). Effects of cutting parameters on tool wear in drilling of polymer composite by Taguchi method. *The International Journal of Advanced Manufacturing Technology*, *58*(9), 915–921. https://doi.org/10.1007/s00170-011-3464-6

Velayudham, A., & Krishnamurthy, R. (2007). Effect of point geometry and their influence on thrust and delamination in drilling of polymeric composites. *Journal of Materials Processing Technology*, *185*(1), 204–209. https://doi.org/10.1016/j.jmatprotec.2006.03.146

Verma, R. K., Singh, V. K., Singh, D. K., & Kharwar, P. K. (2021). Experimental investigation on surface roughness and circularity error during drilling of polymer nanocomposites. *Materials Today: Proceedings*, *44*, 2501–2506. https://doi.org/10.1016/j.matpr.2020.12.597

Vigneshwaran, S., John, K. M., Johnson, R. D. J., Uthayakumar, M., Arumugaprabu, V., & Kumaran, S. T. (2021). Conventional and unconventional machining performance of natural fiber-reinforced polymer composites: A review. *Journal of Reinforced Plastics and Composites*, *40*(15–16), 553–567. https://doi.org/10.1177/0731684420958103

Wan, M., Li, S.-E., Yuan, H., & Zhang, W.-H. (2019). Cutting force modelling in machining of fiber-reinforced polymer matrix composites (PMCs): A review. *Composites Part A: Applied Science and Manufacturing*, *117*, 34–55. https://doi.org/10.1016/j.compositesa.2018.11.003

Wang, Haijin, Sun, J., Li, J., Lu, L., & Li, N. (2016a). Evaluation of cutting force and cutting temperature in milling carbon fiber-reinforced polymer composites. *The International Journal of Advanced Manufacturing Technology*, *82*(9), 1517–1525. https://doi.org/10.1007/s00170-015-7479-2

Wang, Haijin, Sun, J., Zhang, D., Guo, K., & Li, J. (2016b). The effect of cutting temperature in milling of carbon fiber reinforced polymer composites. *Composites Part A: Applied Science and Manufacturing*, *91*, 380–387. https://doi.org/10.1016/j.compositesa. 2016.10.025

Wang, Hongxiao, Zhang, X., & Duan, Y. (2018). Effects of drilling area temperature on drilling of carbon fiber reinforced polymer composites due to temperature-dependent properties. *The International Journal of Advanced Manufacturing Technology*, *96*(5), 2943–2951. https://doi.org/10.1007/s00170-018-1810-7

Wang, Hui, Cong, W., Ning, F., & Hu, Y. (2018). A study on the effects of machining variables in surface grinding of CFRP composites using rotary ultrasonic machining. *The International Journal of Advanced Manufacturing Technology*, *95*(9), 3651–3663. https://doi.org/10.1007/s00170-017-1468-6

Wang, Hui, Hu, Y., Cong, W., & Burks, A. R. (2019). Rotary ultrasonic machining of carbon fiber–reinforced plastic composites: effects of ultrasonic frequency. *The International Journal of Advanced Manufacturing Technology*, *104*(9), 3759–3772. https://doi.org/10. 1007/s00170-019-04084-4

Wang, Hui, Ning, F., Hu, Y., & Cong, W. (2018). Surface grinding of CFRP composites using rotary ultrasonic machining: a comparison of workpiece machining orientations. *The International Journal of Advanced Manufacturing Technology*, *95*(5), 2917–2930. https://doi.org/10.1007/s00170-017-1401-z

Wang, J., Feng, P., Zhang, J., Cai, W., & Shen, H. (2017). Investigations on the critical feed rate guaranteeing the effectiveness of rotary ultrasonic machining. *Ultrasonics*, *74*, 81–88. https://doi.org/10.1016/j.ultras.2016.10.003

Yazdanfar, A., & Shahrajabian, H. (2021). Experimental investigation of multi-wall carbon nanotube added epoxy resin on the EDM performance of epoxy/carbon fiber/MWCNT hybrid composites. *The International Journal of Advanced Manufacturing Technology*, *116*(5), 1801–1817. https://doi.org/10.1007/s00170-021-07593-3

Yildiz, Y., & Sundaram, M. M. (2012). Cryogenic machining of composites. In H. Hocheng (Ed.), *Machining Technology for Composite Materials* (pp. 365–393). Woodhead Publishing. https://doi.org/10.1533/9780857095145.3.365

Zajac, J., Hutyrová, Z., & Orlovský, I. (2014). Investigation of Surface Roughness after Turning of One Kind of the Bio-Material with Thermoplastic Matrix and Natural Fibers. *Materials and Processes Technologies V*, *941*, 275–279. https://doi.org/10.4028/www. scientific.net/AMR.941-944.275

Zenia, S., Ben Ayed, L., Nouari, M., & Delamézière, A. (2015). Numerical analysis of the interaction between the cutting forces, induced cutting damage, and machining parameters of CFRP composites. *The International Journal of Advanced Manufacturing Technology*, *78*(1), 465–480. https://doi.org/10.1007/s00170-014-6600-2

6 Advanced Machining Processes of Polymer Matrix Composites
An Overview

Deepak Kaushik, Binaz Varikkadinmel,
Sandeep Gairola, and Inderdeep Singh

6.1 INTRODUCTION

Composite materials have gained significant attention in the past decades for their contribution to a variety of industries such as aviation, automotive, military, and marine. These materials are composed of two distinct materials (known as matrix and reinforcement) having different physical and chemical characteristics depending on matrix and reinforcement materials, composites can be classified into several categories. In the case of polymer matrix composites (PMCs), polymers (polypropylene, polylactic acid, epoxy, etc.) are taken as matrix materials whereas fibres such as carbon, aramid, glass, and flax are typically used for reinforcement. PMCs have high strength-to-weight ratios; therefore, the application spectrum for these materials is continuously increasing to replace traditional materials. Even though PMCs are fabricated to near net shape, machining is still required to ensure the accurate shape and size of the finished components. However, as PMCs are made of two distinct materials, the machining of PMCs is a difficult task due to their anisotropic and non-homogeneous nature. Delamination, fibre pull-out, matrix fracture, or melt are some of the difficulties that are encountered when machining PMCs. The abrasive nature of fibres can also present challenges, such as tool wear, thermal damage from excessive heat generation, and melting of the polymer matrix. As a result, specialized cutting tools and a careful selection of cutting variables and machining processes are needed to address these difficulties. Cooling and a lubrication system can also be employed to reduce thermal damage and improve chip evacuation during machining.

Additionally, non-conventional or advanced machining techniques can be useful to reduce high cutting force generation, tool wear, thermal-induced damage, matrix melting, surface roughness, and dimensional inaccuracy. Furthermore, the abrasive water jet machining (AWJM) process has benefits in the form of being an environmentally benign machining process compared to traditional machining processes that use cutting fluid or produce a significant amount of waste.

DOI: 10.1201/9781032665375-6

In general, advanced machining processes for PMCs provide specific advantages in terms of accuracy, surface quality, tool wear, and the ability to manufacture complicated geometries. These processes address the issues of traditional machining techniques and offer substitute approaches for the proper and efficient machining of these sophisticated materials.

6.2 AWJM

AWJM is a hybrid machining process that combines mechanical and fluid energy to remove material from the workpiece. In this technique, a high-pressure jet of water combined with abrasive particles is directed towards the workpiece to remove material. Due to the absence of a lubricant, toxic fumes, and less wastage of material, AWJM is also termed a green manufacturing process [1]. Since AWJ machining is neither thermal, chemical, or electrical, the metallurgical and physical properties of the work material remain the same.

6.2.1 Cutting Mechanism

The cutting mechanism AWJM is complex and involves the interaction of the abrasive particles, water jet, and workpiece. Mechanisms of material removal in AWJM are considered on the scale of micro- and macro-processes. Concerning micro-processes, the abrasive particles, along with high-velocity water jet, impact the workpiece and cut it by either microfracture or micro-cutting or ploughing deformation, depending on the type of workpiece material. Brittle materials are machined based on brittle fracture whereas for ductile material ploughing deformation takes place. The impact of a sharp abrasive particle at a sufficiently narrow impact angle causes micro-cutting to occur.

By comparison, macro-processes are concerned with the kerf geometry formation due to jet behaviour and other controlling parameters.

6.2.2 Process Parameters

The quality of machined parts by AWJ is influenced by various parameters as mentioned in Figure 6.1. The quality of AWJ-machined parts is measured in terms of delamination, kerf geometry and surface roughness. In addition, depth of cut and material removal rate (MMR) are also considered as outcomes of the AWJM. MRR and depth of cut are clearly impacted by pressure, orifice diameter, and abrasive flow rate. When combined with an appropriate abrasive flow rate, increasing the water jet pressure, and decreasing the orifice diameter led to an increase in jet speed (momentum energy), which would enable enhanced MMR. Surface quality is strongly associated with mass flow rate and size of abrasive particles whereas the width of cut is a function of mixing tube diameter.

6.2.2.1 Target Material

It was also reported that the performance of AWJM (in terms of surface roughness and kerf taper) of woven glass fibre–reinforced composites was better than coconut

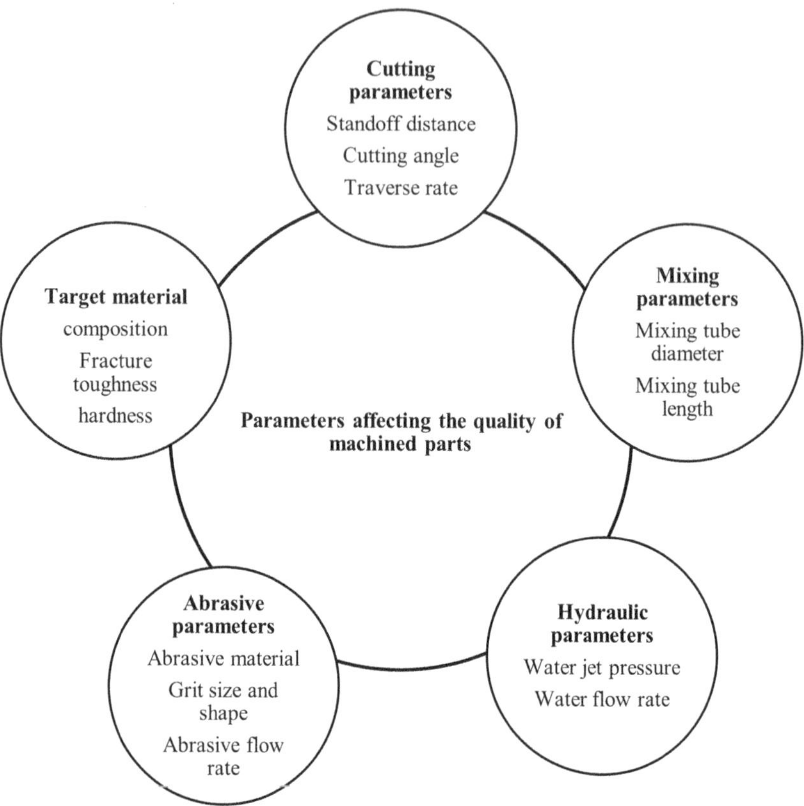

FIGURE 6.1 Parameters affecting the quality of parts machined by AWJ.

sheath composites [2]. More damage to the fibres and matrix degradation were observed for coconut-sheath composites because of poor interfacial adhesion of the fibre and the matrix. Another similar investigation was conducted to compare the AWJ drilling performance of three composites that were reinforced with carbon fibre, flax fibre and hybrid flax/carbon fibre [3]. The surface roughness of carbon fibre composites was reported to have the least, whereas the roughness of flax fibre–based composites had the most. AWJ cutting can perform a variety of machining operations, such as milling, hole cutting, hole drilling, and turning [4].

6.2.2.2 Cutting and Hydraulic Parameters

High water jet pressure, along with the low standoff distance and traverse rate, can be employed to minimize the kerf taper for carbon and glass fibre–reinforced composites [5]. In an investigation on AWJM of aramid fibre composites, water jet pressure was reported as the most effective parameter with the highest contributing rate to the response variables, such as surface roughness, delamination, and kerf angle [6]. An increase in the water jet pressure and a decrease in the standoff distance caused a decrease in kerf angle, delamination factor, and surface roughness values. However, the traverse rate was reported for the most significant parameters, with a 54.61% contribution compared to water jet pressure (30.84%), standoff distance (0.39%), and abrasive mass flow rate (1.54%), affecting the surface roughness of carbon/epoxy composites [7]. High water jet pressure combined with a low traverse rate yielded a good surface finish of AWJ-machined carbon/epoxy composites. The author also suggested optimum values of process parameters (water jet pressure – 260 MPa, abrasive mass flow rate – 800 g/min, standoff distance – 2.0 mm, and traverse rate – 50 mm/min) to obtain the minimum surface roughness.

6.2.2.3 Abrasive and Mixing Parameters

The role of abrasive in water jet drilling cannot be ignored as a higher material removal rate is obtained in the case of AWJM compared to water jet machining (WJM). Thus, a far better surface finish can be obtained using AWJM than WJM [8]. The effect of abrasive type was evaluated on the kerf taper ratio and surface roughness of AWJ-machined glass/epoxy composites [9]. Aluminium oxide abrasive performed better compared to garnet abrasive due to its better hardness. The authors also examined that increasing the water jet pressure and abrasive mass flow rate while reducing the standoff distance and traverse rate may improve the AWJM performance of glass/epoxy composites. However, machining performance was not affected by cutting orientation.

Nozzle design also plays a vital role in the performance of AWJM [10–12]. Nozzle with internal threads along with high water jet pressure, low standoff distance, and low particle size were found as optimal parameters for obtaining the better surface finish of the machined hole in carbon fibre–reinforced polymer composites [13].

Variations in cutting angle or jet impact angle also influence the performance of AWJM. The jet impact angle during AWJM represents the cutting head tilt from the original flowing direction of AWJ. Up to 25% increase in depth of cut was reported as the cutting angle increased from 50° to 80° during AWJM of polymer matrix composite [14]. However, a decrease in the depth of cut was observed beyond an 80° cutting angle. In another investigation, the author found that the surface roughness of an AWJ-machined polymer matrix composite reduced by 50% when the cutting angle was increased from 50° to 70° [15].

6.2.3 Modelling the AWJM Process

Modelling the AWJ cutting process has also been attempted by researchers to predict the optimum values for process parameters so that the performance of AWJM in terms of kerf geometry, depth of cut, and surface roughness can be enhanced. The majority of these models are based on regression analysis of experimental data [16], volume-displacement correlations [17], or other similar concepts.

6.3 LASER MACHINING

Laser beam machining (LBM) is another advanced machining technique of PMCs. LBM removes the material with the help of vaporization and the ablation phenomenon. CO_2 lasers are generally used for the LBM of fibre-reinforced composites.

High-power CO_2 lasers may provide up to 3kW power, and nonmetals can effectively absorb their radiation. However, the solid-state ND:YAG laser is operated in pulse mode, and an average power of 400W can be obtained with peak power up to 7–10 kW. Glass fibres and other organic materials may not successfully absorb ND:YAG laser radiations. Consequently, the LBM of PMCs is mostly performed with CO_2 lasers.

As no contact between tool and workpiece is not needed during LBM, therefore cutting forces, part distortion, and tool wear are not concerns. Furthermore, with LBM being a thermal process, the strength and hardness of the material do not affect its performance. As a result, it works well for machining heterogeneous materials consisting of many phases with different mechanical characteristics. Concerning the difficulties encountered during traditional machining of PMCs such as fibre pull-out, delamination, and the like, LBM eliminates such problems. In addition, the recast layer of matrix material which is formed due to laser intensity serves as a protective barrier to both the fibre and matrix. LBM also provides a high material removal rate, the capacity to cut intricate shapes with flexibility, and a thin kerf width. Nevertheless, the use of LBM for PMCs is constrained by some of its limitations, including work material strength loss caused by heat-affected zones (HAZ), decreased cutting efficiency for thicker materials, and the production of hazardous chemical decomposition products. Moreover, it has been reported that the LBM of aramid/epoxy generates significant amounts of hydrogen cyanide, which might be dangerous to human health [18]. Fumes following the LBM of graphite/epoxy, aramid/epoxy, and glass/epoxy were analysed using mass spectrometry and gas chromatography, which revealed the existence of fragmented powders of fibre materials and significant amounts of CO_2, CO, and low-molecular-weight organic compounds. Many of these issues may be significantly encountered by carefully choosing the process variables for optimum machining performance. Investigations on the LBM of PMCs suggest that to obtain the good quality of cut surface thermal conductivity of both fibre and matrix should be nearer to each other [19].

A summary of findings pertaining to the process parameter selection of LBM is provided in the next section.

6.3.1 Performance of LBM

Figure 6.2 represents the parameters of the LBM technique affecting the quality of machine parts. Performance of the LBM is generally observed in terms of the size of the HAZ, the kerf geometry (kerf taper and kerf width), and surface roughness.

6.3.1.1 HAZ

Polymer and reinforcement materials inside PMCs have different thermal and physical characteristics, thus both act in a different way when exposed to a laser beam of high intensity. During the laser machining of PMCs, the matrix material is exposed first to the laser beam. The thermoplastic matrix material in PMCs is typically cut by shearing a localized melt with a laser beam, whereas for thermoset matrix materials (epoxy), chemical degradation of material takes place when they are exposed to high temperatures compared to thermoplastic materials.

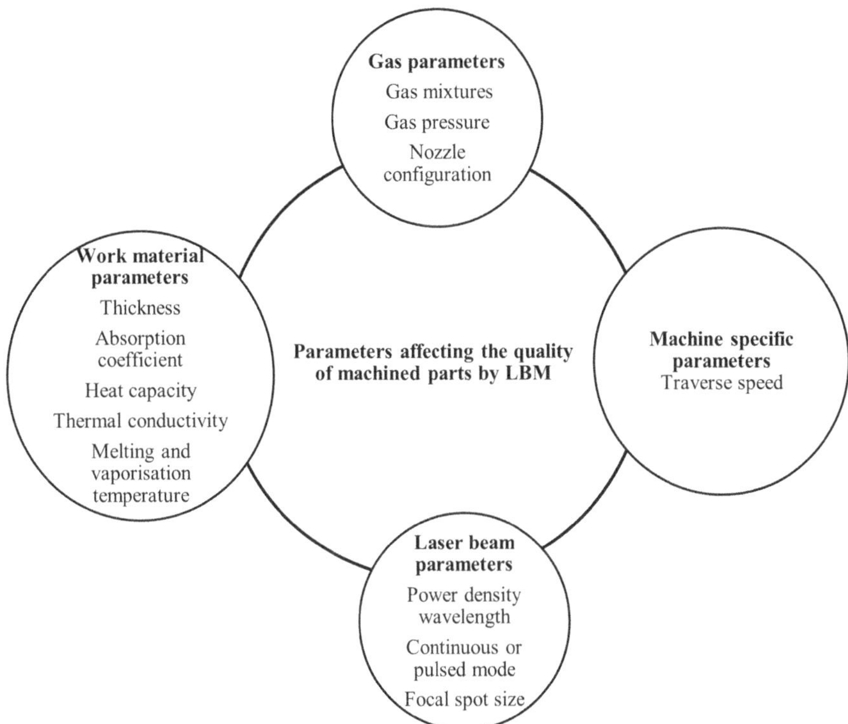

FIGURE 6.2 Parameters of LBM.

If both polymer and reinforced fibres have identical thermal characteristics, then the LBM of such PMCs will be simple. Because of this, aramid fibre–based composites are ideal candidates for LBM. Glass and carbon fibres have high vaporization temperatures from 2300°C to 3300°C. Due to the high thermal conductivity of carbon fibres, there is significant heat dissipation inside the work material, which results in a significant HAZ. Cabron fibre composites are therefore the least appropriate material for high-power CO_2 laser cutting. Yet reasonable cutting quality for carbon fibre composites can be obtained using pulsed Nd:YAG lasers. Pulsed laser beams limit the amount of time that a laser beam is exposed to a composite, hence reducing the HAZ's overall size. The anisotropic heat conduction model might be used to predict the HAZ between cuts [20].

Carbon fibre composites of a 3-mm thickness were machined using CO_2 laser in both continuous and pulse mode [21]. The authors reported a reduction in the HAZ in pulse mode as compared to continuous mode.

6.3.1.2 Kerf Geometry

Mathew et al, [22] analysed the effect of gas pressure, pulse repetition rate, pulse duration, and cutting speed on the kerf width (both top and bottom side), kerf taper, and the HAZ produced during ND:YAG laser cutting of carbon fibre composites. Pulse repetition rate and pulse energy were reported as the most dominating parameters affecting the kerf width. In a similar study, the influence of various laser machining parameters was analysed on the kerf width, surface, and kerf taper of coir/carbon fibre–based hybrid composites [23]. In order to estimate the kerf width and kerf taper based on experimental data from the laser machining of glass fibre composites and aramid fibre composites, Cenna et al. [24] devised a model based on the energy balance equation.

6.3.2 Remarks

Based on the preceding discussion, it can be deduced that optimization of laser cutting parameters is necessary to obtain the good quality of machined surface. Therefore, models of sufficient accuracy are required to predict the performance of laser machining operations. The laser cutting of large, thick composite specimens without causing any machining damage is still a challenging task. The depth of drilling and cutting of thick carbon fibre composites is also constrained by the ablation vapours' flow rate.

6.4 ELECTRIC DISCHARGE MACHINING

PMCs can be cut using unconventional machining methods, including laser, water jet, and ultrasonic; however, there are certain drawbacks, such as some failure modes that exhibit features of impact damage, intralaminar fractures, step-like delamination, and high-density micro failure zones.

Electric discharge machining (EDM) is another advanced machining process for the cutting of PMCs. Among recently created non-traditional machining processes, EDM is a very popular machining process when cutting hard-to-machine materials, such as electrically conductive ceramics, composites [25], and tool steel; the benefits of EDM machining process become particularly clear. During EDM, electric sparks are generated between an electrode (tool) and a workpiece submerged in a dielectric fluid; subsequently, material removal takes place due to the thermal eroding of the work material. The highly strong and localized heat of the spark discharge melts or vapourizes microscopic bits of the working material. Therefore hardness, strength, or toughness of the workpiece has no impact on the rate of material removal in the EDM process since it does not use mechanical energy. Other advantages of EDM include a smooth surface finish, high cutting precision, and the absence of direct contact between the workpiece and electrode (no forces), therefore its ability to process highly delicate workpieces without damaging them. EDM can be easily applied for cavities, slots, shallow-angle tiny-hole drilling, complicated die geometries, and curved surfaces. A process accuracy of 0.0025–0.127 mm and a surface finish ranging from 0.04 to 3.1 mm are simple to attain. Volumetric MRRs of 0.016–1.6 cm^3/hr can be obtained by employing appropriate dielectric fluid, power, and electrode material.

However, EDM offers very low MRRs and, like other thermal techniques, produces HAZ and a recast layer on the cut surface. However, compared to laser cutting, the HAZ produced by EDM is substantially lower [26]. Also, EDM can only be used on composite materials that have electrical conductivity, such as metal matrix composites and composites based on graphite fibres. Due to poor electrical conductivity, Kevlar- and glass fibre–reinforced polymers may not be machinable with EDM. However, since the carbon fibres in composites are electrically conductive, EDM has been used to machine carbon/polymer and carbon/carbon materials.

6.4.1 Process Parameters

The performance of EDM is significantly influenced by process parameters as shown in Figure 6.3 MRR, wear ratio, surface finish, the depth of the recast layer, and the HAZ are some of the key parameters for performance measurement of EDM process. A 120V DC power source is used for common EDM operations, with current values ranging from 0.5 to 10 A [18]. Die-sinking EDM typically uses a pulse duration of 10–100 s, whereas wire EDM often uses pulse durations of less than 2 s. A significant increase in MRR at the expense of surface quality can be obtained by increasing the current while maintaining all other parameters constant. However, a better surface finish can be achieved with the help of an increase in the spark or pulse frequency.

6.4.2 Performance

EDM performance is dependent on process parameters mentioned in the previous section. Researchers have investigated the effect of these process parameters on the performance of EDM of composites in terms of delamination, surface roughness, recast layer, wear ratio and material removal rate. The influence of process parameters investigated by various researchers is summarized in this section.

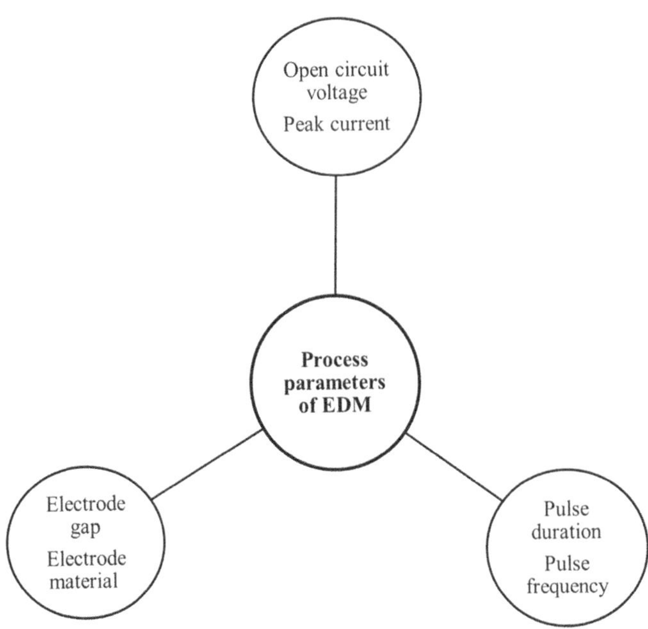

FIGURE 6.3 Process parameter of EDM.

Guu et al. [27] analysed the effect of EDM parameters on the surface roughness, thickness of recast layer and delamination on the carbon/carbon composites, using graphite electrode. The authors reported an increase in the thickness of the recast layer, delamination, and surface roughness with an increase in the power input. A better quality of drilled holes was obtained at low discharge conditions.

Hocheng et al. [28] employed a copper electrode for EDM of carbon/carbon composite materials. The author reported the effect of pulse current and pulse duration on the MRR, the recast layer, and surface roughness. It was noted that melting and vaporization were responsible for material removal. Delamination around the hole was reduced at a smaller value of pulse current (below 0.5A) and pulse on duration. Recast layer and fibre-tear defects were observed at the high values of pulse current. The diffusion of electrode material inside the coated layer was confirmed with the help of energy-dispersive spectroscopic analysis.

Habib et al. [29] reported the increase in MRR and surface roughness with an increase in the open circuit voltage, pulse on time, and peak current during of EDM of CFRP. The outcomes of the study also indicated a high MRR while EDM using a graphite electrode compared to a copper electrode.

6.5 HYBRID MACHINING PROCESSES

6.5.1 Ultrasonic Vibration–assisted Machining

In these types of machining processes ultrasonic vibrations are applied on either workpiece or tool for the machining of polymer composites. Ultrasonic vibration–assisted (UV-A) processes are also known as hybrid machining processes because in these processes ultrasonic vibrations are applied to the tool or workpiece of conventional machining processes such as turning, drilling, and grinding, among others. Ultrasonic machining differs from UV-A machining, as the latter needs abrasive material for machining purposes.

As the tool may not always be in direct contact with the workpiece during UV-A machining, the cutting mechanism is different from that of traditional machining methods. When compared to conventional machining methods, the use of ultrasonic vibrations results in a decrease in cutting force and chip size, which also improves tool life and surface smoothness of machined items [30].

Kim and Lee [31] performed UV-A turning of CFRP using three different tool materials (single-crystal diamond, poly-crystal diamond, and tungsten carbide). The authors found a better surface finish of machined surface compared to conventional machining. Additionally, in the case of UV-A turning, surface roughness was influenced by feed rate and spindle speed rather than depth of cut. Vibration amplitude is considered an additional process variable other than traditional parameters, such as feed rate, cutting speed, and the like.

Ultrasonic vibrations when drilling fibre-reinforced composites possess many advantages, such as better chip evacuation, absence of burr formation, reduced thrust force and torque, an increase in MRR, and a better surface finish to the hole. Research investigations on the UV-A and ultrasonic machining majorly focus on the ceramic matrix composites and metal matrix composites, a few investigations discuss the performance of UV-A machining on the polymer matrix composites.

Wang et al. [32] compared the performance of UV-A drilling with that of conventional drilling on the hemp/vinyl ester composites. The authors reported less drilling-induced damage, a better surface finish, and diminished fibre pull-out and burr formation in the case of UV-A drilling compared to traditional drilling.

Zhang et al. [33] investigated the influence of drill tool parameters (tool material, point angle, diameter), spindle speed and feed rate on the thrust force, and the delamination factor of the hole during ultrasonic-assisted drilling of glass/epoxy composites. The results indicated that a carbide drill with the least point angle (120°) and diameter (4 mm) was suitable for reducing the thrust force and delamination damage.

Compared to the traditional drilling process, a significant reduction (~80%) in thrust force and torque was obtained during ultrasonic drilling of carbon/epoxy composites [34], which subsequently led to better hole quality

Rotary ultrasonic drilling is another cutting-edge hybrid advanced machining technique that makes use of a tool impregnated with diamond abrasives to combine the benefits of ultrasonic machining with diamond grinding. Shard et al. [35] observed the effect of temperature rise was drilling factors, including rotating speed, feed rate, abrasive grit size, and ultrasonic power on the temperature rise during rotary ultrasonic drilling of polyetherimide composites. It was discovered that temperature rises as feed rate, rotational speed, and abrasive grit size increase, whereas the contrary effect was observed for ultrasonic power.

Wang et al. [36] performed surface grinding of CFRP using rotary ultrasonic machining, in order to study the impact of machining factors (tool rotation speed, ultrasonic power, depth of cut, and feed rate) on output variables (surface roughness, cutting force, and torque). The findings showed that employing higher levels of tool rotation, greater levels of ultrasonic power, speed, lower levels of cut depth, or lower levels of feed rate might reduce the cutting force. Reduced ultrasonic power, increased tool rotation speed, decreased feed rate, and reduced depth of cut might all lead to lower surface roughness.

6.5.2 UV-A Laser Machining

As discussed earlier, laser machining of polymer matrix composites possesses certain limitations such as large HAZ, matrix shrinkage, and fibre pullout, among others. However, by effectively cooling the processing area, thermal damage could be minimized, and machining quality could be improved. Water and ethanol can be used as a liquid cooling medium for laser processing of polymer composites [37, 38]. However cavitation bubbles are induced during water-assisted laser processing of fibre-reinforced composites, which lowers machining quality and efficiency. Zhou et al. [39] utilised ultrasonic vibrations to reduce the negative effect of cavitation bubbles during laser processing of carbon fibre composites. Ultrasonic vibrations help in exploding the bubble, and therefore, the interference of bubbles with the laser beam is reduced.

To enhance the performance of laser machining of a polymer matrix, ultrasonic vibrations may be used to create hybrid machining methods like UV-A laser machining and UV-A EDM.

Ghavidel et al. [40] utilized UV-A laser machining to fabricate the ultrasensitive strain gauge from nanocomposites (carbon nanotubes/polymethyl methacrylate). The findings demonstrated that laser cutting alone can reduce surface electrical resistance down to 3 kΩ under ideal conditions, while laser cutting with ultrasonic assistance can lower it to 2.5 kΩ.

6.5.3 UV-A EDM

EDM is typically thought of as a slow process, and it might be difficult to get accurate components. Through the generation of efficient sparking, UV-A EDM corrects the flaws of the EDM process. First, they help in breaking the dielectric fluid film around the electrode and workpiece, which is necessary for the electrical discharge to occur. This enhances the flushing of debris and improves the efficiency of material removal. Second, the ultrasonic vibrations agitate the molten material, causing it to be expelled from the machining zone more effectively. This reduces the chances of recast layer formation and improves the surface finish of the machined part. Furthermore, the ultrasonic vibrations promote the re-solidification of the molten material in a uniform manner, leading to a refined microstructure and reduced occurrence of cracks or defects. The enhanced re-solidification also improves the dimensional accuracy of the machined part. Using a graph theory approach, Kumar et al. [41] established a systematic model for the UV-A EDM process and discovered that the thickness of recast layers was less than that of the EDM process. Kumaran et al. [42] applied UV-A EDM process for deburring of CFRP and suggested optimum values of parameters to improve the deburring and minimize the tool wear rate.

It is important to note that ultrasonic-assisted EDM requires specialized equipment capable of generating and controlling the ultrasonic vibrations. This typically involves incorporating piezoelectric transducers or magneto-strictive actuators into the EDM setup. In conclusion, ultrasonic-assisted EDM is an advanced machining process that combines the benefits of both EDM and ultrasonic vibrations.

6.6 SUMMARY

The chapter equips the readers with the unique characteristics of PMCs and the associated challenges when machining PMCs. The role of advanced machining processes such as EDM, ultrasonic machining, LBM, and AWJM to tackle these challenges have been discussed in detail. The chapter provided a comprehensive understanding of the cutting mechanisms of various advanced manufacturing processes and hybrid machining processes. Also, to further enhance the quality of machined components, the concept of hybrid machining techniques such as UV-A, UV-A laser, UV-A EDM has been presented. Additionally, the chapter also highlighted the influence of process parameters of each technique on the performance parameters such as surface roughness, delamination, kerf width, heat affected zone, and others.

6.7 FUTURE SCOPE

Although researchers have addressed some of the difficulties produced during the traditional machining process, with the help of advanced and hybrid machining processes outlined earlier. However, various research investigations are still needed to improve the machining efficiency and quality of PMCs. Some potential areas of future research in the field of advanced machining of PMCs include developing new cutting tool material with improved thermal stability and wear resistance, use of soft computing techniques for cutting parameters selection, adaptive machining strategies, and others. Adaptive machining strategy involves the real-time monitoring of surface finish, cutting forces and temperature, and modifying the cutting parameters accordingly. This can assist to compensate for intrinsic material property changes and increase accuracy while machining PMCs. Researchers may also focus on developing hybrid machining techniques particularly tailored for PMCs.

REFERENCES

[1] Jagadish; Bhowmik, S.; Ray, A. Prediction of Surface Roughness Quality of Green Abrasive Water Jet Machining: A Soft Computing Approach. *J. Intell. Manuf.*, 2019, *30* (8), 2965–2979. https://doi.org/10.1007/s10845-015-1169-7

[2] Kalirasu, S.; Rajini, N.; Winowlin Jappes, J. T.; Uthayakumar, M.; Rajesh, S. Mechanical and Machining Performance of Glass and Coconut Sheath Fibre Polyester Composites Using AWJM. *J. Reinf. Plast. Compos.*, 2015, *34* (7), 564–580. https://doi.org/10.1177/0731684415574870

[3] Dhakal, H. N.; Ismail, S. O.; Ojo, S. O.; Paggi, M.; Smith, J. R. Abrasive Water Jet Drilling of Advanced Sustainable Bio-Fibre-Reinforced Polymer/Hybrid Composites: A Comprehensive Analysis of Machining-Induced Damage Responses. *Int. J. Adv. Manuf. Technol.*, 2018, *99* (9–12), 2833–2847. https://doi.org/10.1007/s00170-018-2670-x

[4] Thakur, R. K.; Singh, K. K. Abrasive Waterjet Machining of Fiber-Reinforced Composites: A State-of-the-Art Review. *J. Brazilian Soc. Mech. Sci. Eng.*, 2020, *42* (7), 1–25. https://doi.org/10.1007/s40430-020-02463-7

[5] Shanmugam, D. K.; Masood, S. H. An Investigation on Kerf Characteristics in Abrasive Waterjet Cutting of Layered Composites. *J. Mater. Process. Technol.*, 2009, *209* (8), 3887–3893. https://doi.org/10.1016/j.jmatprotec.2008.09.001

[6] Akıncıoğlu, S. Investigation of Effect of Abrasive Water Jet (AWJ) Machining Parameters on Aramid Fiber-Reinforced Polymer (AFRP) Composite Materials. *Aircr. Eng. Aerosp. Technol.*, 2021, *93* (4), 615–628. https://doi.org/10.1108/AEAT-11-2020-0249

[7] Dhanawade, A.; Kumar, S. Study on Carbon Epoxy Composite Surfaces Machined by Abrasive Water Jet Machining. *J. Compos. Mater.*, 2019, *53* (20), 2909–2924. https://doi.org/10.1177/0021998318807278

[8] Ramulu, M.; Arola, D. Water Jet and Abrasive Water Jet Cutting of Unidirectional Graphite/Epoxy Composite. *Composites*, 1993, *24* (4), 299–308. https://doi.org/10.1016/0010-4361(93)90040-F

[9] Azmir, M. A.; Ahsan, A. K. A Study of Abrasive Water Jet Machining Process on Glass/Epoxy Composite Laminate. *J. Mater. Process. Technol.*, 2009, *209* (20), 6168–6173. https://doi.org/10.1016/j.jmatprotec.2009.08.011

[10] Balasubramanian, M.; Madhu, S. Evaluation of Delamination Damage in Carbon Epoxy Composites under Swirling Abrasives Made by Modified Internal Threaded Nozzle. *J. Compos. Mater.*, 2019, *53* (6), 819–833. https://doi.org/10.1177/0021998318791340

[11] Zhang, Y.; Wang, L.; Yao, F.; Wang, H.; Men, X.; Ye, B. Experimental Study on the Influence of Nozzle Diameter on Abrasive Jet Cutting Performance. *Adv. Mater. Res.*, 2011, *337*, 466–469. https://doi.org/10.4028/www.scientific.net/AMR.337.466

[12] Madhu, S.; Balasubramanian, M. Effect of Swirling Abrasives Induced by a Novel Threaded Nozzle in Machining of CFRP Composites. *Int. J. Adv. Manuf. Technol.*, 2018, *95* (9–12), 4175–4189. https://doi.org/10.1007/s00170-017-1488-2

[13] Madhu, S.; Balasubramanian, M. Influence of Nozzle Design and Process Parameters on Surface Roughness of CFRP Machined by Abrasive Jet. *Mater. Manuf. Process.*, 2017, *32* (9), 1011–1018. https://doi.org/10.1080/10426914.2016.1257132

[14] Wang, J.; Kuriyagawa, T.; Huang, C. Z. An Experimental Study to Enhance the Cutting Performance in Abrasive Waterjet Machining. *Mach. Sci. Technol.*, 2003, *7* (2), 191–207. https://doi.org/10.1081/MST-120022777

[15] Wang, J. Abrasive Waterjet Machining of Polymer Matrix Composites - Cutting Performance, Erosive Process and Predictive Models. *Int. J. Adv. Manuf. Technol.*, 1999, *15* (10), 757–768. https://doi.org/10.1007/s001700050129

[16] Arola, D.; Ramulu, M. A Study of Kerf Characteristics in Abrasive Waterjet Machining of Graphite/Epoxy Composite. *J. Eng. Mater. Technol. Trans. ASME*, 1996, *118* (2), 256–265. https://doi.org/10.1115/1.2804897

[17] Abrasive-Waterjets, I.; Awjs, W.; Awj, T. A Model for Abrasive-Waterjet (AWJ) Machining. *J. Eng. Mater. Technol. Trans. ASME*, 1989, *111* (2), 154–162.

[18] Sheikh-Ahmad, J. Y. *Machining of Polymer Composites*; 2009. https://doi.org/10.1007/978-0-387-68619-6

[19] Tagliaferri, V.; Ilio, A. D. I.; Visconti, I. C. Laser Cutting of Fibre-Reinforced Polyesters. *Composites*, 1985, *16* (4), 317–325.

[20] Pan, C. T.; Hocheng, H. Prediction of Extent of Heat Affected Zone in Laser Grooving of Unidirectional Fiber-Reinforced Plastics. *J. Eng. Mater. Technol. Trans. ASME*, 1998, *120* (4), 321–327. https://doi.org/10.1115/1.2807021

[21] Riveiro, A.; Quintero, F.; Lusquiños, F.; Del Val, J.; Comesaña, R.; Boutinguiza, M.; Pou, J. Experimental Study on the CO_2 Laser Cutting of Carbon Fiber Reinforced Plastic Composite. *Compos. Part A Appl. Sci. Manuf.*, 2012, *43* (8), 1400–1409. https://doi.org/10.1016/j.compositesa.2012.02.012

[22] Mathew, J.; Goswami, G. L.; Ramakrishnan, N.; Naik, N. K. Parametric Studies on Pulsed Nd:YAG Laser Cutting of Carbon Fibre Reinforced Plastic Composites. *J. Mater. Process. Technol.*, 1999, *89–90*, 198–203. https://doi.org/10.1016/S0924-0136(99)00011-4

[23] Singh, Y.; Singh, J.; Sharma, S.; Aggarwal, V.; Pruncu, C. I. Multi-Objective Optimization of Kerf-Taper and Surface-Roughness Quality Characteristics for Cutting-Operation On Coir and Carbon Fibre Reinforced Epoxy Hybrid Polymeric Composites During CO_2-Pulsed Laser-Cutting Using RSM. *Lasers Manuf. Mater. Process.*, 2021, *8* (2), 157–182. https://doi.org/10.1007/s40516-021-00142-6

[24] Cenna, A. A.; Mathew, P. Analysis and Prediction of Laser Cutting Parameters of Fibre Reinforced Plastics (FRP) Composite Materials. *Int. J. Mach. Tools Manuf.*, 2002, *42* (1), 105–113. https://doi.org/10.1016/S0890-6955(01)00090-6

[25] Mohan, B.; Rajadurai, A.; Satyanarayana, K. G. Electric Discharge Machining of Al-SiC Metal Matrix Composites Using Rotary Tube Electrode. *J. Mater. Process. Technol.*, 2004, *153–154* (1–3), 978–985. https://doi.org/10.1016/j.jmatprotec.2004.04.347

[26] Lau, W. S.; Lee, W. B. A comparison between edm wire-cut and laser cutting of carbon fibre composite materials. *Materials and Manufacturing Processes*, 2007, *6* (2), 331–342. https://doi.org/10.1080/10426919108934760

[27] Guu, Y. H.; Hocheng, H.; Tai, N. H.; Liu, S. Y. Effect of Electrical Discharge Machining on the Characteristics of Carbon Fiber Reinforced Carbon Composites. *J. Mater. Sci.*, 2001, *36* (8), 2037–2043. https://doi.org/10.1023/A:1017539100832

[28] Hocheng, H.; Guu, Y. H.; Tai, N. H. The Feasibility Analysis of Electrical-Discharge Machining of Carbon-Carbon Composites. *Mater. Manuf. Process.*, 1998, *13* (1), 117–132. https://doi.org/10.1080/10426919808935223

[29] Habib, S.; Okada, A. Influence of Electrical Discharge Machining Parameters on Cutting Parameters of Carbon Fiber-Reinforced Plastic. *Mach. Sci. Technol.*, 2016, *20* (1), 99–114. https://doi.org/10.1080/10910344.2015.1133914

[30] Brehl, D. E.; Dow, T. A. Review of Vibration-Assisted Machining. *Precis. Eng.*, 2008, *32* (3), 153–172. https://doi.org/10.1016/j.precisioneng.2007.08.003

[31] Kim, J. Du; Lee, E. S. A Study of Ultrasonic Vibration Cutting of Carbon Fibre Reinforced Plastics. *Int. J. Adv. Manuf. Technol.*, 1996, *12* (2), 78–86. https://doi.org/10.1007/BF01178947

[32] Wang, D.; Onawumi, P. Y.; Ismail, S. O.; Dhakal, H. N.; Popov, I.; Silberschmidt, V. V.; Roy, A. Machinability of Natural-Fibre-Reinforced Polymer Composites: Conventional vs Ultrasonically-Assisted Machining. *Compos. Part A Appl. Sci. Manuf.*, 2019, *119* (2018), 188–195. https://doi.org/10.1016/j.compositesa.2019.01.028

[33] Zhang, Y.; Chen, T.; Li, H.; Duan, Z.; Li, H. Influence of Tool Parameters on Ultrasonic Assisted Drilling of GFRP Composites. *Adv. Compos. Mater.*, 2022, *32* (3), 303–323. https://doi.org/10.1080/09243046.2022.2085646

[34] Makhdum, F.; Phadnis, V. A.; Roy, A.; Silberschmidt, V. V. Effect of Ultrasonically-Assisted Drilling on Carbon-Fibre-Reinforced Plastics. *J. Sound Vib.*, 2014, *333* (23), 5939–5952. https://doi.org/10.1016/j.jsv.2014.05.042

[35] Shard, A.; Agarwal, R.; Garg, M. P.; Gupta, V. In-Situ Temperature Monitoring during Rotary Ultrasonic-Assisted Drilling of Fiber-Reinforced Composites. *J. Thermoplast. Compos. Mater.*, 2022, *0* (0), 1–24. https://doi.org/10.1177/08927057221142854

[36] Wang, H.; Cong, W.; Ning, F.; Hu, Y. A Study on the Effects of Machining Variables in Surface Grinding of CFRP Composites Using Rotary Ultrasonic Machining. *Int. J. Adv. Manuf. Technol.*, 2018, *95* (9–12), 3651–3663. https://doi.org/10.1007/s00170-017-1468-6

[37] Tangwarodomnukun, V.; Chen, H. Y. Laser Ablation of PMMA in Air, Water, and Ethanol Environments. *Mater. Manuf. Process.*, 2015, *30* (5), 685–691. https://doi.org/10.1080/10426914.2014.994774

[38] Cao, X. W.; Chen, Q. D.; Fan, H.; Zhang, L.; Juodkazis, S.; Sun, H. B. Liquid-Assisted Femtosecond Laser Precision-Machining of Silica. *Nanomaterials*, 2018, *8* (5), 1–8. https://doi.org/10.3390/nano8050287

[39] Zhou, L.; Jiao, H.; Qin, T.; Huang, P.; Zhang, G.; Huang, Y.; Zhou, J.; Long, Y. Study on the Mechanism of Ultrasonic-Assisted Laser Processing Carbon Fiber Reinforced Plastics in Ethanol Solution. *J. Laser Appl.*, 2022, *34* (4), 042012. https://doi.org/10.2351/7.0000811

[40] Karimzad Ghavidel, A.; Zadshakoyan, M.; Kiani, G.; Lawrence, J.; Moradi, M. Innovative Approach Using Ultrasonic-Assisted Laser Beam Machining for the Fabrication of Ultrasensitive Carbon Nanotubes-Based Strain Gauges. *Opt. Lasers Eng.*, 2023, *161* (2022), 107325. https://doi.org/10.1016/j.optlaseng.2022.107325

[41] Kumar, S.; Grover, S.; Walia, R. S. Analyzing and Modeling the Performance Index of Ultrasonic Vibration Assisted EDM Using Graph Theory and Matrix Approach. *Int. J. Interact. Des. Manuf.*, 2018, *12* (1), 225–242. https://doi.org/10.1007/s12008-016-0355-y

[42] Kumaran, S. T.; Ko, T. J.; Kurniawan, R. Grey Fuzzy Optimization of Ultrasonic-Assisted EDM Process Parameters for Deburring CFRP Composites. *Meas. J. Int. Meas. Confed.*, 2018, *123* (March), 203–212. https://doi.org/10.1016/j.measurement.2018.03.076

7 Eccentric Sleeve Grinding

Progressive-Intermittent Machining Strategy for Fibre-Reinforced Polymer Matrix Composites

V. S. Sooraj and Danish Handa

7.1 INTRODUCTION

Fibre-reinforced polymer (FRP) composites are highly popular in various engineering applications, especially in the aerospace domain, due to their superior properties, such as strength-to-weight ratio, stiffness, corrosion resistance, impact strength and others, combined with low coefficient of thermal expansion (Gilpin 2009; W. Xu and Zhang 2019). FRPs with different combinations of fibre and matrix have been employed successfully in key applications, such as launch vehicles, satellite systems, thermal protection systems, civil construction and maintenance (Krishnadas Nair and Srinivasan 2008), electromagnetic shielding, energy storage devices, structural health monitoring systems and others (Forintos and Czigany 2019).

According to the functional application, FRP components are generally fabricated using primary manufacturing sequences including pultrusion, vacuum infusion, filament winding, pre-preg preparation, compression moulding, resin transfer moulding and others. Innovative choices and methodologies of using matrix, fibres, prepregs and other additives could yield better outcomes in the aforementioned primary operations. However, precision requirements such as assembly fit, geometrical tolerances and accuracies, surface integrity and others, shall invite additional requirements of *secondary material removal/machining operations*. Various machining approaches, like drilling, milling and turning, and others, under controlled cutting environments using traditional and non-traditional methods are already reported in the literature (Caggiano 2018; Uhlmann et al. 2016; Song et al. 2022). Although these methods are attempted in various forms, anisotropy and heterogeneity of fibre-matrix systems in FRP composites make them unfriendly for fine removal of materials. Differences in the mechanical behaviour and strength of fibre and matrix cause undesired damages at their interfaces, making barriers to the surface integrity of machined FRP composites.

DOI: 10.1201/9781032665375-7

In accordance with Hui Wang, Pei and Cong (2020a), damages in relatively weak matrix regions, in the form of cracks during the interaction of cutting edges and subsequent initiation of delamination, fibre bending and breakage, intra-laminar and inter-laminar failures, fibre pull-out, fibre-matrix debonding and others, are the serious bottlenecks during FRP machining. Following research by Chen et al. (2018) and Hui Wang et al. (2018), fibre volume fraction in the matrix, orientation of fibres, selection of process variables and other factors are found to be critical while addressing the performance of FRPs after machining. Type of matrix, fibre material, cutting tool material, cutting edge rake angle and other variables are also reported to have a prevalent impact on aforementioned surface/sub-surface damages during the machining of FRP composites (Ben Soussia, Mkaddem, and El Mansori 2014; Gao et al. 2021; W. Xu and Zhang 2014).

Cutting temperature is another serious concern while machining FRPs, due to its low thermal conductivity and anisotropic material characteristics. If the cutting temperature exceeds the glass transition temperature of the polymer (generally epoxy resin) matrix, it will be softened, and the reinforced fibres will be pulled out while cutting action continues in this softened layer (Haijin Wang et al. 2016). Due to the weak bonding of fibre and matrix, undesired effects such as delamination and fibre pull-out may occur during machining (W. Xu and Zhang 2019). Such flaws are not preferred for functional FRP components, especially in aerospace applications. Although there is a conventional practice of cutting fluids to deal with temperature in metal cutting, it is not widely practised during the machining of FRP composites. The formation of agglomerates from the machining swarfs (fine dust) that may cause hindrances to the cutting action, and the possibilities of cutting fluid entrapment into the fibre–matrix system are the reasons perceived for the same. In typical aerospace/space applications, it is also observed that a vacuum-sucking system for dust collection under a dry-cutting environment is practised (Ahmad 2009; Ramulu and Abdullah 2021).

In accordance with the results reported by (Su et al. 2017), depth of cut is identified as one of the critical process variables contributing to defect generation while machining FRPs. Cutting forces and cutting edge-induced damages are proven to be more sensitive to depth of cut while machining the surfaces of a fibre–matrix system, as stated by (L. Zhang et al. 2019). Higher levels of depth of cut exhibited enhanced crack propagation and acute sub-surface damage in the fibre–matrix system. The accumulation of such flaws during machining may trigger undesired failure modes in FRP components, indulging the benefits obtained via novel procedures in composition and ingredients (W. Xu and Zhang 2016; Rao, Mahajan, and Bhatnagar 2007; Abena, Kahwash, and Essa 2021).

In this context, grinding is projected to be one of the promising options for surface machining of FRPs, due to the involvement of fine cutting edges of the order smaller than the diameter of the fibre, as reported by Hui Wang, Pei, and Cong (2020b). Contemplating the possibility of fine removal of material, various researchers have revealed encouraging demonstrations regarding the possibility of investigating grinding strategies for FRPs (W. Xu, Zhang, and Wu 2013; Hu and Zhang 2003). Although grinding has been suggested as a potential machining option for FRPs in literature (Soo et al. 2012), it is not found popular for end-use applications in industries and research and development organizations, typically for aerospace/space applications. A detailed literature review and preliminary surface grinding experiments revealed the following rationales for the same.

The removal of material while grinding FRPs appears to be influenced by an unrestrained, intermittent brittle fracture caused by heterogeneous, anisotropic fibre-matrix interfaces (Sheikh-Ahmad 2009). Conventional dry grinding of FRPs showed relatively elevated grinding force owing to the continuous interaction of multiple-abrasive cutting edges (Gao et al. 2022). As the fibres are continuously loaded against the polymer matrix, generation and propagation of defects may get triggered in the matrix zone, which is relatively the weaker element in the system. As the loading continues, the fibres will lose their backup support from the matrix, and the damages at the interface will be enhanced. As the situation of continuous loading increases, the fibres may undergo severe bending action, which will damage the matrix further. As this action continues, fibre may break or get pulled out of the damaged matrix, which will get reflected as surface/sub-surface damages that will seriously affect the surface integrity of ground FRPs.

As the cutting action is continuous, the heat accumulation at the interface of grinding also becomes high. Due to the lower thermal conductivity for the fibre–matrix system, the dissipation of this grinding heat is also difficult. Such an accumulation of force and temperature may trigger damage to the fibre–matrix system. The continuous action of abrasive grains and debris generation may invite serious wheel loading, which initiates extreme rubbing, wheel wear, higher cutting force and accumulation of heat at the grinding zone (Yuan, Gao, and Liang 2010). Eventually, all these may again lead to the generation of unsuitable flaws for instance matrix smearing, fibre bending, fibre–matrix debonding, matrix degradation and so on (Altin Karataş and Gökkaya 2018; Hu and Zhang 2003). Typical defects observed during preliminary trials of conventional surface grinding on CFRP have been presented in Figure 7.1.

One of the interesting observations related to the grinding performance of FRPs was their high sensitivity to depth of cut (Hu and Zhang 2004). It has been reported that the sub-surface damage on ground FRPs is susceptible to the depth of cut, irrespective of the fibre orientations. The defects, both force and temperature-driven damages on fibre matrix systems, were observed to be severe during continuous interaction of abrasive cutting edges, especially at elevated depths of cut (Hu and Zhang 2003, 2004).

Vibration-assisted cutting is one of the innovative hybrid techniques reported to resolve the surface integrity issues of FRP composites (W. Xu, Zhang, and Wu 2013; CAI et al. 2022). According to F. D. Ning et al. (2016), when the cutting tool is given ultrasonic vibration along and normal to the direction of cutting, the cutting force was observed to be significantly lowered. Fibre fracture in such elliptical vibration-assisted (EVA) cutting has been concentrated near the cutting edge, with minimum fibre deformation and debonding depth, providing better surface integrity. High-frequency vibration normal to cutting direction in EVA cutting will impart supplementary tensile stress to the surface of fibre, along with the dynamic vibrational effect in the cutting direction. Such an intermittent and high-frequency cutting action

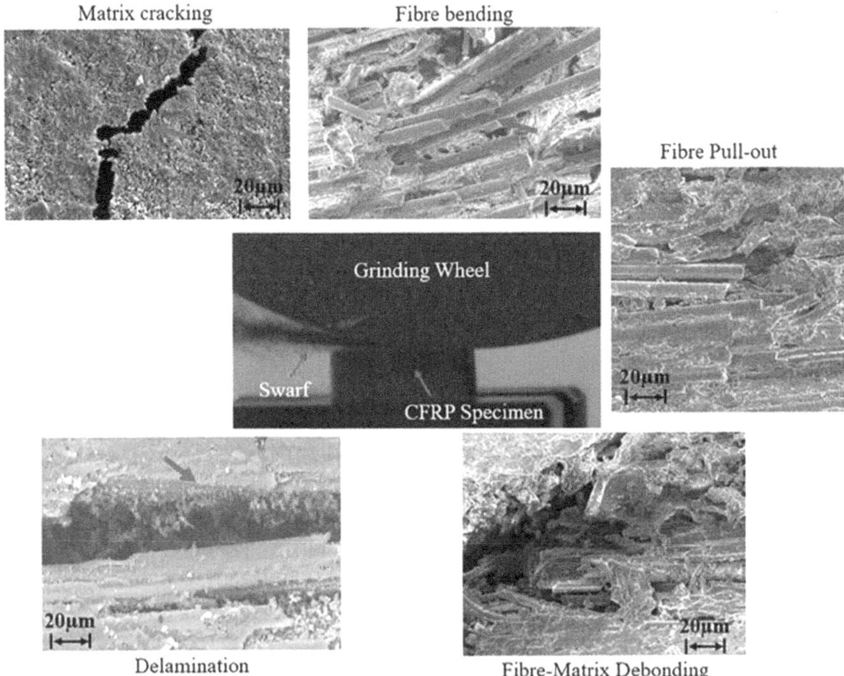

FIGURE 7.1 Undesired flaws during conventional surface grinding of CFRP samples.

is proven to be suitable for accelerating the fibre fracture with minimum fibre deflection. W. X. Xu, Zhang and Wu (2012) observed that the depth of fibre-matrix debonding is significantly lower in the vibration-assisted cutting process, contributing to a lower order of subsurface defects in the fibre matrix system. Furthermore, the vibration of cutting tool at an ultrasonic frequency, considerably greater than the natural frequency of cutting system, is proven to be effective for the increase in dynamic stiffness of cutting tool that facilitates easy fracture of fibres under smaller levels of cutting forces. Such intermittency in cutting introduced via ultrasonic vibration also could improve the dissipation of heat at the interface of grinding, resulting in the depletion of grinding temperature and force, as illustrated by Chen et al. (2018). Orbital drilling and vibration-assisted drilling are the two classical examples in this category of machining operations for FRPs. The thermomechanical advantages of intermittency in cutting reported for vibration-assisted cutting and the improvement in surface integrity achieved via such intermittent cutting action are the inspiring outcomes noticed in these research attempts. Approximately 40% reduction in cutting force and a 50% reduction in cutting temperature were observed during vibration-assisted drilling (VAD) of holes in FRPs, in comparison with traditional drilling (C. Zhang and Lu 2018). In addition, improvements in machining efficiency and delamination-free holes were also attained using VAD (Cao et al. 2022). Raster fly cutting is another example in this category, where a cutting edge (typically diamond) flies with spindle rotation and removes the material via discontinuous/intermittent cutting trajectory (S. J. Zhang et al. 2016; G. Zhang et al. 2016). This method is also reported to be flexible for the fabrication of FRP surfaces with good surface quality and reduced roughness. However, the cost associated with vibration setup and control over the trajectory of tool are some of the complexities involved in vibration-assisted cutting.

Intermittent grinding using segmented wheels is another interesting methodology reported in the literature for difficult-to-machine materials (Tawakoli and Azarhoushang 2011). Segmented wheels dedicatedly manufactured for reducing the continuous engagement of cutting edge via passive and active areas of abrasives (Tawakoli and Azarhoushang 2011) could reduce the specific cutting energy and average cutting force. As the passive areas could dissipate heat, the grinding temperature was also observed to be significantly lower in segmented grinding, typically during the machining of low thermal conducting material like titanium alloys (Handa et al. 2021; X. Xu, Yu, and Huang 2003). In this context, the major research queries are the following:

- Is it possible to impart progressive cutting and thus the control over the sensitivity of depth of cut and effects of cutting load accumulation?
- Is it possible to combine the merits of intermittent cutting along with the aforementioned progressiveness for effective thermal management at the grinding zone?
- How such a cutting strategy that incorporates the capabilities of intermittency and progressiveness can be devised to achieve minimum damage–high surface quality machining of FRP composites?
- Is it possible to achieve all the preceding features using a traditional surface grinding machine and wheels to make it more acceptable to a wide range of end users?

A possible solution for all these queries has been presented in this chapter, and the technique is called "eccentric sleeve grinding" (ESG).

7.2 ESG: CONCEPT AND GEOMETRIC CONFIGURATION

Eccentric sleeve grinding process (ESG can be adopted from a very simple and reconfigurable amendment of the traditional surface grinding system. The typical arrangement for the grinding wheel mounting system in conventional surface grinding (CSG) is illustrated in Figure 7.2. Here, the grinding wheel's geometric centre is aligned with the spindle axis of the machine to maintain a concentric rotation. As the mode shifts to ESG, the mounting of the wheel is modified using an additional eccentric sleeve (Part 3 in Figure 7.3) between the wheel and the drive flange. As the sleeve is introduced, the geometric centre of grinding wheel will establish an eccentricity between its geometric centre and spindle axis (which is the axis of rotation of wheel). Typical wheel assembly in ESG configuration is shown in Figure 7.4 (Handa and Sooraj 2019).

A non-circular profile of rotation is expected in ESG, which will establish a minimum and maximum locus analogous to the movement of a circular cam. The point

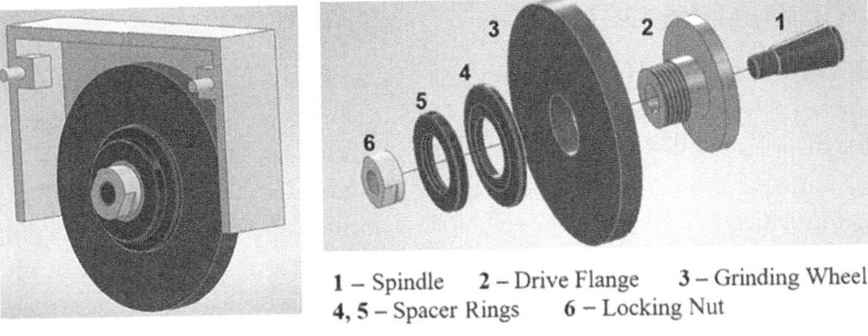

1 – Spindle   2 – Drive Flange   3 – Grinding Wheel
4, 5 – Spacer Rings   6 – Locking Nut

FIGURE 7.2 Typical grinding wheel mounting configuration for CSG.

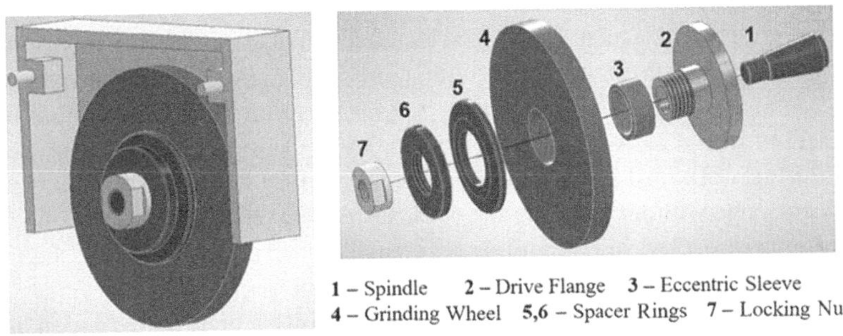

1 – Spindle   2 – Drive Flange   3 – Eccentric Sleeve
4 – Grinding Wheel   5,6 – Spacer Rings   7 – Locking Nut

FIGURE 7.3 Wheel mounting configuration for eccentric sleeve grinding.

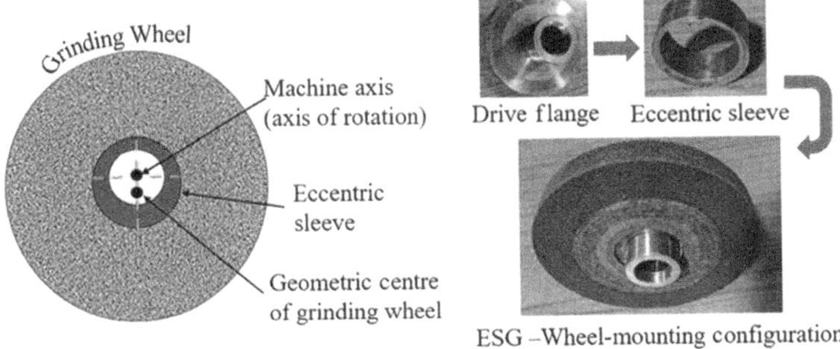

FIGURE 7.4 Grinding wheel assembly in ESG configuration.

on the wheel periphery that has the smallest diameter with reference to the eccentric centre of rotation generates the minimum locus, whereas the one with the largest diameter produces the maximum locus. As the grinding wheel is rotated over a workpiece in this eccentric mode, the abrasives near maximum locus of the wheel will establish contact with work surface. In contrast, the remaining abrasives near the minimum locus will be passive (in a dwell state). The segment establishing contact with work surface can be referred as 'cutting zone of grinding wheel in ESG', and the latter part is denoted as 'non-cutting zone'.

7.3 KINEMATICS OF ESG

During conventional surface grinding, as the work piece is fed against a grinding wheel rotating concentric to spindle axis at a depth of cut 'd', all active grains on the wheel will pose similar depth of engagement and trajectory as presented in Figure 7.5. All abrasive grains on the periphery of the wheel are expected to follow identical cutting trajectory in this non-eccentric grinding configuration. It is completely accepted that there may be micro-scale deviations due to randomness in the shape and size of abrasive grains and its protrusion height over the wheel periphery. Apart from this, the initial depth of engagement and trajectory followed by each individual abrasive grain on wheel periphery remains identical in the case of conventional surface grinding, as presented in Figure 7.5 (Handa and Sooraj 2019).

The situation becomes different when the configuration is modified to ESG. As the wheel rotates with eccentricity at a selected depth of cut and fed against a linearly moving workpiece, cutting and non-cutting zones will appear as described earlier. As a representative example, typical cutting zone (maximum locus region) in an eccentric wheel is shown as A–B–C in Figure 7.6, which will participate in the grinding action during work feed and remaining portion of wheel C–D–A will follow a minimum locus zone without touching the work surface. Now let us consider some abrasive grains G1 to G5 within the cutting segment of wheel periphery, where the contact with work piece during its linear feed happens in the sequence G1 to G5. A closer look at the cutting trajectory of G1 to G5 clearly shows a progressive action, as presented in Figure 7.6. The depth of engagement (depth of penetration) increases progressively from d_1 to d_3 for grain G1 to G3 and reduces back to d_2 and then to d_3 for grain G4 to G5. Correspondingly, the cutting trajectory generated by the rotation of wheel and the translation of work piece also varies progressively from G1 to G5 as presented in Figure 7.6 (Handa and Sooraj 2019). The penetration depth and the trajectory for individual grain depend on its radial position on the periphery of the wheel, with reference to the eccentric centre of rotation.

From the earlier illustration, it is apparent that the cutting strategy established by ESG is completely different from CSG. In CSG, all the abrasive grains will follow an identical cutting trajectory, starting its cutting motion from the same penetration

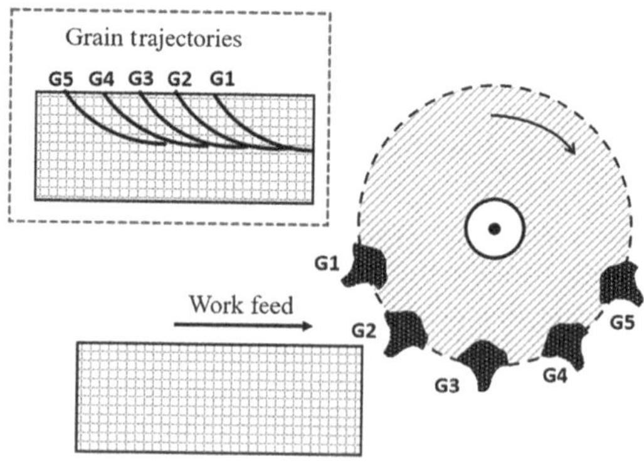

FIGURE 7.5   Abrasive grain trajectories expected during CSG.

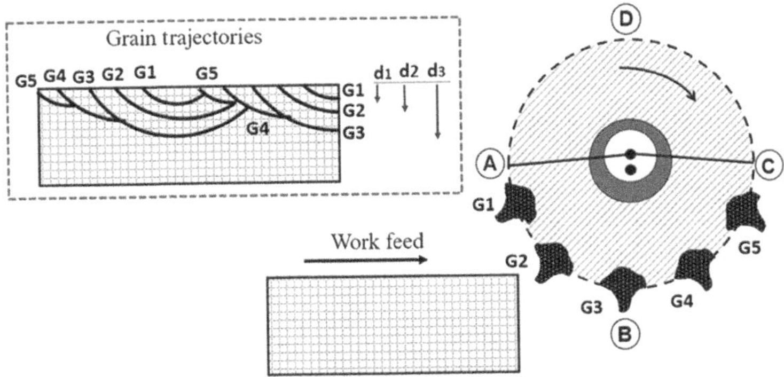

FIGURE 7.6 Abrasive grain trajectories during ESG.

depth (which is equal to the maximum value d_3), as presented in Figure 7.5. In the case of ESG, the depth of engagement will be d_3 only for the abrasive grains (Type G3 grains in location B across the width of wheel) located at maximum distance from the eccentric centre of rotation, whereas it will vary progressively for other grains (Type G1, G2, G4, and G5 grains) depending on its radial position with reference to center of rotation. Thus, the cutting strategy in ESG is fully progressive, unlike the continuous/identical cutting trajectory in CSG.

In addition to progressiveness, intermittency in cutting action is another highlight of ESG, unlike the conventional scheme. Intermittency brings a number of advantages while machining heterogeneous-anisotropic material like FRP, which is already proven via vibration-assisted cutting, segmented grinding, and other techniques. Such details and results for ESG have been illustrated in the subsequent sections of this chapter.

There may be apprehensions for a reader regarding the adverse effects of intermittent cutting scheme on the surface integrity of FRP surfaces after ESG. However, the control of the grinding parameters, typically the eccentricity level and the feed rate of work piece per revolution of grinding wheel showed very promising results in the minimization of uncut regions expected from intermittency, which is covered in Section 7.6.

7.4 MECHANICS OF ESG

While dealing with the machining of materials like FRP, the main elements under the cutting action include (a) reinforcement of fibres, (b) polymer matrix segment, and (c) fibre–matrix interface. While a hard cutting edge interacts with these heterogeneous anisotropic elements, crack/fracture may get generated in fibre, matrix, and/or its interfaces. These defects are mainly contributed by the cutting load (cutting force) that imparts localized stress/strain on the work material. Besides the cutting force magnitude, the loading rate and thermal effects induced by the cutting load may also contribute to the machining-induced damage to the FRP system. Step-by-step/progressive material removal, with non-continuous cutting load accumulation, is one of the promising options to reduce undesired defects in this context.

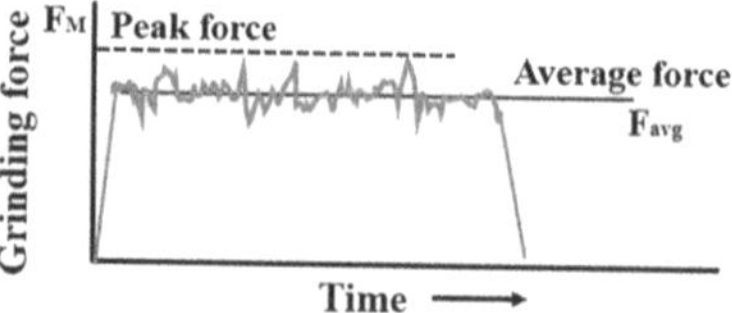

FIGURE 7.7 Typical grinding force signal expected during conventional surface grinding.

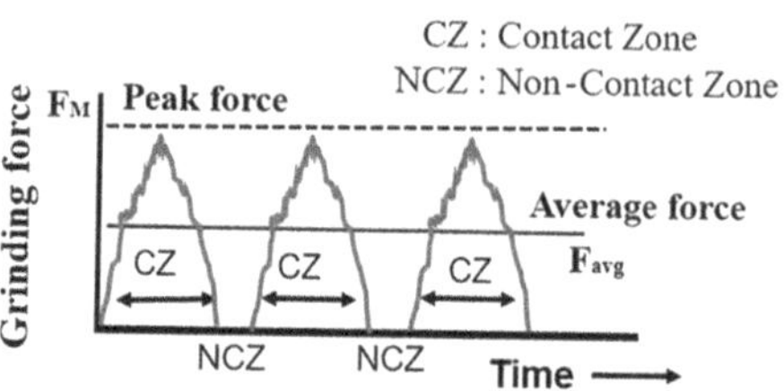

FIGURE 7.8 Typical grinding force signal expected during eccentric sleeve grinding.

Since a unique cutting scheme combining the merits of progressiveness and intermittency is followed in ESG, the primary interest was to see its impact on cutting force during the machining of FRP. Typical trends of grinding force expected during the CSG, as well as ESG, mode are compared in Figures 7.7 and 7.8 (Handa and Sooraj 2019).

In the case of CSG, the grinding force was maintained almost at the same level as expected in Figure 7.7, owing to the successive abrasive grains cutting action with identical cutting trajectory and cut thickness. Approximately constant force signal is observed during entire grinding time, except minor variations due to heterogeneity in material behaviour, randomness in abrasive shape and size, rake angle of cutting edge, and the like. Average magnitude of grinding force was near the peak value (F_M), due to the similarity in the cutting trajectory, as well as the penetration depth of successive abrasive grains.

During ESG, the force signal will be of a pulsed nature with an ON–OFF behaviour as expected in Figure 7.8. As the abrasive grains engage with FRP in a progressive manner with step-by-step increase in the level of depth, the magnitude of grinding force also increases progressively and then reduces back to zero in each rotation. The maximum force signal corresponds to the contribution of grains at the maximum locus region (Type G3 grains). As the work feed (feed/revolution of FRP) is lowered, the frequency of ON–OFF pulse was increased with a significant reduction in non-cutting zones (NCZs) for selected workpiece length and depth of cut. Compared to a conventional cutting scheme, the average grinding force was observed to be substantially lower during ESG.

7.5 MODELLING OF CUTTING MECHANICS IN ESG

A typical approach attempted for modelling grinding force and material removal in ESG is discussed in this section, which can be extended/improved further to other

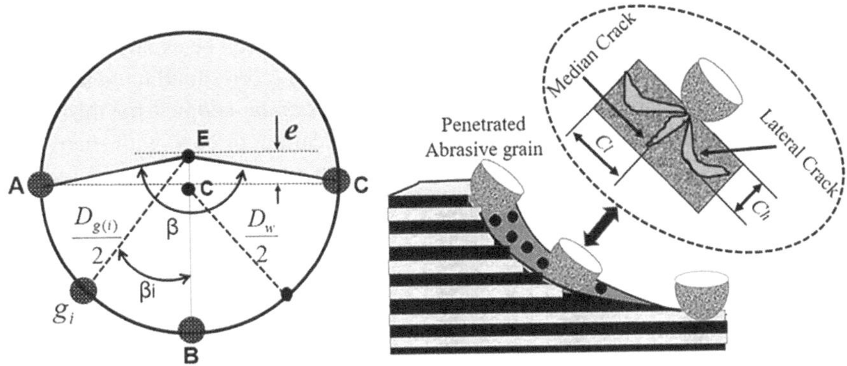

FIGURE 7.9 Kinematics and Cutting dynamics of ESG – a schematic view

methodologies. In the first phase, the idea was to incorporate the process variables, geometric configuration, grinding wheel specification and kinematic aspects such as cutting trajectory into the proposed model. For this, the cutting trajectory of an individual grain (*i*th grain g_i) in the grinding zone was modelled as shown in Figure 7.9.

When such a wheel of diameter 'D_w' rotates at 'N' rpm and interacts with a linearly moving work piece of feed rate 'F' mm/min, at a set depth of 'd' mm from the work surface, the depth of engagement d_i (the starting depth of trajectory analogous to d_1, d_2, and so on in Figure 7.6) for *i*th grain can be written as in Equation 7.1 (Handa and Sooraj 2019).

$$d_i = \left(\frac{D_{g(i)}}{2} \right) - \left[\frac{D_w}{2} + e - d \right] \tag{7.1}$$

The length of cutting path (trajectory length) covered by this grain g_i can be expressed as in Equation (7.2).

$$l_{c(i)} = \frac{F}{N} \frac{\left(\beta - \beta_i \right)}{360} + \left[\frac{D_{g(i)}}{2} * \cos^{-1} \left[\frac{\left(\frac{D_{g(i)}}{2} \right) - d_i}{\frac{D_{g(i)}}{2}} \right] \right] \tag{7.2}$$

A detailed derivation of these equations is provided in the article by Handa and Sooraj (2019).

In ESG, the starting depth of engagement, as well as the length of trajectory for each individual grain, will vary progressively based on its radial location from the eccentric centre of rotation. Because of the combined effect of work feed and wheel rotation, the width of cut and depth of penetration (uncut thickness) of each individual grain also varies from zero-tomaximum as represented in Figure 7.9 (Handa and Sooraj 2019). Depending on the shape of abrasive grain under consideration, the average area of cut (cross-sectional area) by an individual grain g_i can be evaluated. Due to the progressive nature, this area of cut also will vary progressively for distinct grains located from A to C in the cutting segment of eccentric wheel. For selected shape of an ith grain, the average cross-sectional area $A_{c(i)}$ can be evaluated nearly as half of its maximum value at the exit of grain trajectory (corresponding to maximum uncut thickness). Various possible shapes of grains can be adopted for this model, typically spherical, truncated cone with hemispherical tip, or cone with sharp tip as suggested in the tribological studies by (Marinescu et al. 2004). Once the area of cross section and length of cut are known, the material removal volume by an ith grain ($V_{g(i)}$) can be illustrated by Equation 7.3 (Handa and Sooraj 2019).

$$V_{g(i)} = A_{c(i)} * l_{c(i)} \tag{7.3}$$

In the second phase, the objective was to incorporate the material properties (Fibre-Matrix characteristics) and mode of material removal into the model. For the convenience of modelling, the FRP system was approximated as an equivalent bulk material whose properties were evaluated from fibre–matrix volume fraction as described by (F. Ning et al. 2017). In such a bulk material with brittle characteristics, the material removal volume by a single abrasive grain can be estimated from the dimensions of lateral cracks (C_l as C_h illustrated in Figure 7.9) generated during the cutting action (Li et al. 2017; F. Ning et al. 2017). From the aforementioned hypothesis, material removal volume by ith grain is as per Equation 7.4 (Handa and Sooraj 2019):

$$V_{g(i)} = 2\, C_l\, C_h\, l_{c(i)}, \tag{7.4}$$

where C_l and C_h can be expressed in terms of normal grinding force induced by ith grain ($F_{N(i)}$), the hardness of FRP composite (H_{FRP}), the stress intensity factor of FRP composite (which is the product of elastic modulus [E_{FRP}] and fracture toughness [G_{FRP}]), as presented in Equation 7.5 and Equation 7.6, respectively (Handa and Sooraj 2019).

$$C_l = k_l \left[\frac{F_{N(i)}}{\sqrt{E_{FRP}\, G_{FRP}}} \right]^{\frac{3}{4}} \tag{7.5}$$

$$C_h = k_h \left[\frac{F_{N(i)}}{H} \right]^{\frac{1}{2}} \tag{7.6}$$

Since both Equations 7.3 and 7.4 provide the same estimation of $V_{g(i)}$, they can be equated to get an approximate prediction of cutting force in ESG as a function of progressive-intermittent cutting kinematics, cutting parameters, work material properties, and grinding wheel configuration.

Total number of active abrasive grains in the grinding zone of ESG can be evaluated from the grinding wheel designation (such as grain size, structure number, etc.) and the area of cutting zone over the periphery of grinding wheel (which is a function of its diameter, width, eccentricity level, and overall depth of cut). As the grinding force varies progressively for each individual grain in ESG, an appropriate averaging procedure can be adopted to get the average grinding force per revolution of wheel.

A detailed description of the aforementioned averaging procedures and cumulative estimation of material removal volume and grinding force is illustrated by Handa and Sooraj (2019).

7.5.1 Case Study 1: Comparison of Grinding Force during Eccentric Sleeve Grinding of Unidirectional CFRP Composites with Conventional Surface Grinding

Grinding of carbon fibre–reinforced polymer composite (CFRP) samples were performed in a traditional surface grinding machine both in CSG and ESG mode, following the cutting conditions depicted in Table 7.1. (Handa and Sooraj 2019). Unidirectional CFRP (carbon fabric/epoxy resin) blocks of 20 × 20 × 10 mm were fabricated through a professional composite manufacturing entity. Grinding force signals were captured using a KISTLER make piezoelectric dynamometer. Eccentricity was maintained at 250 μm, and all the experiments were performed using freshly dressed and balanced grinding wheel. The details of experimental system have been presented in Figure 7.10.

TABLE 7.1
ESG Experimental Condition Used for Case Study 1

Grinding Wheel	**Al_2O_3 A80L5V10**
Fibre Orientation	0/90° Bidirectional
Grinding wheel Size	ϕ200 × 20 mm
Wheel rotation speed	300 RPM
CFRP sample feed rate	5 m/min
Wheel depth of cut	10100 μm
Eccentricity	250 μm for ESG and 0 μm for CSG
Coolant	Nil (Dry Grinding)

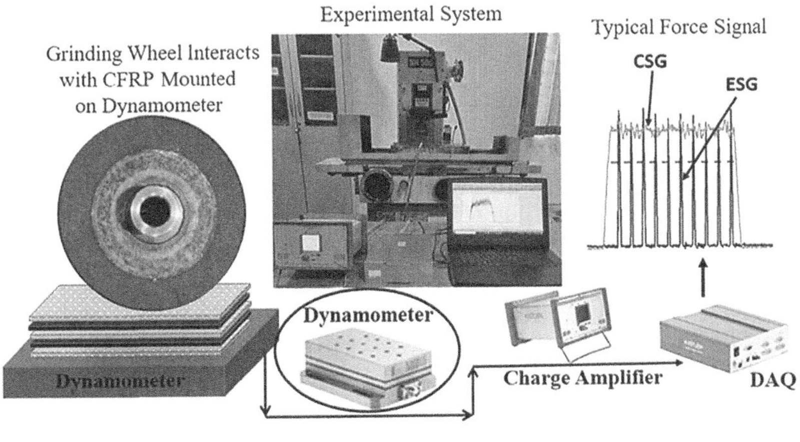

FIGURE 7.10 Experimental Configuration for case study 1.

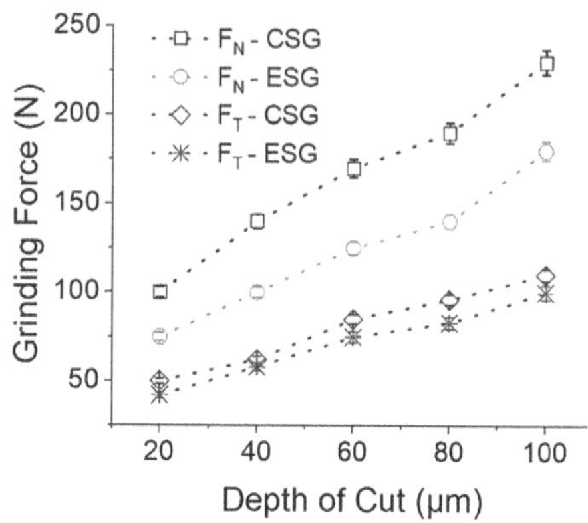

FIGURE 7.11 Grinding force versus depth of cut: Comparison of ESG and CSG.

The comparison of grinding forces (normal as well as tangential) after ESG and CSG have been plotted in Figure 7.11. Significant reduction in grinding force was visible during ESG which is attributed to the unique progressive-intermittent cutting scheme adopted to facilitate step-by-step removal of fibre–matrix system without the hassles of continuous loading of cutting edges. Further details of these experiments can be found in Handa and Sooraj's (2019) article.

7.5.2 Case Study 2: Assessment of Grinding Efficiency of ESG in Terms of Force Ratio in Bidirectional CFRP Composites

The outcome of previous case study exhibited the ability of ESG to decrease the grinding forces during the machining of unidirectional CFRPs. However, it was found interesting to explore the role of fibre orientation in such an experimental situation. As bidirectional and multidirectional CFRPs are typical choices in engineering/aerospace applications (Krishnadas Nair and Srinivasan 2008), another case study was conducted on ESG of 0/90° bidirectional composites. ESG experiments were performed using bidirectional CFRP specimens (Handa and Sooraj 2020), considering the same operating conditions identified for previous case study on unidirectional CFRPs. The performance index under comparison was the ratio of tangential grinding force to normal grinding force, which is defined as 'grinding force ratio' (Table 7.2). This force ratio (tangential to normal grinding force) is stated as an indirect indication of grinding efficiency (Rowe 2014), where higher levels of force ratio indicate a smooth cutting action without an undesired loading of abrasive grains in the normal direction. As the ratio becomes less, it is an indication of higher normal loading of abrasive grains into the work material, in comparison with its tangential cutting motion, which may lead to deeper sub-surface damages and undesired damages that hinder smooth surface generation.

At all levels of depth (between 10 and 100 μm), nearly 14% increase in force ratio was noted in ESG, compared to conventional cutting scheme. It is again validating the capability of progressive-intermittent cutting scheme offered by the eccentric rotation of wheel.

TABLE 7.2

Performance of ESG and CSG on Bidirectional CFRP – Observations of Case Study 2

Depth of cut	Average Normal Grinding Force (N)		Average Tangential Grinding Force (N)		Force Ratio (Tangential to Normal force)	
	CSG	ESG	CSG	ESG	CSG	ESG
10	60	38	32	23	0.53	0.60
50	100	55	51	32	0.51	0.58
100	180	130	90	74	0.5	0.56

Details of these experiments can be seen in the article by Handa and Sooraj (2020).

7.6 PROCESS VARIABLES IN ESG

On the basis of geometric configuration and cutting scheme, major process variables that may influence the surface generation, grinding performance, and machining defects during ESG can be listed as shown in the Fishbone diagram (Figure 7.12).

While selecting the operating condition for a given grinding wheel–FRP combination, the major parameters to be accounted for are (a) wheel rotation, (b) workpiece feed rate, (c) depth of cut set for the wheel, and (d) the level of eccentricity.

7.6.1 SELECTION OF WORK FEED PER WHEEL REVOLUTION

Since the cutting scheme in ESG is intermittent, one of the apprehensions has been the generation of uncut material scallops during the non-cutting segment of eccentric wheel, as visible in the overlaps of cutting trajectories presented in Figure 7.13. The simplest way to minimize this uncut waviness on the ground surface is to select an appropriate level of work feed for the FRP sample, per unit revolution of grinding wheel. A detailed theoretical investigation on cutting trajectory and subsequent experiments of ESG by Handa and Sooraj (2022a) proposed a critical feed per

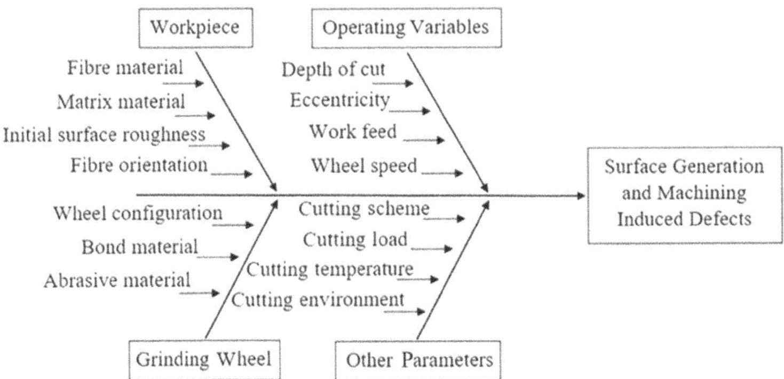

FIGURE 7.12 Major process variables of ESG – A fishbone view.

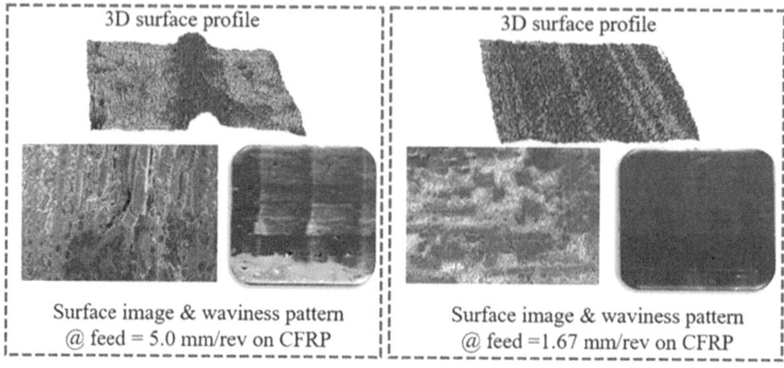

FIGURE 7.13   Surface generation on CFRP during ESG above and below the critical feed.

revolution for minimizing the undesired effect of uncut scallops that may appear due to intermittency of eccentric wheel. According to this, the selection of FRP feed (work feed) per grinding wheel revolution in ESG is preferred below the critical limit f_{cr} stated in Equation 7.7 (Handa and Sooraj 2022a).

$$f_{cr} = \left(\frac{D_w}{2} + e\right)\sin\beta, \tag{7.7}$$

where β is the included angle set by the cutting trajectory with reference to the eccentric centre of rotation, and the same can be evaluated as is Equation (7.8).

$$\beta = \cos^{-1}\left[\frac{\dfrac{D_w}{2} + e - d}{\dfrac{D_w}{2} + e}\right] \tag{7.8}$$

Furthermore, the height of uncut scallop Z_p expected to be generated on the surface by the overlaps of intermittent cutting trajectory in ESG can be expressed as in Equation 7.9 (Handa and Sooraj 2022a).

$$Z_p = \left(\frac{D_w}{2} + e\right) - \sqrt{\left[\frac{D_w}{2} + e\right]^2 - \left[\frac{[F/N]}{2}\right]^2} \tag{7.9}$$

According to a set limit of Z_p, the FRP feed per wheel revolution can be selected during ESG using Equation 7.7.

A detailed derivation of the preceding equations can be found in Handa and Sooraj's (2022a) article.

7.6.2   CASE STUDY 3: SELECTION OF WORKPIECE FEED RATE FOR MINIMUM–SCALLOP SURFACE GENERATION DURING ESG OF CFRP

ESG experiments were performed on CFRP composite specimens at various levels of feed per revolution, specifically chosen above and below the critical limit f_{cr} stated in Equation 7 (Handa and Sooraj 2022a).

For a selected grinding wheel A80L5V10 of diameter 200 mm rotating at 3000 rpm at a depth of cut of 100 µm and eccentricity of 250 µm, the critical feed was estimated to be 4.48 mm/rev as per Equation 7.7. The feed rate of CFRP samples was selected as 15 m/min and 5 m/min to accommodate the feed levels above and below the critical limit, 5 mm/ rev and 1.67 mm/rev, respectively.

Unidirectional CFRP samples for the experiments were prepared using epoxy resin, hardener, and carbon fibre prepreg, following a systematic sequence of vacuum degassing, layup, compression moulding under 5-bar pressure, and curing cycle for 24 hours. Samples having minimal defects in the cross-sectional micrographs, with initial surface waviness measured in the range of 3.5 to 4.0 µm were screened to maintain uniformity in the surface condition before grinding experiments. A three-dimensional (3D) non-contact profilometer, Zeiss make scanning electron microscope, and a skidless type TALYSURF surface tester were used for the surface inspection of samples. Appropriate dressing and balancing procedures were followed for each experiment, and the runs were replicated three times to verify the observations.

The waviness of ground CFRP surface (average value) measured using a surface tester, with a cut-off wavelength 0.8 mm, was of the order 10.11 µm along the grinding direction and 10.16 µm cross the grinding direction for the feed value of 5.0 mm/ rev. As the workpiece feed rate reduced below the critical limit to 1.67 mm/rev, the mean waviness was reduced to 7.3 µm and 7.0 µm, respectively, along and across the cutting direction.

The scanning electron micrographs and 3D surface images, as presented in Figure 7.13, were clearly indicating the minimization of scallops (uncut regions) induced by intermittency and related waviness effects on machined CFRP surfaces.

Details of experiments in this case study can be seen in the publication by (Handa and Sooraj 2022a).

7.6.3 Selection of Eccentricity

The right selection of eccentricity level is an essential requirement in ESG, as the eccentric rotation and related runout effects may not be favorable all the time in a grinding situation. To decide an appropriate level of eccentricity, a detailed experimental study was performed by Handa and Sooraj (2023), where a number of performance indices, *such as (a) grinding force variation, (b) defect estimation via X-ray computed tomography, (c) surface defect assessment via scanning electron microscopy, (d) residual stress pattern via Raman Spectra, (e) surface topography via 3D surface profiler, and others*, were compared at various levels of eccentricity. A typical case study on CFRP to decide the level of eccentricity for ESG is presented as Case Study 4.

7.6.4 Case Study 4: Selection of eccentricity level in ESG of CFRP Composites

Grinding experiments were performed on CFRP samples using traditional Al_2O_3 wheel (A80L5V10 of diameter 200 mm) on a hydraulic-surface grinding machine. Different sleeves were manufactured using a precision turn-mill center, with an eccentricity of 5 to 500 μm. Benchmark experiments were also conducted using conventional mode with zero eccentricity for comparison. Unidirectional carbon fibre–reinforced epoxy specimens, screened appropriately via scanning electron microscopy (SEM) and computed X-ray tomography, were selected for the experiments. Performance indices such as grinding force, changes in residual stress patterns, surface/sub-surface defects and surface roughness were considered for deciding the optimum electricity level of ESG (Handa and Sooraj 2023). A freshly dressed and balanced grinding wheel rotating at 3000 rpm was fed against CFRP samples that move with a linear speed below critical feed limit described in Case Study 3. In addition to surface assessment using SEM, maximum volume of defect per unit volume was also used for the screening of initial samples. Through X-ray tomography, specimens with maximum volume of defect less than 6 mm^3 (which is less than 0.15% of specimen volume) were only chosen as the initial sample for experimental benchmarks.

While analyzing the grinding force, high impulsive force patterns were visible at elevated eccentricity levels, typically for e = 300 μm and above. Peak normal force was observed to be of the order 300 N, with grinding force ratio of nearly 0.5 for eccentricity greater than 300 μm. It was highly undesirable, as the conventional (zero eccentricity) situation of a similar specimen produced only 200 N peak force with a force ratio of 0.51. As the eccentricity was chosen between 5 and 20 μm, peak normal force was of the order 198 N, with a force ratio of 0.52, which was almost similar to conventional mode. Since eccentricity induces progressiveness in cutting, the magnitude of average grinding forces in tangential and normal directions was significantly reduced in ESG.

X-ray tomography outcomes also validated the aforementioned observations on force pulses. Maximum volume of defects generated after grinding CFRP specimens at various eccentricity levels has been presented in Figure 7.14.

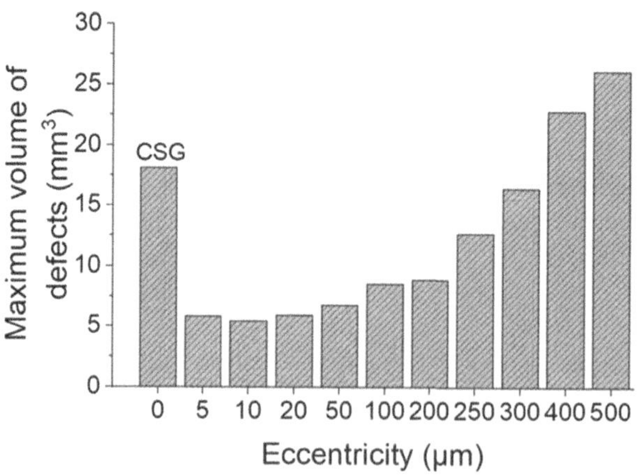

FIGURE 7.14 Maximum volume of defects generated on CFRP at various wheel eccentricities.

It is very evident from Figure 7.14 that the damages were extremely triggered by the impulsive force pulses at higher eccentricity levels (typically above e = 300), even goes worse than conventional (e = 0) situation. Whereas at lower levels of eccentricity (5–20 μm), the maximum volume of damage was considerably reduced to an acceptable level in comparison with initial and zero-eccentricity situations. The magnitudes of surface roughness observed at lower levels of eccentricity were also promising; for example, e = 10 μm generated CFRP ground surface of Sz = 5.8 μm and Sa = 260 nm, in comparison with its values in conventional non-eccentric situation (Sz = 24.23 μm and Sa = 1.35 μm). The impact of reduced average grinding force achieved via progressive-intermittent cutting scheme was clearly evident at low eccentric settings. The trends of residual stress evaluated using Raman Spectroscopy were moving towards more compressive (less tensile) side at lower eccentricities, indicating reducing possibility of median crack propagation, lateral crack expansion, and subsequent accumulation of defects.

Details of experiments presented in this case study can be seen in Handa and Sooraj's (2023) article.

7.7 THERMAL PERFORMANCE OF ESG

In CSG, as stated earlier, the depth of engagement and cutting trajectories for all the successive abrasive grains (active grains) on the wheel will be identical, except the micro-level deviations caused by grain shape and size. Thus, the parameters such as cutting arc length, grinding force, and corresponding heat flux density by successive grains will be similar in CSG. It may produce continuous heat flux accumulation in the cutting zone, and such an accumulation may result in the formation of undesirable thermal damage on resin (polymer) matrix and fibre–matrix interfaces. Conversely, when the configuration shifts to ESG, the length of cutting trajectory, chip thickness, and cutting loads induced by successive grains in the grinding zone varies progressively. Owing to this, the heat flux produced due to the cutting action of consecutive grains also will vary progressively. Another interesting fact associated with ESG is the intermittency per rotation of wheel, where accumulated heat in the cutting zone gets some room to dissipate while it reaches the non-cutting segment. This may repeat in each rotation of wheel, and thus, the accumulated heat during cutting may get transferred more effectively in comparison with CSG.

A detailed thermal investigation of ESG can be found in Handa and Sooraj's (2022b) article, where the theoretical modelling of cutting temperature in ESG is presented with a progressive heat flux distribution in the grinding zone.

7.7.1 CASE STUDY 5: THERMAL PERFORMANCE OF ESG IN UNIDIRECTIONAL CFRP COMPOSITES

Dry grinding experiments were performed using Al_2O_3 grinding wheel on 0° unidirectional CFRP specimens at an eccentricity of 10 μm, for the assessment of thermal performance during ESG (Handa and Sooraj 2022b, 2022c). A K-type thermocouple calibrated using a hot plate and pyrometer is used for the measurement of grinding temperature, with the aid of NI PXIe-1071 unit and LabVIEW software.

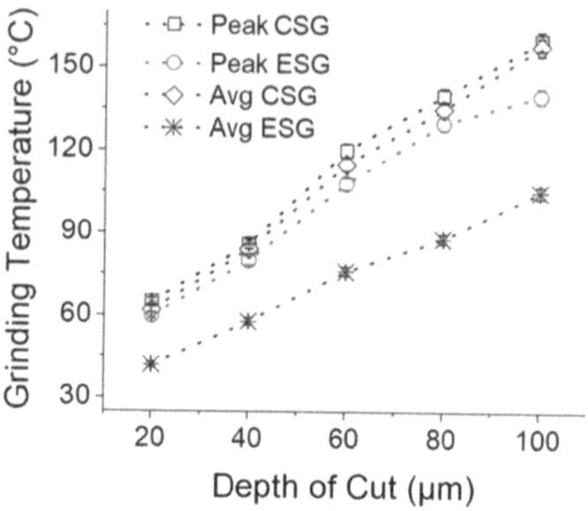

FIGURE 7.15 Grinding temperature on CFRP vs. depth of cut: Comparison of ESG and CSG.

The thermocouple was embedded inside the CFRP sample using the methodology described in prior work (Shen et al. 2008). The depth of cut was varied from 20 to 100 µm, where the grinding wheel was rotated at 3000 rpm against a work feed of 5 m/min (below the critical feed limit). The peak and average (root mean square) values of cutting temperature observed are plotted in Figure 7.15, which clearly demonstrates the effectiveness of progressive-intermittent cutting scheme in thermal management of FRP grinding. As the grinding temperature reduces, the defects associated with thermal accumulation on the polymer matrix and corresponding damages on fibre–matrix interfaces will also be controlled. Significantly lower peak and average temperature in ESG, in comparison with CSG, even with the use of cutting fluids (wet CSG), clearly demonstrate this capability. Surface images of ground CFRP were also promising while the operation was in eccentric mode.

Further details of these experiments in the case study may be found in Handa and Sooraj's (2022b, 2022c) articles.

7.8 SURFACE ASPECTS OF ESG

According to the earlier case studies and discussions, grinding force and temperature were considerably lower in ESG owing to the unique cutting strategy in progressive-intermittent mode. Furthermore, the X-ray tomography results clearly exhibited the effectiveness of such a cutting scheme to minimize the surface/subsurface defects. As a combined result of all these, the surface finish appears to be significantly improved via ESG. Typical roughness values obtained during ESG experiments on CFRP sample using Al_2O_3 wheel, at an eccentricity level of 10 µm, depth of cut of 50 to 100 µm, and a work feed of 1.67 mm/rev (below critical limit) have been presented in Table 7.3 (Handa and Sooraj 2022a, 2023).

From the results presented in Table 7.3, it is evident that the roughness of CFRP surface showed a significant reduction after progressive-intermittent cutting scheme in comparison with CSG. From the SEM images captured before and after grinding (both in ESG and CSG), presented in Figures 7.16 to 7.18, the aforementioned improvement in surface finish is clearly visible.

TABLE 7.3

Roughness of CFRP after grinding: Comparison of ESG and CSG

Roughness parameter	Depth of cut = 50 μm		Depth of cut = 100 μm	
	CSG	ESG	CSG	ESG
Ra	1.4	0.75	1.98	0.84
Rq	2.1	1.1	2.4	1.26
Rt	16	8.5	28.5	12
Rz	9	5	11.5	5.8
Sa	1.37	0.61	1.76	0.75
Sz	11.55	6.23	14.6	6.57

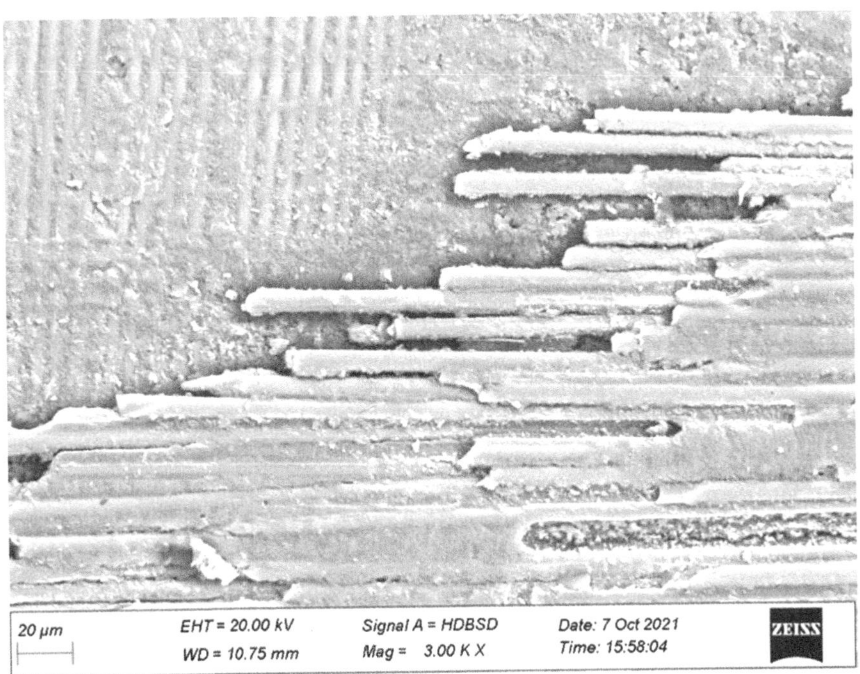

FIGURE 7.16   Surface of CFRP before grinding.

7.9   ESG USING SUPER ABRASIVES

After illustrating the capability of ESG using traditional abrasive wheels, experiments were performed using electroplated diamond wheel on unidirectional CFRP composites. The wheel speed was maintained at 3000 rpm against a work feed of 1.67 mm/rev at a depth of cut of 100 μm and eccentricity of 10 μm. The performance indices of ESG using a super-abrasive wheel are summarized in Table 7.4.

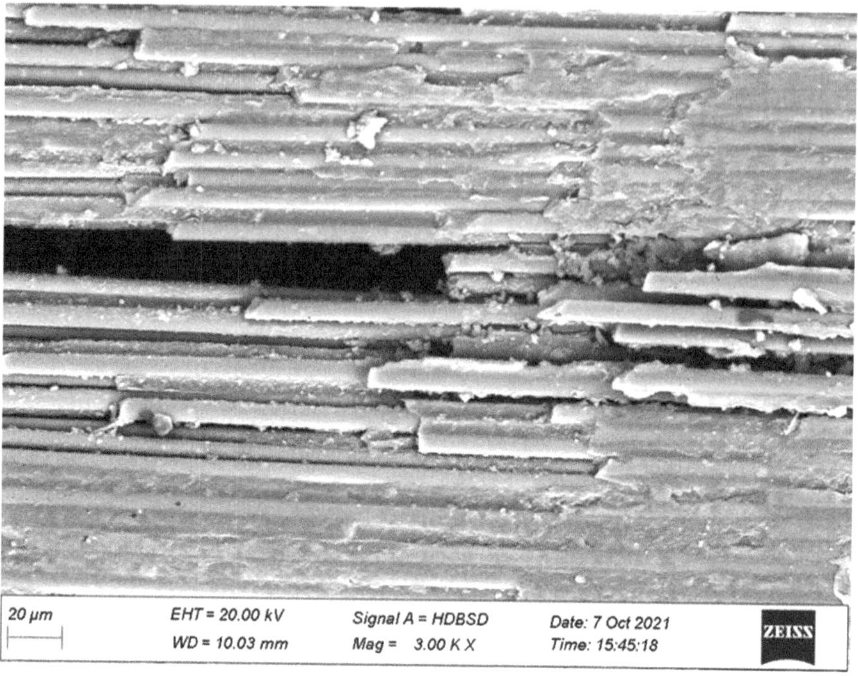

FIGURE 7.17 Surface of CFRP after CSG.

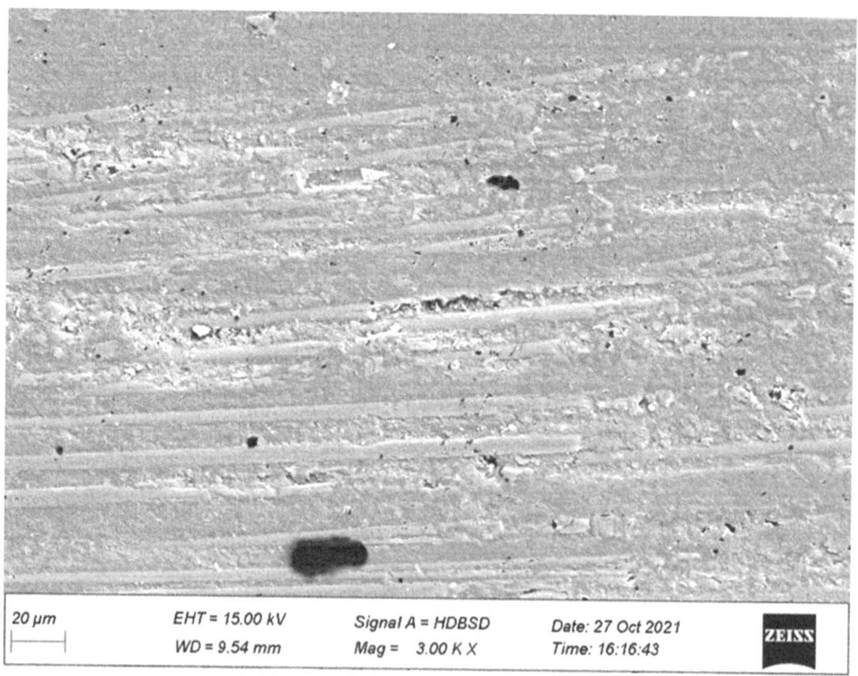

FIGURE 7.18 Surface of CFRP after ESG.

TABLE 7.4

Performance indices of ESG using super abrasive wheels during ESG and CSG.

	Grinding Force (N)		Temperature (°C)		Roughness (μm)(Cut off 0.8)	
	Normal	Tangential	Peak	Average	Ra	Rz
ESG	31.53 + 0.94	16.65 ± 0.53	70.33 ± 2.1	48.15 ± 2.1.4	0.51 ± 0.01	4.11 ± 0.13
CSG	42.63 ± 1.3	23.36 ± 0.71	75.51 ± 2.2	71.24 ± 2.1	0.69 ± 0.02	7.79 ± 0.23

7.10 SUMMARY

This chapter covered various aspects of a new technique, referred as "Eccentric Sleeve Grinding ESG, for surface machining of fibre-reinforced polymer composites. ESG adopts a progressive-intermittent cutting scheme for grinding heterogeneous-anisotropic fibre–matrix system via controlled eccentricity in grinding wheel rotation. ESG is easy to adapt in a traditional surface grinding machine by introducing a precisely machined eccentric sleeve into the wheel mounting system.

ESG exhibited superior performance while machining fibre-reinforced polymer composites such as CFRPs, through a significant reduction of grinding force, grinding temperature, and surface defects, with substantial improvement in surface finish. The progressive (step-by-step) cutting trajectory in ESG could relax the continuous accumulation of cutting load, thereby reducing the associated damage on FRPs. The progressive cutting scheme, along with the intermittency, is also proven to be effective in the reduction of thermal damage in fibre–matrix systems. ESG could demonstrate better surface integrity on machined CFRPs, with reduced defects and roughness, in comparison with CSG, for the selected experimental cases.

7.11 FUTURE PROSPECTIVES

The content of this chapter, incorporated with specific case studies, could demonstrate the possibility of adopting a progressive-intermittent cutting scheme for minimal damage machining of anisotropic-heterogeneous fibre–matrix systems in FRPs. Although a typical methodology referred as 'eccentric sleeve grinding (ESG) and its fundamental aspects have been covered in this chapter, the thoughts can be extended further to the following aspects in future:

- Long-term dynamic effects of eccentric wheel rotation on a machine spindle
- Dynamic balancing aspects of various wheel configurations in eccentric mode
- Investigations on the grinding wheel wear and related tribological aspects of ESG
- Comparison of ESG with vibration-assisted grinding and segmented grinding

- Case studies on ESG for various combinations of fibres, matrices, and fibre orientations
- Cost-effective modifications and hybridization of ESG for optimum machining performance on FRPs

REFERENCES

Abena, Alessandro, Fadi Kahwash, and Khamis Essa. 2021. "Modelling Machining of FRP Composites BT - Advances in Machining of Composite Materials: Conventional and Non-Conventional Processes." In, edited by Islam Shyha and Dehong Huo, 101–27. Cham: Springer International Publishing. https://doi.org/10.1007/978-3-030-71438-3_5

Ahmad, Jamal. 2009. "Health and Safety Aspects in Machining FRPs BT - Machining of Polymer Composites." In, edited by Jamal Ahmad, 293–307. Boston, MA: Springer US. https://doi.org/10.1007/978-0-387-68619-6_7

Altin Karataş, Meltem, and Hasan Gökkaya. 2018. "A Review on Machinability of Carbon Fiber Reinforced Polymer (CFRP) and Glass Fiber Reinforced Polymer (GFRP) Composite Materials." *Defence Technology* 14 (4): 318–26. https://doi.org/10.1016/j.dt.2018.02.001

Caggiano, Alessandra. 2018. "Machining of Fibre Reinforced Plastic Composite Materials." *Materials (Basel, Switzerland)* 11 (3): 442. https://doi.org/10.3390/ma11030442

Cai, Chongyan, Dang, Jiaqiang, An, Qinglong, Ming, Weiwei, and Chen, Ming. 2022. "Surface Morphology Characterization of Unidirectional Carbon Fibre Reinforced Plastic Machined by Peripheral Milling." *Chinese Journal of Aeronautics* 35 (2): 361–75. https://doi.org/10.1016/j.cja.2021.04.024

Cao, Shiyu, Hao Nan Li, Wenjian Huang, Qi Zhou, Ting Lei, and Chaoqun Wu. 2022. "A Delamination Prediction Model in Ultrasonic Vibration Assisted Drilling of CFRP Composites." *Journal of Materials Processing Technology* 302: 117480. https://doi.org/10.1016/j.jmatprotec.2021.117480

Chen, Yan, Yuhong Liang, Jiuhua Xu, and Andong Hu. 2018. "Ultrasonic Vibration Assisted Grinding of CFRP Composites: Effect of Fiber Orientation and Vibration Velocity on Grinding Forces and Surface Quality." *International Journal of Lightweight Materials and Manufacture* 1 (3): 189–96. https://doi.org/10.1016/j.ijlmm.2018.08.003

Forintos, N., and T. Czigany. 2019. "Multifunctional Application of Carbon Fiber Reinforced Polymer Composites: Electrical Properties of the Reinforcing Carbon Fibers – A Short Review." *Composites Part B: Engineering* 162: 331–43. https://doi.org/10.1016/j.compositesb.2018.10.098

Gao, Teng, Changhe Li, Min Yang, Yanbin Zhang, Dongzhou Jia, Wenfeng Ding, Sujan Debnath, Tianbiao Yu, Zafar Said, and Jun Wang. 2021. "Mechanics Analysis and Predictive Force Models for the Single-Diamond Grain Grinding of Carbon Fiber Reinforced Polymers Using CNT Nano-Lubricant." *Journal of Materials Processing Technology* 290: 116976. https://doi.org/10.1016/j.jmatprotec.2020.116976

Gao, Teng, Yanbin Zhang, Changhe Li, Yiqi Wang, Yun Chen, Qinglong An, Song Zhang, et al. 2022. "Fiber-Reinforced Composites in Milling and Grinding: Machining Bottlenecks and Advanced Strategies." *Frontiers of Mechanical Engineering* 17 (2): 24. https://doi.org/10.1007/s11465-022-0680-8

Gilpin, Andrew. 2009. "Tool Solutions for Machining Composites." *Reinforced Plastics* 53 (6): 30–33. https://doi.org/10.1016/S0034-3617(09)70260-7

Handa, Danish, Shankar Kumar, Sarath Babu Thekkoot Surendran, and V. S. Sooraj. 2021. "Simulation of Intermittent Grinding for Ti-6Al-4V with Segmented Wheel." *Materials Today: Proceedings*. https://doi.org/10.1016/j.matpr.2020.12.626

Handa, Danish, and V. S. Sooraj. 2022a. "Generation of Scallop Free Machined Surfaces in CFRPs with Minimum Waviness and Defects Using Eccentric Sleeve Grinding." *Journal of Materials Processing Technology* 301 (March): 117431. https://doi.org/10.1016/J.JMATPROTEC.2021.117431

Handa, Danish, and V. S. Sooraj. 2022b. "Eccentric Sleeve Grinding for Thermal Management and Dry Grinding of Carbon Fibre Reinforced Composites." *Journal of Reinforced Plastics and Composites*, September, 07316844221122073. https://doi.org/10.1177/07316844221122073

Handa, Danish, and V. S. Sooraj. 2022c. "Thermal Investigations on Eccentric Sleeve Grinding of Fibre Reinforced Composites." *Journal of Manufacturing Processes* 84 (December): 1404–27. https://doi.org/10.1016/J.JMAPRO.2022.11.002

Handa, Danish, and V. S. Sooraj. 2023. "Some Constructive Observations on the Effect of Wheel Eccentricity during Grinding of CFRP." *Journal of Manufacturing Processes* 97: 62–75. https://doi.org/10.1016/j.jmapro.2023.04.066

Handa, Danish, and V.S. Sooraj. 2019. "An Eccentric Sleeve Grinding Strategy for Fibre-Reinforced Composites." *Composites Part B: Engineering* 176 (November): 107332. https://doi.org/10.1016/J.COMPOSITESB.2019.107332

Handa, Danish, and V.S. Sooraj. 2020. "Performance Assessment of a Hybrid Intermittent-Progressive Grinding Strategy for Bi-Directional Carbon Fibre Reinforced Composites." *Materials Today: Proceedings*, February. https://doi.org/10.1016/j.matpr.2019.12.314

Hu, N.S., and L.C. Zhang. 2003. "A Study on the Grindability of Multidirectional Carbon Fibre-Reinforced Plastics." *Journal of Materials Processing Technology* 140 (1–3): 152–56. https://doi.org/10.1016/S0924-0136(03)00704-0

Hu, N.S., and L.C. Zhang. 2004. "Some Observations in Grinding Unidirectional Carbon Fibre-Reinforced Plastics." *Journal of Materials Processing Technology* 152 (3): 333–38. https://doi.org/10.1016/J.JMATPROTEC.2004.04.374

Krishnadas Nair, C G, and R Srinivasan. 2008. *Materials and Fabrication Technology for Satellite and Launch Vehicle*. Bangalore: Harishree Publishing Company.

Li, Chen, Feihu Zhang, Binbin Meng, Lifei Liu, and Xiaoshuang Rao. 2017. "Material Removal Mechanism and Grinding Force Modelling of Ultrasonic Vibration Assisted Grinding for SiC Ceramics." *Ceramics International* 43 (3): 2981–93. https://doi.org/10.1016/j.ceramint.2016.11.066

Marinescu, Ioan D., W. Brian Rowe, Boris Dimitrov, and Ichiro Inasaki. 2004. *Tribology of Abrasive Machining Processes. Tribology of Abrasive Machining Processes*. Elsevier Inc. https://doi.org/10.1115/1.1819313

Ning, F D, W L Cong, Z J Pei, and C Treadwell. 2016. "Rotary Ultrasonic Machining of CFRP: A Comparison with Grinding." *Ultrasonics* 66: 125–32. https://doi.org/10.1016/j.ultras.2015.11.002

Ning, Fuda, Weilong Cong, Hui Wang, Yingbin Hu, Zhonglue Hu, and Zhijian Pei. 2017. "Surface Grinding of CFRP Composites with Rotary Ultrasonic Machining: A Mechanistic Model on Cutting Force in the Feed Direction." *International Journal of Advanced Manufacturing Technology* 92 (1–4): 1217–29. https://doi.org/10.1007/s00170-017-0149-9

Ramulu, Mamidala, and Mohammad Sayem Bin Abdullah. 2021. "Health and Safety Considerations in Machining of Composites BT - Advances in Machining of Composite Materials: Conventional and Non-Conventional Processes." In, edited by Islam Shyha and Dehong Huo, 517–25. Cham: Springer International Publishing. https://doi.org/10.1007/978-3-030-71438-3_19

Rao, G. Venu Gopala, P. Mahajan, and N. Bhatnagar. 2007. "Micro-Mechanical Modeling of Machining of FRP Composites – Cutting Force Analysis." *Composites Science and Technology* 67 (3): 579–93. https://doi.org/10.1016/j.compscitech.2006.08.010

Rowe, Brian William. 2014. *Principles of Modern Grinding Technology: Vol-2*. 2nd ed. Waltham, MA: Williams Andrew Publishing. https://doi.org/10.1016/C2013-0-06952-6

Sheikh-Ahmad, Jamal Y. 2009. "Conventional Machining Operations BT - Machining of Polymer Composites." In, edited by Jamal Ahmad, 37–62. Boston, MA: Springer US. https://doi.org/10.1007/978-0-387-68619-6_2

Shen, Bin, Guoxian Xiao, Changsheng Guo, Stephen Malkin, and Albert J Shih. 2008. "Thermocouple Fixation Method for Grinding Temperature Measurement." *Journal of Manufacturing Science and Engineering* 130 (5). https://doi.org/10.1115/1.2976142

Song, Yang, Huajun Cao, Wei Zheng, Da Qu, Lei Liu, and Chunping Yan. 2022. "Cutting Force Modeling of Machining Carbon Fiber Reinforced Polymer (CFRP) Composites: A Review." *Composite Structures* 299: 116096. https://doi.org/10.1016/j.compstruct.2022.116096

Soo, Sein Leung, Islam S. Shyha, Tom Barnett, David K. Aspinwall, and Wei Ming Sim. 2012. "Grinding Performance and Workpiece Integrity When Superabrasive Edge Routing Carbon Fibre Reinforced Plastic (CFRP) Composites." *CIRP Annals - Manufacturing Technology* 61 (1): 295–98. https://doi.org/10.1016/j.cirp.2012.03.042

Soussia, Aymen Ben, Ali Mkaddem, and Mohamed El Mansori. 2014. "Rigorous Treatment of Dry Cutting of FRP – Interface Consumption Concept: A Review." *International Journal of Mechanical Sciences* 83: 1–29. https://doi.org/10.1016/j.ijmecsci.2014.03.017

Su, Youliang, Zhenyuan Jia, Bin Niu, and Guangjian Bi. 2017. "Size Effect of Depth of Cut on Chip Formation Mechanism in Machining of CFRP." *Composite Structures*. https://doi.org/10.1016/j.compstruct.2016.11.044

Tawakoli, Taghi, and Bahman Azarhoushang. 2011. "Intermittent Grinding of Ceramic Matrix Composites (CMCs) Utilizing a Developed Segmented Wheel." *International Journal of Machine Tools and Manufacture* 51 (2): 112–19. https://doi.org/10.1016/J.IJMACHTOOLS.2010.11.002

Uhlmann, E., F. Sammler, S. Richarz, G. Reucher, R. Hufschmied, A. Frank, B. Stawiszynski, and F. Protz. 2016. "Machining of Carbon and Glass Fibre Reinforced Composites." *Procedia CIRP* 46 (January): 63–66. https://doi.org/10.1016/J.PROCIR.2016.03.197

Wang, Haijin, Jie Sun, Jianfeng Li, Laixiao Lu, and Nan Li. 2016. "Evaluation of Cutting Force and Cutting Temperature in Milling Carbon Fiber-Reinforced Polymer Composites." *International Journal of Advanced Manufacturing Technology* 82 (9–12): 1517–25. https://doi.org/10.1007/s00170-015-7479-2

Wang, Hui, Weilong Cong, Fuda Ning, and Yingbin Hu. 2018. "A Study on the Effects of Machining Variables in Surface Grinding of CFRP Composites Using Rotary Ultrasonic Machining." *The International Journal of Advanced Manufacturing Technology* 95 (9): 3651–63. https://doi.org/10.1007/s00170-017-1468-6

Wang, Hui, Z J Pei, and Weilong Cong. 2020a. "A Feeding-Directional Cutting Force Model for End Surface Grinding of CFRP Composites Using Rotary Ultrasonic Machining with Elliptical Ultrasonic Vibration." *International Journal of Machine Tools and Manufacture* 152: 103540. https://doi.org/10.1016/j.ijmachtools.2020.103540

Wang, Hui, Z J Pei, and Weilong Cong. 2020b. "A Mechanistic Cutting Force Model Based on Ductile and Brittle Fracture Material Removal Modes for Edge Surface Grinding of CFRP Composites Using Rotary Ultrasonic Machining." *International Journal of Mechanical Sciences* 176: 105551. https://doi.org/10.1016/j.ijmecsci.2020.105551

Xu, Wei Xing, Liang Chi Zhang, and Yong Bo Wu. 2012. "Micromechanical Modelling of Elliptic Vibration-Assisted Cutting of Unidirectional FRP Composites." *Advanced Materials Research* 591–593: 531–34. https://doi.org/10.4028/www.scientific.net/AMR.591-593.531

Xu, Weixing, and L C Zhang. 2014. "On the Mechanics and Material Removal Mechanisms of Vibration-Assisted Cutting of Unidirectional Fibre-Reinforced Polymer Composites." *International Journal of Machine Tools and Manufacture* 80–81: 1–10. https://doi.org/10.1016/j.ijmachtools.2014.02.004

Xu, Weixing, L C Zhang, and Yongbo Wu. 2013. "Elliptic Vibration-Assisted Cutting of Fibre-Reinforced Polymer Composites: Understanding the Material Removal Mechanisms Elliptic Vibration-Assisted Cutting of Fibre-Reinforced Polymer Composites: Understanding the Material Removal Mechanisms." *Composites Science and Technology* 92 (January): 103–11. https://doi.org/10.1016/j.compscitech.2013.12.011

Xu, Weixing, and Liangchi Zhang. 2016. "Mechanics of Fibre Deformation and Fracture in Vibration-Assisted Cutting of Unidirectional Fibre-Reinforced Polymer Composites." *International Journal of Machine Tools and Manufacture* 103: 40–52. https://doi.org/10.1016/j.ijmachtools.2016.01.002

Xu, Weixing, and Liangchi Zhang. 2019. "Heat Effect on the Material Removal in the Machining of Fibre-Reinforced Polymer Composites." *International Journal of Machine Tools and Manufacture* 140 (January): 1–11. https://doi.org/10.1016/j.ijmachtools.2019.01.005

Xu, Xipeng, Yiqing Yu, and Hui Huang. 2003. "Mechanisms of Abrasive Wear in the Grinding of Titanium (TC4) and Nickel (K417) Alloys." *Wear* 255 (7): 1421–26. https://doi.org/10.1016/S0043-1648(03)00163-7

Yuan, H P, H Gao, and Y D Liang. 2010. "Fabrication of a New-Type Electroplated Wheel with Controlled Abrasive Cluster and Its Application in Dry Grinding of CFRP." *International Journal of Abrasive Technology* 3 (4): 299–315. https://doi.org/10.1504/IJAT.2010.036963

Zhang, Chen, and Ming Lu. 2018. "A Novel Variable-Dimensional Vibration-Assisted Actuator for Drilling CFRP." *The International Journal of Advanced Manufacturing Technology* 99 (9): 3049–63. https://doi.org/10.1007/s00170-018-2680-8

Zhang, Guoqing, Suet To, Shaojian Zhang, and Zhiwei Zhu. 2016. "Case Study of Surface Micro-Waves in Ultra-Precision Raster Fly Cutting." *Precision Engineering* 46 (October): 393–98. https://doi.org/10.1016/J.PRECISIONENG.2016.06.009

Zhang, Lifeng, Sheng Wang, Zhan Li, Weilin Qiao, Ying Wang, and Tao Wang. 2019. "Influence Factors on Grinding Force in Surface Grinding of Unidirectional C/SiC Composites." *Applied Composite Materials* 26 (3): 1073–85. https://doi.org/10.1007/s10443-019-09767-5

Zhang, S.J., S. To, Z.W. Zhu, and G.Q. Zhang. 2016. "A Review of Fly Cutting Applied to Surface Generation in Ultra-Precision Machining." *International Journal of Machine Tools and Manufacture* 103 (April): 13–27. https://doi.org/10.1016/J.IJMACHTOOLS.2016.01.001

8 Drilling Characteristics of Fibre-reinforced Polylactic Acid Composites

A Study on Himalayacalamus falconeri Fibre-based Composites

*Mayank Pokhriyal, Pawan Kumar Rakesh,
Rajesh Kumar, Shaurya Bhatt, Vivek Bahuguna,
and Hitesh Sharma*

8.1 INTRODUCTION

As per the strict government policy regarding saving the environment and ecosystem, synthetic fibres are being replaced fully or partially by natural fibres, namely, flax fibre, neetle fibre, jute fibre, hemp fibre, and others, which are light in weight, renewable, fully biodegradable and compatible to the environment. Keeping this in mind, natural fibre is reinforced with polylactic acid for the fabrication of biodegradable composites to fulfil the demands of today's needs, such as household items, electronic racks for helicopters, deck panels, footwear, helmet shells, mobile phone casing, and orthopaedic implants, among others [1, 2, 3]. Biocomposites are well recognized for compatible strength, extraordinary specific stiffness, vibration and sound absorption properties, corrosion resistance, and dimensional stability [4, 46]. The parameters like fibre selection, fibre properties, fibre dispersion, laminate orientation, and interfacial strength, influencing the properties of biocomposites are mentioned in Table 8.1. The manufacturing of biocomposites can be done by hand lay-up, spray-up moulding, prepreg lay-up (vacuum bag and autoclave moulding), compression moulding, resin injection moulding, filament winding, and the pultrusion process [5]. The major limitation of manufacturing processes is their inability to make complex natural fiber reinforced polymer composite (NFRPC) products. The secondary manufacturing of NFRPCs, that is, conventional and unconventional machining, is most frequently practised due to the need for the assembly of parts

DOI: 10.1201/9781032665375-8
"

TABLE 8.1

Factors Influencing Natural Fibre–based Composites

Parameters	Factors
Natural fibre collection	Type of fibre, plant geography related to availability, harvesting time, length, and diameter
Natural fibre properties	Specific yield strength, density, coefficient of thermal expansion, and aspect ratio
Natural fibre dispersion	Hydrophobic and hydrophilic nature
Natural fibre orientation	Aligned fibres yarn and inclined orientation
Interfacial strength	Fibre matrix interfacial strength, bonding, cohesion, and structure

to get final products. Due to the distinct constituents of composite materials, it is very challenging to produce damage-free holes on biocomposites [6–9]. In order to decrease/optimize the drilling-induced damage in the biocomposites, investigators have undertaken studies on the drilling parameters [10, 11], drilling forces prediction [12–15], cutting mechanism [16], and control of cutting parameters [17–19]. Several other research papers have been published that focus on specific issues, such as multifaceted drills [20, 21], drill wear [22, 23], dynamics and control [24, 25], and mechanical behaviour of drilled holes [26–28].

An extensive literature survey focused on cutting parameters and their effects on the forces and power consumption with hole quality. The drilling-induced damage could be correlated directly or indirectly through drilling forces. A conventional method to avoid delamination/damage, such as the helical feed method, oscillatory assisted drilling, and woodpecker-type drilling cycle, has been reported. Scanning microscopy was used to evaluate surface textures in terms of average roughness height, and measured surface roughness profiles. Rakesh et al. [28] carried out drilling on unidirectional glass fibre–reinforced plastics (UD-GFRP) laminates and analysed the unidirectional strength of drilled hole laminates. The tensile strength declined because of the accumulation of the damaged area around the drilled hole, which is contrariwise proportional to the damage zone. The following points need to be deliberated while drilling:

- Material of the laminates and different orientations of fibres
- Manufacturing conditions (temperature and pressure)
- Drill material and drill geometry
- Drilling parameters

Generally, drilling parameters (drill materials), laminate properties (fibre orientation, fibre concentration, number of layers, types of matrix materials), and output parameters are considered. Damage during drilling can be controlled by optimizing the feed rate and drill geometry, as shown in Figure 8.1. This chapter aims to present the feasibility study on the hole-making process in *Himalayacalamus falconeri* fibre–reinforced polylactic composites.

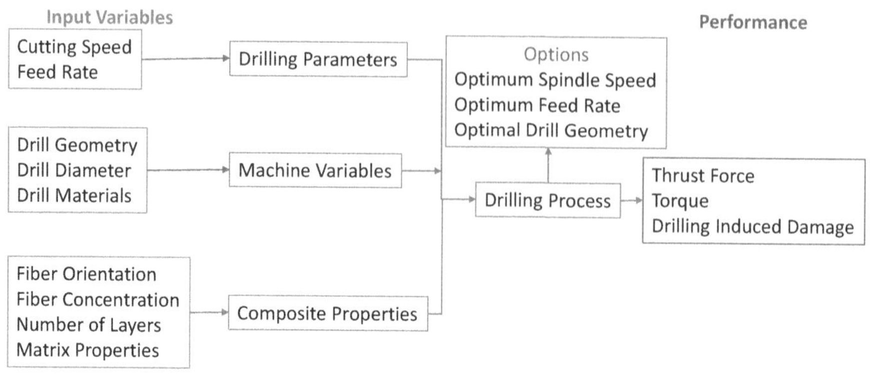

FIGURE 8.1   Input/output drilling parameters.

8.2   MODELLING OF DRILLING OF FIBRE-REINFORCED PLASTIC COMPOSITES

8.2.1   ANALYTICAL MODELLING

An analytical model was established to envisage the drilling forces and damage. Bhatnagar et al. [29] proposed a dimensional analysis technique to correlate the thrust force with the hole diameter and the thickness of laminates. A regression model has been developed for the thrust forces and the burr height for the process. The thrust force was least affected by the spindle speed; however, it was affected by the time of cut during the drilling process. Langella et al. [30] presented an automatous classic for envisaging drilling forces in fibre-reinforced plastic composites, which was dedicated to drill geometry and drilling parameters. The predicted equations can be used as the basis for formulating an economic model for cost optimization of the drilling operation.

8.2.2   NUMERICAL MODELLING

Rakesh et al. [31] modelled the drilling process in ABAQUS software to envisage the damage. The ratio of the damaged area (D_{max}) to the hole area (D) is termed the delamination factor. Three drill point geometries, namely, trepanning tool, conventional twist, and JO drill, were deliberated in the investigation. It was reported that the delamination factor was maximum in the case of the twist drill (4.43) compared to the trepanning drill (3.08), and the JO drill (3.79). It was also suggested that finite element (FE) modelling can be used to predict the thrust forces and torque. Singh et al. [32] studied the finite element modelling of the drilling of FRP materials. The FE drilling was made in UD-GFRP lamina with different drill point angles (118°, 104°, 90°). The feed rate considered was 0.3 mm/min and constant speed (375 rpm) for all the cases. A Tsai Wu failure criterion was implemented to predict the delamination factor. It was observed that the drill point geometry having minimum point angle (90°) contributes less delamination factor as compared to the other two drill point angles (118°, 104°) [33].

8.2.3 NEURAL NETWORK

A neural network-based controller was initiated to model the drilling dynamic forces and to make the correlation between drilling forces and drilling-induced damage. Al-Jarrah and Al-Oqla [34] utilized artificial neural networks (ANNs) to prepare estimates and categorize the mechanical properties of renewable natural fibres to augment their executions in green composites. The estimation and categorization were based on the chemical composition and mechanical properties of natural fibres. The model was established based on the experimental data available for different types of cellulosic fibres. The complete process was made in two steps: (a) back propagation neural network and (b) shallow neural networks. Twenty hidden layer sizes and 11 numbers of neuron output were considered. The model was able to predict the tensile strength, the percentage of elongation, and Young's modulus, as well as cellulose and moisture content. Alhaddad et al. [35] developed a machine learning–based model on experimental data to predict the quality parameters (tensile strength, layer thickness, volume fraction, fibre orientation and fibre form, and infill pattern). The ANN model was verified by the parametric data. Furthermore, the ultimate tensile strength of composites was optimized by using an artificial bee colony [36].

8.3 UNCONVENTIONAL METHODS

Experimental studies were carried out on the laminates to characterize the drilling process and to understand the various stages as the drill advances through the laminate. It was proposed to develop a prototype drill machine on the basis of the control strategy and test it for machining damage-free holes. An unconventional machining process for material removal such as ultrasonic machining, water jet and abrasive water jet, and laser surface treatment, has been studied for biocomposites.

8.3.1 ULTRASONIC DRILLING

An ultrasonically assisted vibration drilling (37.9 kHz with peak-to-peak vibration amplitude of 32.3 μm) of hemp fibre–reinforced vinyl ester composites was deliberated by Wang et al. [37]. A Harrison lathe machine with a spindle speed of 2500 rpm was selected for the drilling. A twist drill (point angle of 135° and helix angle of 28°) was used for experiments. A Kistler 9345b two-component dynamometer was used to measure the thrust force. It was indicated that ultrasonically assisted vibration drilling is more proficient than conventional drilling in terms of thrust force, surface finish and hole quality including burr formation and fibre pull-outs. They demonstrated that ultrasonically assisted vibration drilling is a worthwhile machining method for the value-added machinability of heterogeneous NFRP materials.

8.3.2 WATER JET AND ABRASIVE WATER JET DRILLING

Water jets and abrasive water jets are multipurpose cutting processes for nonhomogeneous materials. The abrasive water jet machining of pineapple fibre–reinforced epoxy composites is investigated by Karthikeyan et al. [38]. The abrasive particle size of 80 mesh, unthreaded nozzle at 90°, is injected with water. The optimized input parameters, that is, a water pressure of 3000 bar, an abrasive flow rate of 350 gm/min, and a transverse speed of 235 mm/min, were recommended for an extreme material removal rate of 39.23 gm/min based on selected parameters. The input parameters are hydraulic pressure (water jet orifice), abrasive shape and size (length, diameter, grit size), and cutting parameters (angle of attack), among others, for the abrasive machining of the composite materials. The output parameters (kerf width, kerf taper, surface morphology, dimensional accuracy, material removal rate) are generally considered for composite machining [39, 40]. The input parameters significantly affect the quality of composite machining. In comparison to water jet machining, the abrasive water jet is more suitable for laminated composites in terms of kerf measurement.

8.3.3 LASER SURFACE TREATMENT

The Lambda-Physik LPX210i and Lambda-Physik EMG102MSC lasers were used for fibre treatment under ambient conditions. The duration of the laser pulse varies with the gas used (ArF: 20ns, KrF: 23ns, XeCl: 14ns) at a frequency of 1 Hz. A concavo-convex lens is used for focus and fluence adjustment, measured with a joule meter and oscilloscope. The sample is placed in a holder between the lens and the joule meter. The irradiation fluence (100–500 mJ/cm^2) and 0–100 pulses can be analysed to study the surface structure of a fibre over an irradiation area of 0.24–1.20 cm^2 [41].

8.4 DRILLING BEHAVIOR OF *H. FALCONERI* FIBRE–REINFORCED COMPOSITES: A CASE STUDY

H. falconeri fibre is extracted from *H. falconeri* culms by the water retting method [42, 43]. Extracted *H. falconeri* fibre is dried in an electric oven at the temperature of 90°C to remove moisture from the fibre surface. The extracted *H. falconeri* fibre is reinforced into polylactic acid to fabricate biocomposites via an injection moulding machine [44–46]. The fibre reinforcement of 15% is considered in the present investigation. Two different drill geometries (Figure 8.2), namely, twist drill and JO

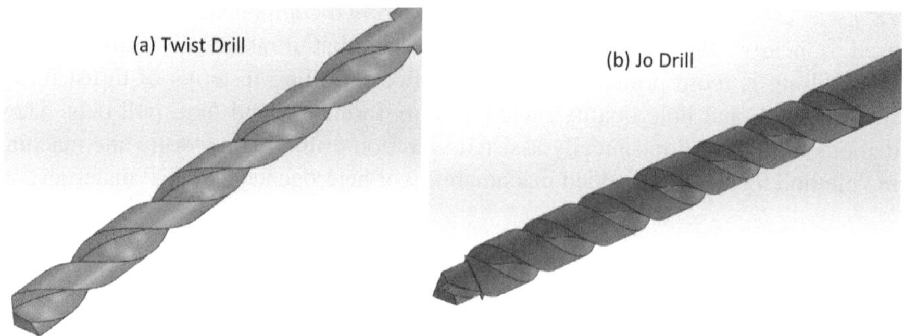

FIGURE 8.2 Drill point geometry. (a) Twist drill, (b) JO drill.

drill, with cutting speed of 560 rpm and a feed rate of 0.3 and 0.19 mm/rev, respectively, are selected for the drilling process. The twist drill point angle is 118°. The JO drill point has two two-step cutting mechanisms, which can provide better cutting as compared to the twist drill in laminated composites. The drill materials are considered solid carbide. The drill diameter is used as 4 mm. Thrust force value of 110 N and torque of 0.28 Nm with the twist drill was recorded via a Kistler dynamometer (9272) for the 15% reinforcement of fibres into polylactic acid–based composites, as shown in Figure 8.3a. A thrust force value of 55 N and torque of 0.18 Nm with the JO drill were recorded, as shown in Figure 3b. The JO drill cuts in two steps; therefore, the force value is reduced to 50%. Similarly, the torque was reduced by 36%. It is a well-known fact that lower thrust force and torque values mean the minimum value of the delamination factor, which is required in the case of composite materials. It was observed that the variation in thrust force and torque was due to their orthotropic material properties. The hard particles of extracted fibres came in front of drill geometry, resulting in increasing the thrust forces and torque on the deformation of the polylactic acid (PLA) composites.

8.5 SUMMARY

The drilling of *H. falconeri* fibre–based PLA composites has been done successfully. The conventional twist drill is already being used for drilling in metals/ceramics material, which is not appropriate for drilling in fibre-reinforced plastic (FRP) laminates, due to its different material constituents, a high amount of drilling-induced damage occurs around the hole boundary. It is suggested to use JO drill point geometry (which has a two-step cutting mechanism) for FRP composite materials, which gives the best results compared to twist drill geometry. Apart from the drilling process, different conventional and unconventional processes have been successful in hole-making in FRP laminates. Unconventional machining is more efficient for making holes in FRP laminates, but it is not economical. As per the market strategy, the fabrication cost of any product must be low, and then the profit margin can be maximized by making the laminated product. Ultimately, conventional drilling is only one of the options to make holes in the FRP laminates. However, further exhaustive investigation can be performed to suggest the specific drill geometry and feed rate for the drilling of FRP laminates.

8.6 FUTURE SCOPE

The fabrication of fibre–reinforced composites can be done by various manufacturing methods. For joining plates, various applications are needed for the hole-making process. The hole making in FRP laminates can be done by optimizing the process parameters and drill point geometry. The different drill geometry can be used to analyse the induced forces by dynamometers or strain gauge-based systems. It was already a fact that spindle speed is one of the parameters that has less influence in drilling damage but that an increased spindle speed enhances the smooth surface finish. Drill geometry and feed rate are the most influencing factors for the drilling damage. The drill tool life also influences the performance of the drilled hole. The

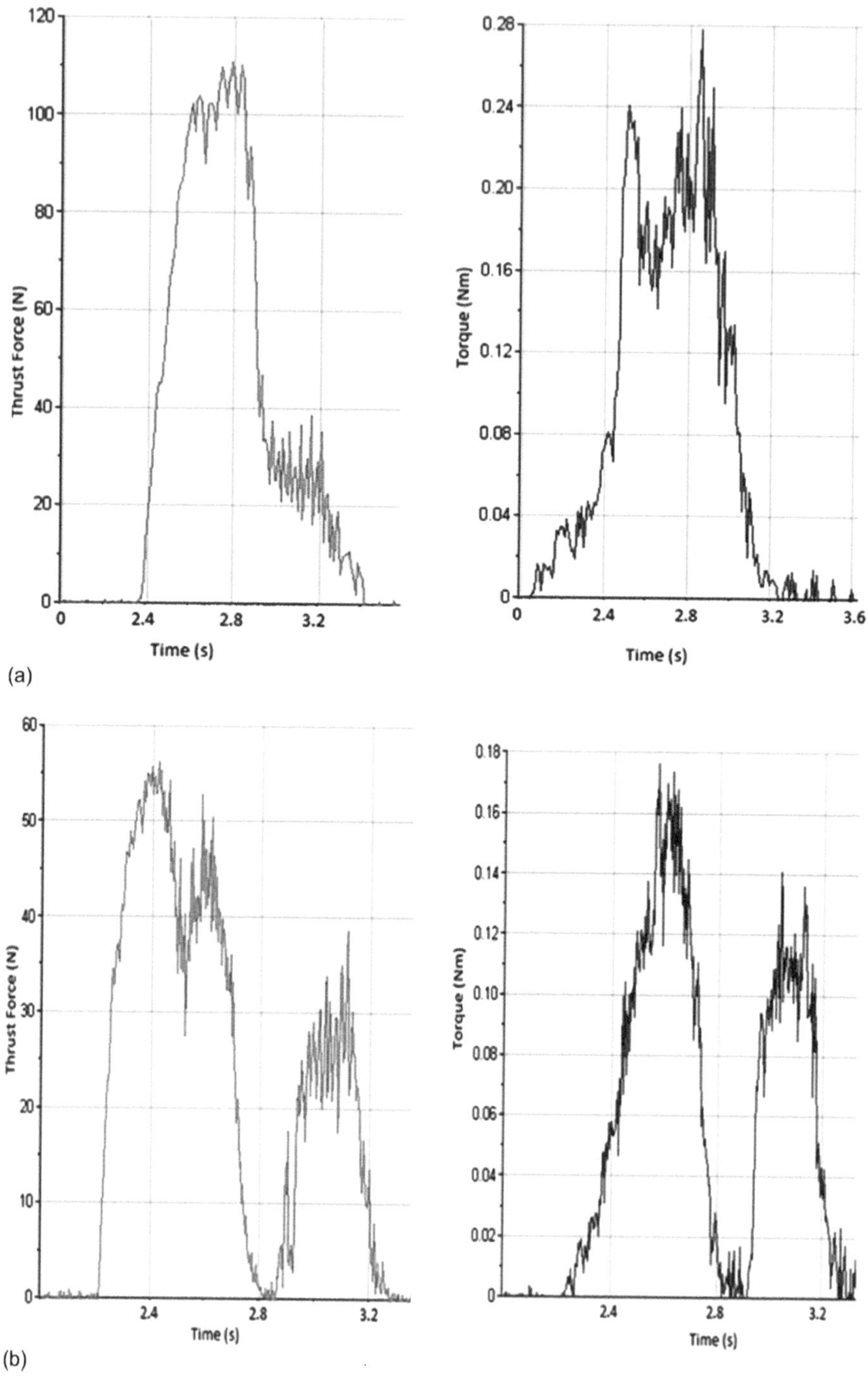

FIGURE 8.3 Thrust force and torque of *Himalayacalamus falconeri* fibre–reinforced poly-lactic acid composites. (a) Twist drill geometry (560 rpm, 0.3 mm/rev.), (b) JO drill geometry (560 rpm, 0.19 mm/rev.).

authors already tried different drill materials like sintered carbide, high-speed steel, and coating by various materials, for reducing tool wear for the drilling process. Still, the selection of drill materials and geometry is under investigation stage. No such tools and techniques are available to select the drill geometry and materials for the drilling of biocomposites.

FUNDING

This funding was supported by Indian Council of Medical Research, New Delhi (Grant No. 2021-14663). The author is greatly acknowledged the supported of Design Innovation Center, NIT Uttarakhand, India.

REFERENCES

1. Chaturvedi A, Ranakoti L, Rakesh PK, Mishra NK, Experimental investigations on mechanical properties of walnut shell and pine needle ash polylactic acid biocomposites, *Compos Theory Pract*, 21:3, 2021, 114–120.
2. Kishore RA, Tiwari R, Dvivedi A, Singh I, Taguchi analysis of the residual tensile strength after drilling in glass fiber reinforced epoxy composites, *Mater Des*, 2009, 30, 2186–2190.
3. Davim JP, Rubio JC, Abrao AM, A novel approach based on digital image analysis to evaluate the delamination factor after drilling composite laminates, *Compos Sci Technol*, 2007, 67, 1939–1945.
4. Rakesh PK, Davim JP, Cellulose Composites: Processing and Characterization, ISBN: 978-3-11-076869-5, 2022, De Gruyter, Germany.
5. Rakesh PK, Singh I, *Processing of Green Composites*, Springer, ISBN 978-981-13-6019-0, 2018.
6. Varatharajan R, Malhotra SK, Vijayaraghavan L, Krishnamurthy R, Mechanical and machining characteristics of GF/PP and GF/Polyester composites, *Mater Sci Eng*, 2006, 132 B, 134–137.
7. Rubio JC, Abrao AM, Faria PE, Correia AE, Davim JP, Effects of high speed in the drilling of glass fiber reinforced plastic: Evaluation of the delamination factor, *Int J Mach Drills Manuf*, 2008, 48, 715–720.
8. Kishore RA, Tiwari R, Singh AP, Singh I, A knowledge based drill for damage free drilling of fiber reinforced plastics, *Manuf Technol Today*, 2008, 9(7), 3–10.
9. Rakesh PK, Singh I, Kumar D. Drilling of composite laminates with solid and hollow drill point geometries, *J Compos Mater*, 2012, 46, 3173–3180.
10. Kumar S, Chauhan SR, Rakesh PK, Singh I, Davim JP, Drilling of glass fiber/vinyl ester composites with fillers, *Mater Manuf Processes*, 2012, 27(3), 314–319.
11. Kishore RA, Tiwari R, Rakesh PK, Singh I, Bhatnagar N, Investigation of drilling in fiber reinforced plastics using response surface methodology. *Proc Inst Mech Eng Part B*, 2011, 225(3), 453–457.
12. Inderdeep S, Rakesh PK, Malik J, *Soft Computing Techniques for Predicting Forces and Damage in Drilling of Polymer Composites, Mechatronics and Manufacturing Engineering: Research and Development*, Woodhead Publishing, UK. ISBN: 978 0 85709 150 5, 2012. 227–255.
13. Mishra R, Malik J, Singh I, Davim JP, Neural network approach for estimating the residual tensile strength after drilling in uni-directional glass fiber reinforced plastic laminates, *Mater Des*, 31(6), 2010, 2790–2795.

14. Tsao CC, Chen W, Prediction of the location of delamination in the drilling of composite laminates. *J Mater Process Technol*, 1997, 70, 185–189.
15. Zhang LB, Wang LJ, Liu XY, A mechanical model for predicting critical thrust forces in drilling composite laminates. *Pi Mech Eng B-J Eng*, 2001, 215(B), 135–146.
16. Ilio AD, Tagliaferri V, Veniali F, Cutting mechanisms in drilling of aramid composites. *Int J Mach Drills Manuf*, 1991, 31(2), 155–165.
17. Kilickap E, Optimization of cutting parameters on delamination based on taguchi method during drilling of GFRP composite, *Expert Syst Appl*, 2010, 37, 6116–6122.
18. Abrao AM, Rubio JCC, Faria PE, Davim JP, The effect of cutting drill geometry on thrust force and delamination when drilling glass fiber reinforced plastic composite, *Mater Des*, 2008, 29, 508–513.
19. Shyha I, Soo SL, Aspinwall D, Bradley S, Effect of laminate configuration and feed rate on cutting performance when drilling holes in carbon fiber reinforced plastic composites, *J Mater Process Technol*, 2010, 210, 1023–1034.
20. Singh RVS, Latha B, Senthilkumar VS, Modeling and analysis of thrust force and torque in drilling GFRP composites by multi-facet drill using fuzzy logic, *Int J Recent Trend Eng*, 2009, 1(5), 66–70.
21. Sardinas RQ, Reis P, Davim JP, Multi-objective optimization of cutting parameters for drilling laminate composite materials by using genetic algorithms, *Compos Sci Technol*, 2006, 66, 3083–3088.
22. Inoue H, Aoyama E, Hirogaki T, Ogawa K, Matushita H, Kitahara Y, Katayama T, Influence of drill wear on internal damage in small diameter drilling in GFRP. *Compos Struct* 39, 55–62, 1997.
23. Miller SF, Blau PJ, Shih AJ, Drill wear in friction drilling, *Int J Mach Drills Manuf*, 2007, 47, 1636–1645.
24. Arul S, Vijayaraghavan L, Malhotra SK, Online monitoring of acoustic emission for quality control in drilling of polymeric composites, *J Mater Process Technol*, 2007, 185, 184–190.
25. Singh AP, Sharma M, Singh I, A review of modeling and control during drilling of fiber reinforced plastic composites, *Compos B Eng*, 47, 2013, 118–125.
26. Rakesh PK, Singh I, Kumar D. Flexural behavior of glass fiber reinforced plastic laminates with drilled hole, *Proc Inst Mechan Eng Part L*, 2012, 226 (2), 149–158.
27. Rakesh PK, Singh I, Kumar D. Compressive behavior of glass fiber reinforced plastic laminates with drilled hole, *Adv Mater Res*, 2012, 410, 349–352.
28. Rakesh PK, Singh I, Kumar D, Failure prediction in glass fiber reinforced plastics laminates with drilled hole under uni-axial loading, *Mater Des*, 2010, 31(6), 3002–3007.
29. Bhatnagar N, Jalutharia MK, Singh I, Prediction of thrust force and torque when drilling composite materials, *Int J Mater Prod Technol*, 2008, 32(2/3), 213–224.
30. Langella A, Nele L, Maio A, A torque and thrust prediction model for drilling of composite materials, *Composites Part A*, 2005, 36, 83–93.
31. Rakesh PK, Sharma V, Singh I, Kumar D, Delamination in fiber reinforced plastics: A finite element approach, *Engineering*, 2011, 3, 249–254.
32. Singh I, Bhatnagar N, Viswanath P, Drilling of uni-directional glass fiber reinforced plastics: experimental and finite element study, *Mater Des*, 2008, 29(2), 546–553.
33. Budan DA, Vijayarangan S. Quality assessment and delamination force evaluation in drilling glass fiber-reinforced plastic laminates- a finite element analysis and linear elastic fracture mechanics approach. *Pi Mech Eng B-J Eng*, 2002, 216, 173–182.
34. Rami A-J, Al-Oqla FM, A novel integrated BPNN/SNN artificial neural network for predicting the mechanical performance of green fibers for better composite manufacturing, *Compos Struct*, 2022, 289, 115475.

35. Wael A, Minjuan H, Yahia H, Almajhali KYM, Optimizing the material and printing parameters of the additively manufactured fiber-reinforced polymer composites using an artificial neural network model and artificial bee colony algorithm, *Structures*, 2022, 46, 1781–1795.

36. Nwosu-Obieogu K, Dzarma GW, Christian G, Nonso UC, Awele AM, Anozie OO, Physico-chemical/mechanical properties of treated groundnut shell fibre; response surface methodology and artificial neural network performance evaluation and optimization, *Cleaner Waste Syst*, 2022, 2, 100017.

37. Wang D, Onawumi PY, Ismail SO, Dhakal HN, Popov I, Silberschmidt VV, Roy A, Machinability of natural fiber reinforced polymer composites: Conventional vs ultrasonically-assisted machining, *Compos Part A Appl Sci Manuf*, 2019, 119, 188–195.

38. Karthikeyan R, Prasad G L, Saraswathy R, Singh DP, Rajkamal MD, Madhu S, Abrasive water jet machining of pineapple fibre epoxy composite at high transverse speed, *Mater Today: Proc*, 2023. https://doi.org/10.1016/j.matpr.2023.02.321

39. Madival AS, Doreswamy D, Shetty R, A review on the physical and mechanical properties of natural fiber reinforced composites and its machinability using abrasive waterjet machining, *Mater Today: Proc*, 2023. https://doi.org/10.1016/j.matpr.2023.04.519

40. Ramraji K, Rajkumar K, Dhananchezian M, Sabarinathan P, Key experimental investigations of cutting dimensionality by abrasive water jet machining on basalt fiber/fly ash reinforced polymer composite, *Mater Today: Proc*, 2023, 22(4), 1351–1359.

41. Mizoguchi K, Ishikawa M, Ohkubo S, et al. Laser surface treatment of regenerated cellulose fiber. *Compos Interfaces*, 2001, 7, 497–509. https://doi.org/10.1163/156855400750262978

42. Pokhriyal M, Rakesh PK, Rangappa SM, Siengchin S, Effect of alkali treatment on novel natural fiber extracted from Himalayacalamus falconeri culms for polymer composite applications, *Biomass Convers Biorefin*, 2023, 23, 1–17.

43. Pokhriyal M, Rakesh PK, Mechanical and Microstructural behaviour of NaOH treated Himalayacalamus falconeri fibers as biodegradable reinforcing material in polymer based composites, *Mater Today: Proc*, 2022, 62(2), 1078–1082.

44. Pokhriyal M, Prasad L, Rakesh PK, Raturi HP, Influence of fiber loading on physical and mechanical properties of Himalayan nettle fabric reinforced polyester composite, *Mater Today: Proc*, 5(9), 2018, 16973–16982.

45. Pokhriyal M, Rakesh PK, Processing and characterization of novel Himalayacalamus falconeri fiber reinforced biodegradable composites, *Biomass Convers Biorefin*, 2023, 23, 1–16.

46. Faruk O, Bledzki AK, Fink HP, Sain M Progress report on natural fiber reinforced composites. *Macromol Mater Eng*, 2014, 299, 9–26.

9 Water Jet Machining of Hybrid Fiber-Reinforced Polymer Composites

Khushi Ram and Pramendra Kumar Bajpai

9.1 INTRODUCTION

Today, composites are widely used in every field such as the aviation industry, the automobile sector, the infrastructure industry, sports equipment, defense applications and the health care industry. Due to the broad spectrum of applications different types of composite materials are available based on the type of matrix and reinforcement (Bajpai, Singh, and Madaan 2014). The main function of matrix material in composite is to transfer the load and maintain the structural integrity of composite material. Based on the matrix, three types of composites are available such as metal matrix composites, polymer matrix composites and ceramics matrix composites. Reinforcement plays an important role in a composite; it provides strength and stiffness and load-bearing components in composites (Ram and Bajpai 2017; Bajpai et al. 2018; Ram and Bajpai 2023). Based on the reinforcement material composites are classified in various forms, which are shown in Figure 9.1. The extensive use of composites is restricted due to their complexity and the high cost involved in the fabrication of desired shape and size products with good surface finish and dimensional tolerances. Composites are a combination of two or more materials at the microscopic level and having different chemical phases which is why they are anisotropic, less ductile and heterogeneous at the microscopic level (Kaw 2005). Due to these properties, composites are difficult to machine, unlike conventional materials. For the machining of composites drilling, sawing and milling are the conventional machining processes. Conventional machining methods create a few problems while machining composites, which makes them difficult to use. First, this process creates delamination of the fibers and the matrix due to which composite strength is reduced. They weaken the interface between fiber and matrix and generate the cracks initiation point for the failure of composites. Second, in conventional machining, close dimensional tolerances are not achievable, and the surface finish is not so good. To get exact dimensions and surface finish, other secondary operations are required, which further increases the cost and some other issues. Third, some reinforcement agents are abrasive in nature. Due to their abrasive nature, they can distort the profile of the tool. This is why special tools are required to machine the composites,

DOI: 10.1201/9781032665375-9

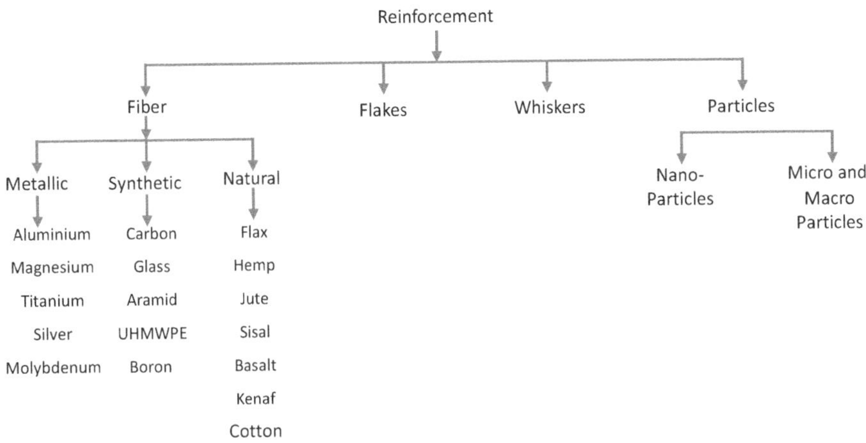

FIGURE 9.1   Reinforcement materials.

which further creates lots of problems. Fourth, during the machining of composites dust particles are developed, which cause serious health issues. Keeping all these problems in mind, for the machining of composites, some unconventional machining processes are required to get the exact shape, dimensions and surface finish. Some of the unconventional machining processes which are used now a days for the machining of composite materials are laser beam machining, ultrasonic machining, water jet machining (WJM) and abrasive water jet machining (AWJM) processes (Deshmukh, Shrivastava, and Thakar 2022; Chenrayan et al. 2022). Table 9.1 compares the unconventional and conventional machining processes. In the literature, machining polymer composites using a WJM machine is missing. In this present study, authors compare machining using WJM and using conventional drilling of the hybrid polymer composite. In this chapter, the effect of process parameters of WJM and AWJM on the machining of hybrid fiber–reinforced polymer composites are discussed to understand the machining behavior of these materials and their applications in different fields to increase the use of unconventional materials in order to save energy and reduce the weight of the component.

9.2   WJM PROCESS

Figure 9.2 shows the schematic diagram of WJM. First, for machining, it is required that water should be pure and free from any salts and chemicals. The pH of the water should be normal. The first water purifier unit receives water from the tank and purifies it, and from the purifier, water is supplied to a chiller that cools the water to normal environment temperature. From the chiller water is supplied to the compressor where with the help of a piston and cylinder the water is pressurized up to the required pressure for machining. Then water is directly transferred to the water jet nozzle through a controlled valve. The workpiece is mounted on a water jet tank and

TABLE 9.1

Comparison of the Machining Process for Composites

Machining Process	Surface Quality	Heat Affected Zone	Tool Wear	Dimensional Accuracy	Delamination
Conventional Drilling	Poor Surface	Generation of Heat	Some reinforcement material is abrasive in nature that distort tool	Less accuracy	Delamination occurs
Milling	Poor surface	Depends on the type of composite	Some reinforcement material is abrasive in nature that distort tool	Low accuracy	Delamination and matrix cracking occur
Sawing	Poor surface	negligible	Some reinforcement material is abrasive in nature that distort tool	Low accuracy	Delamination and matrix cracking occur
WJM	High surface finish	No heat generation	No tool wear	Better dimensional accuracy	No delamination
AWJM	Good surface finish	No heat generation	No tool wear	Lower compared to WJM	No delamination
Laser Cutting	Poor surface finish	Cutting due to burning of fiber and matrix	No tool wear	Low	Burning of fibers

Souces: Nele, Caggiano, and Improta (2021); Abrate and Walton (1992).

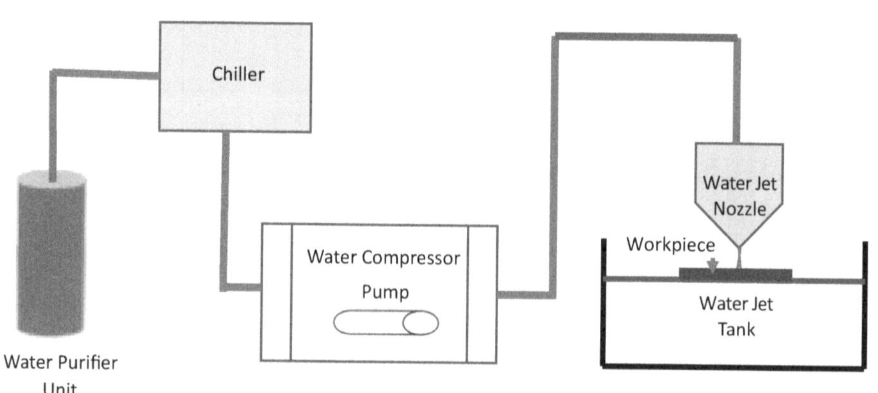

FIGURE 9.2   Schematic diagram of WJM.

the tool motion is according to a computer numerically controlled (CNC) program to cut the specific geometry. The water jet tank is partially filled by normal water so that the pressurized jet does not hit directly the water tank surface. In addition to this, for AWJM, a separate tank containing the abrasive is supplied to the water jet nozzle in a controlled manner. The flow of water and abrasive flow rate is controlled by the CNC program.

9.2.1　Process Parameters of WJM

Process parameters of WJM are shown in Figure 9.3. It includes three parameters: first, the input parameters; second, the type of workpiece; and, third, the desired output parameters. Three types of input parameters on which the cutting of the workpiece depends and the desired accuracy of cutting obtained: First is water pressure, which is the most important parameter of WJM because for every material and certain thickness a required water pressure is required to pierce the workpiece. To cut any gematrical shape, piercing is required, and if the water pressure is insufficient, piercing does not happen and the cutting process will not start. The second input parameter is the machining parameter, which includes tool speed and standoff distance. Tool speed is the rate at which the cutting tool moves over the workpiece. The standoff distance in WJM is the distance between the workpiece surface and the tooltip. The standoff distance should not so little that it can touch the workpiece or not so long that the jet of water disperses and the required dimensions are not obtained. The third input parameter is the water jet nozzle. The nozzle material and its diameter affect the cutting process. If the nozzle diameter is large, the kerf width increases, which will lead to an inaccuracy in the dimension. Workpiece material type also affects the functionality of the WJM process ductile materials are easy to cut while brittle materials are difficult to cut. To cut brittle materials like ceramics special arrangements are required.

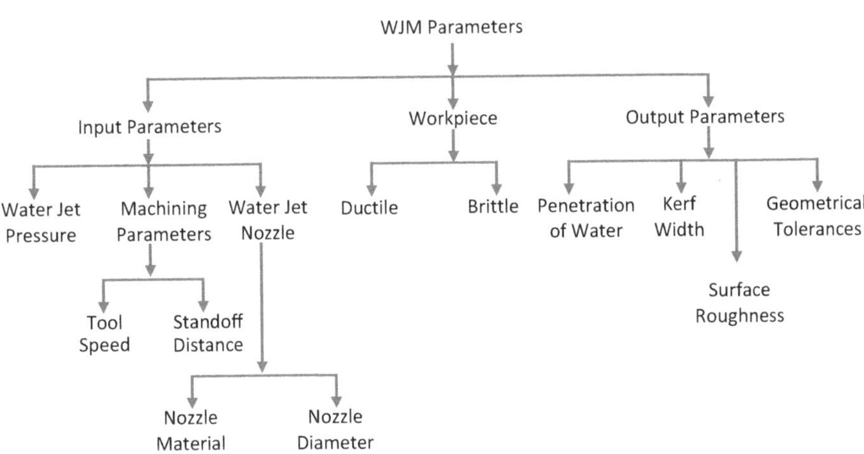

FIGURE 9.3　Process parameters of WJM.

9.3 CUTTING OF A HYBRID POLYMER COMPOSITE BY CONVENTIONAL DRILLING AND WJM

Hoekstra et al. (2022) compared the hole surface of carbon/flax-reinforced epoxy composite machined by conventional drilling process and by AWJM. The authors concluded that AWJM showed peel-off and push-out of the fibers while conventional drilling showed only push-out delamination of the fibers. Since the machining of composites by AWJM is available in literature and machining by WJM is missing this is why the authors compare the machining of a hybrid composite using WJM and a conventional drilling process.

9.3.1 CASE STUDY

A hybrid composite laminate of unidirectional sisal and ramie fiber–reinforced polyester is fabricated by hand lay-up process. Two circles of a 10-mm diameter are cut using WJM and conventional drilling. The cutting surface of the laminate by both techniques is observed visually and under a microscope. Figures 9.4a and 9.4b show the 10-mm circle cut by the WJM and the conventional drilling machine. As shown in Figure 9.4, the hole cut by WJM is smooth, and no surface cracks are observed. However, the hole cut by conventional drilling shows surface cracks, and its soft composite material gets stuck in the drill 2–3 times during the drilling operation. The drill

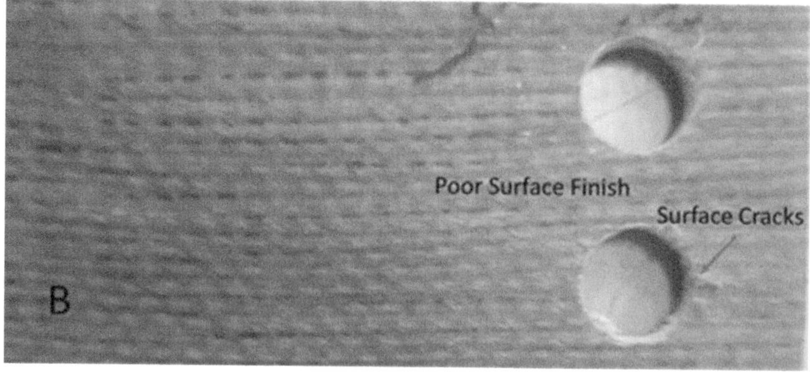

FIGURE 9.4 (a) 10-mm circle cut by WJM (b) 10-mm circle by conventional drilling machine.

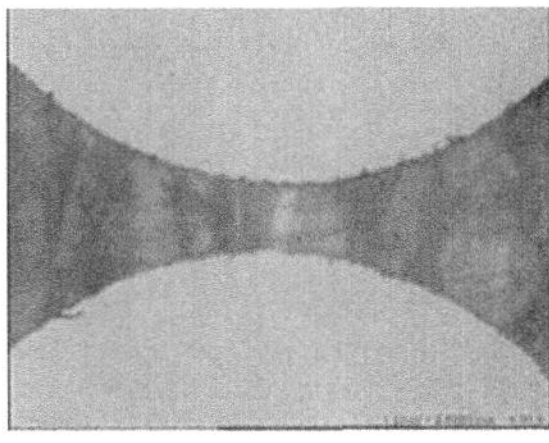 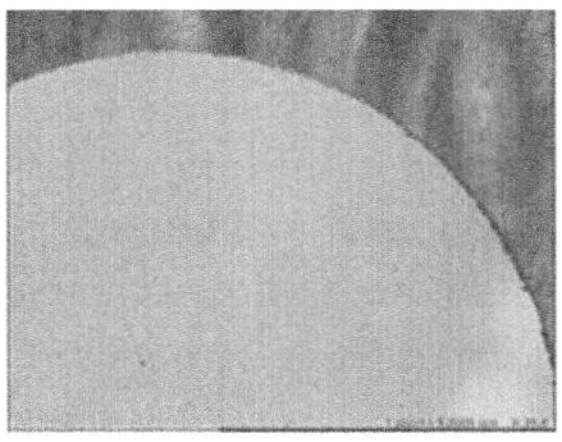 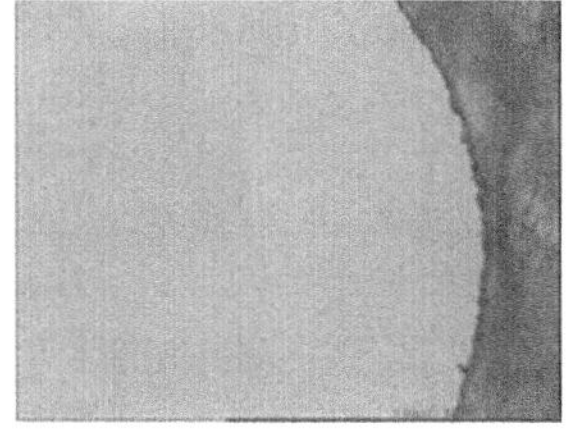

FIGURE 9.5 Cutting-edge surface of a machined hole using the WJM technique.

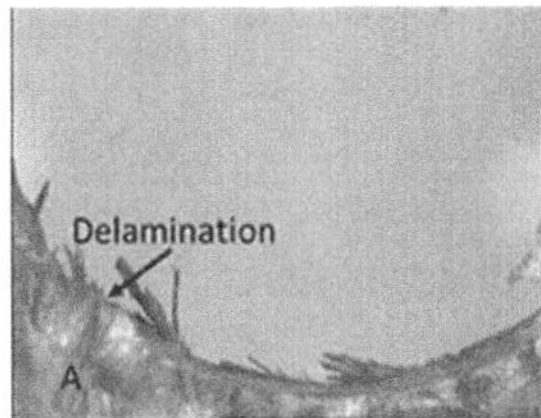

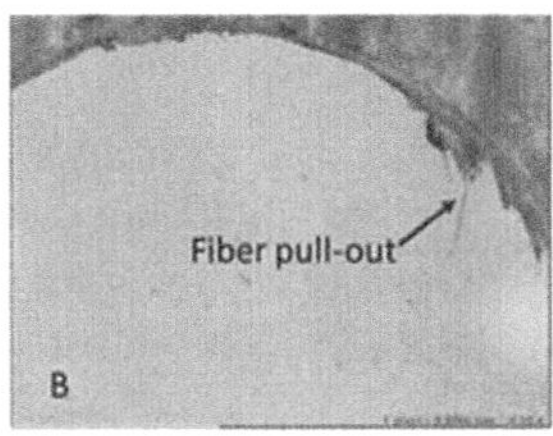

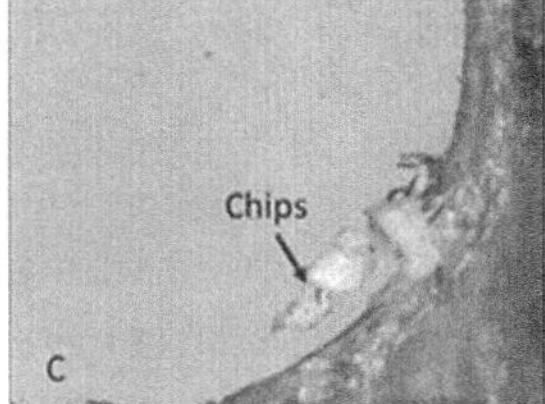

FIGURE 9.6 Cutting-edge surface of the drilled hole by the conventional drilling technique.

bit is stuck during drilling by conventional drilling as the fibers are stuck between the flutes of the drill bit and in the case of WJM no chip or burr formation takes place only a circle of 10-mm cut out from the laminate like a blank. Figure 9.5 shows the cutting-edge surface of the machined hole using the WJM technique, and Figure 9.6 shows the cutting-edge surface of the drilled hole using conventional drilling technique. As shown in Figure 9.5, the cutting-surface edge of the WJM hole in which small fibers are shown, only no fiber pullout or delamination occurs. This is due to the water jet pressure directly hitting the fiber at high pressure without pulling the fibers. As shown in Figure 9.6, for the hole surface when using conventional drilling, delamination of the fibers occurs, and fiber pullout also happened because the tool has physical contact with composite laminate. In the conventional drilling process also, chip formation takes place, and these chips are discontinuous.

9.3.2 Effect of Process Parameters on Cutting of Hybrid Composites

Composites are difficult to machine due to some restrictions, such as they are heat-sensitive, some fibres are abrasive in nature and because of their surface properties. In this part, the effect of WJM process parameters on the machining aspect of hybrid composites is discussed.

Thakur, Singh and Ramkumar (2020a) analyzed the effect of process parameters of AWJM on hybrid glass/carbon-reinforced epoxy composites. They drill holes at two jet pressures, 96 and 304 MPa, and three standoff distances, 1, 2 and 3 mm, along with three tool transverse speeds, 72, 144 and 216 mm/min. The results of the study showed that minimum surface roughness and kerf taper were achieved at a high jet pressure, that is, 304 MPa, and at a low standoff distance and low transverse speed. The authors suggest that the standoff distance is an important parameter in machining composites; otherwise, improper cutting and water absorption occur in laminates due to delamination. Demiral et al. (2022) studied the effect of abrasive particle speed and impact angle of AWJM on carbon-fiber composite experimentally and numerically. Abaqus was used to numerically study the machining behavior of composites, and the Hashin three-dimensional (3D) damage model was used to study the delamination and damage of composite. Study results concluded that when the speed of abrasive particles increased from 300 to 600 m/s, the delamination of fibers reduced from 6.44% to 5.69%. The authors also compared the behavior of AWJM and WJM processes for cutting the same laminate and found out the surface quality of WJM piece was better because fibers were cut properly. The numerical study does not provide much information, but it can be suitable for a process parameter study. Ramraji et al. (2019) studied the behavior of basalt fiber–reinforced fly ash–vinyl ester composites machining by AWJM. The authors studied the effect of fly ash content on the kerf width and the surface quality of machined composites. Fly ash content up to 10% is favorable for cutting parameters, but above 10% fly ash, a deviation in the kerf width and the surface quality is observed. The higher the water jet pressure, the better the surface finish, and no fiber pullouts were observed. As the fly ash content increased, the surface roughness changed from 2 to 5 micrometers. The surface roughness of fly ash–reinforced composite is better than basalt fiber–reinforced vinyl ester polyester composite and 20% fly ash–reinforced composite gives a worse surface finish. The kerf taper width increased as increased in fly ash content and maximum for basalt fiber–reinforced vinyl ester composite. Tripathi et al. (2019) studied the effect of cutting speed and abrasive particle flow rate on the material removal rate of glass fiber polymer composite by applying the Taguchi L_9 array method. The cutting speeds were 50, 75 and 100 cm/min, and the abrasive flow rates were 200, 250 and 300 gm/min. The maximum material removal rate obtained is 5589.23 mm^3/min, with a 100-cm/min cutting speed and a 300-gm/min abrasive flow rate. The authors finally concluded that cutting speed affects the material removal rate, and the predicted S/N ratio is better than the computed one. Thakur, Singh, and Ramkumar (2020b) studied the effect of water pressure, standoff distance and tool speed on surface quality of carbon/glass-reinforced epoxy composite machined using AWJM and applied the analysis of variance (ANOVA) technique to find out the best machining parameters to predict better surface quality. The water pressures were 96 and 302 MPa; the standoff distances were 1, 2 and 3 mm; and tool speeds of 72, 144 and 216 mm/min were considered in this study. The authors defined a delamination factor (F_d), which is the ratio of the maximum width of the cut to the actual width of the cut. The maximum values of F_d were 1.8475 at entry point and 1.4168 at the exit of the hole at 96 MPa, 3-mm standoff distance and 216-mm/min tool speed. The authors concluded that F_ds at exit and entry of the hole increase with the increase in standoff distance and tool speed while decreasing with the increase in water pressure. Manivannan et al. (2021) studied the effect of AWJM process parameters on turkey fiber–reinforced polyester composites. Multi-objective optimization by ratio analysis (MOORA) was applied to optimize the machining parameters. In the MOORA technique, the decision matrix, based on the input parameters, is first determined, followed by calculating the normalized weight and then normalizing the decision matrix. The last step is to determine the normalized performance of possible input parameters. The authors found out the best suitable machining parameters for turkey fiber–reinforced polyester composite are 250 kPa water pressure, 2-mm standoff distance and 30-mm/s feed rate. Gous et al. (2022) reviewed the machining process for composite materials. They reviewed the electric discharge machining (EDM) process and the drilling process for machining of composites. Drill geometry affects the machining and the fibers stuck between the flute of the drill were observed. The EDM process is suitable for electrically conductive materials, so metal matrix composites can be machined using the EDM process. Gulia and Nargundkar (2022) studied the effect of process parameters on the machining of glass fiber composites and optimized the process parameters by the ANOVA method. The authors also tried to develop a mathematical model to optimize the process parameters and apply artificial intelligence methods to analyze the better surface properties via machining.

9.4   CONCLUSION

The following conclusions are drawn based on the present study:

1. WJM process parameters greatly affect the machining of composite laminates in terms of the surface of the workpiece, dimensional accuracy and the cutting geometry.
2. In comparison to conventional drilling the cutting of holes by WJM produces better holes without delamination or peel-out of fibers. Also, no surface cracks are generated during cutting by WJM.
3. The standoff distance has a great effect on dimensional accuracy and the kerf width of composite laminates; the lower the standoff distance, the better the dimensional accuracy and surface obtained.
4. Water jet pressure also greatly affects the machining of composite laminates in terms of fiber pullout; the lower the jet pressure, the more the fiber pullout and there may be chances of improper cutting of composites.
5. Unlike the conventional machining process, the unconventional machining of composite laminates is an easy process and environmentally friendly and does not create serious health issues.

9.5   FUTURE SCOPE

The use of composites is increasing day-by-day in every field of engineering and medicine therefore, it requires their machining without any loss in terms of strength and shape. WJM composites are still being explored because limited work was done by researchers on the machining of polymer composites by WJM and laser machining. These two processes are more in terms of dimensional accuracy and less damage to composite materials. The process parameters optimization of these two processes for different types of polymer composites still needs research.

REFERENCES

Abrate, S., and D. A. Walton. 1992. "Machining of Composite Materials. Part I: Traditional Methods." *Composites Manufacturing* 3 (2): 75–83. doi:10.1016/0956-7143(92)90119-F

Bajpai, Pramendra Kumar, Khushi Ram, Lokesh Kumar Gahlot, and Vivek Kumar Jha. 2018. "Fabrication of Glass/Jute/Epoxy Composite Based Industrial Safety Helmet." *Materials Today: Proceedings* 5 (2): 8699–8706. doi:10.1016/j.matpr.2017.12.296

Bajpai, Pramendra Kumar, Inderdeep Singh, and Jitendra Madaan. 2014. "Development and Characterization of PLA-Based Green Composites: A Review." *Journal of Thermoplastic Composite Materials* 27 (1): 52–81. doi:10.1177/0892705712439571

Chenrayan, Venkatesh, Chandru Manivannan, Kiran Shahapurkar, Girmachew Ashegiri Zewdu, N. Maniselvam, Ibrahim M. Alarifi, Khalid Alblalaihid, Vineet Tirth, and Ali Algahtani. 2022. "An Experimental and Empirical Assessment of Machining Damage of Hybrid Glass-Carbon FRP Composite during Abrasive Water Jet Machining." *Journal of Materials Research and Technology* 19: 1148–1161. doi:10.1016/j.jmrt.2022.05.042

Demiral, Murat, Fethi Abbassi, Tamer Saracyakupoglu, and Mohamed Habibi. 2022. "Damage Analysis of a CFRP Cross-Ply Laminate Subjected to Abrasive Water Jet Cutting." *Alexandria Engineering Journal* 61 (10): 7669–7684. doi:10.1016/j.aej.2022.01.018

Deshmukh, Suhas P., Ramakant Shrivastava, and Chetan M. Thakar. 2022. "Machining of Composite Materials through Advance Machining Process." *Materials Today: Proceedings* 52: 1078–1081. doi:10.1016/j.matpr.2021.10.495

Gous, Mahmad Ziya, Anand Pandey, Shaikh Sarfaraj, and Shahid Tamboli. 2022. "Fabrication and Machining of Fiber Based Composite Materials Using Advance Machining Process, A Review." *Materials Today: Proceedings* 56: 3617–3622. doi:10.1016/j.matpr.2021.12.070

Gulia, Vikas, and Aniket Nargundkar. 2022. "Experimental Investigations of Abrasive Water Jet Machining on Hybrid Composites." *Materials Today: Proceedings* 65: 3191–3196. doi:10.1016/j.matpr.2022.05.372

Hoekstra, Braedon, Atefeh Shekarian, Kamal Kolasangiani, Donatus C.D. Oguamanam, Redouane Zitoune, and Habiba Bougherara. 2022. "Effect of Machining Processes on the Damage Response and Surface Quality of Open Hole Hybrid Carbon/Flax Composites: An Experimental Study." *Composite Structures* 285 (January): 115244. doi:10.1016/j.compstruct.2022.115244

Kaw, Autar K., 2005. *Mechanics of Composite Materials*, CRC Press, Boca Raton, doi: 10.1201/9781420058291

Manivannan, J., S. Rajesh, K. Mayandi, N. Rajini, and Nadir Ayrilmis. 2021. "Investigation of Abrasive Water Jet Machining Parameters on Turkey Fibre Reinforced Polyester Composites." *Materials Today: Proceedings* 45: 8000–8005. doi:10.1016/j.matpr.2020.12.1059

Nele, Luigi, Alessandra Caggiano, and Ilaria Improta. 2021. "Machining of Composite Materials." *Fiber Reinforced Composites: Constituents, Compatibility, Perspectives and Applications*, 83–111. doi:10.1016/B978-0-12-821090-1.00021-1

Ram, Khushi, and Pramendra Kumar Bajpai. 2017. "FEM Analysis of Glass/Epoxy Composite Based Industrial Safety Helmet." *IOP Conference Series: Materials Science and Engineering* 225: 012174. doi:10.1088/1757-899x/225/1/012174

Ram, Khushi, and Pramendra Kumar Bajpai. 2023. "Performance Analysis of Hybrid Bio-Composites Developed through Different Processing Routes for Automobile Application." *Proceedings of the Institution of Mechanical Engineers, Part C: Journal of Mechanical Engineering Science.* doi:10.1177/09544062231156514

Ramraji, K., K. Rajkumar, M. Dhananchezian, and P. Sabarinathan. 2019. "Key Experimental Investigations of Cutting Dimensionality by Abrasive Water Jet Machining on Basalt Fiber/Fly Ash Reinforced Polymer Composite." *Materials Today: Proceedings* 22: 1351–1359. doi:10.1016/j.matpr.2020.01.428

Thakur, R. K., K. K. Singh, and J. Ramkumar. 2020a. "Experimental Investigation of Abrasive Waterjet Hole Cutting on Hybrid Carbon/Glass Composite." *Materials Today: Proceedings* 21: 1551–1558. doi:10.1016/j.matpr.2019.11.085

Thakur, R. K., K. K. Singh, and J. Ramkumar. 2020b. "Delamination Analysis and Hole Quality of Hybrid FRP Composite Using Abrasive Water Jet Machining." *Materials Today: Proceedings* 33: 5653–5658. doi:10.1016/j.matpr.2020.04.056

Tripathi, Dharmagna R., Krupang H. Vachhani, Soni Kumari, Dinbandhu, and Kumar Abhishek. 2019. "Experimental Investigation on Material Removal Rate during Abrasive Water Jet Machining of GFRP Composites." *Materials Today: Proceedings* 26: 1389392. doi:10.1016/j.matpr.2020.02.280

10 Abrasive Water Jet Machining of Sustainable Composites

A Case Study on Hole Making in Basalt-Reinforced PLA and PBS Composites

Binaz Varikkadinmel, Deepak Kaushik, and Inderdeep Singh

10.1 INTRODUCTION

Fiber-reinforced polymer composites offer lightweight, strong alternatives to traditional materials, fueling their adoption across industries [1, 2]. However, environmental concerns drive a shift towards sustainable, biodegradable, and recyclable composites. Over the past four decades, sustainable development has gained prominence, emerging from the environmental movement of the late 20th century. Key sustainability factors encompass water, energy, and chemical usage, greenhouse gas emissions, and waste generation throughout supply chains. With more than 40,000 industrial composite products spanning multiple sectors and increasing usage, addressing the environmental impact of composites is crucial in the face of rising awareness, waste management challenges, and crude oil prices, from both regulatory and consumer perspectives [3, 4]. Utilizing renewable reinforcement materials and biodegradable/compostable polymer matrices to make sustainable composites can mitigate environmental impacts, while prioritizing sustainability and performance in composite manufacturing. Nonetheless, the general characteristics of these composites have yet to match the superior performance of traditional glass/carbon fiber–reinforced counterparts. As such, striking equilibrium between composite capabilities and biodegradability is essential for developing environmentally conscious composites [5]. Ongoing research endeavors are focused on enhancing the resilience and robustness of biocomposites for efficient dynamic applications [6]. The adaptable properties of sustainable composites render them perfect for a wide array of uses, from everyday items to specialized end products. Indeed, sustainable and natural

DOI: 10.1201/9781032665375-10

TABLE 10.1
Applications of Sustainable Composites

Product/Applications	Materials	References
Automotive field: Door panels, aircraft interiors, boat body, dashboards, noise insulation panels, headliners, seat backs, boot lining, decking.	Polymer composites reinforced with coconut fibers, palm fibers, hemp fibers, jute fibers	[7–12]
Building, home decor and furniture: Tables, windows, kitchen equipment, door frames, floor-ceiling work.	Biodegradable composites of PLA–bamboo, jute	[13–15]
Military field: Personal body armor (second-layer protection)	Pineapple leaf fibers, kenaf fiber, ramie fiber, coconut sheath	[16]
Sports equipment: Hockey sticks, fishing gears, tennis racquets, bicycle frames, archery, and snowboards.	Polymer composites using flax fiber, oil palm fruit fibers	[17–19]
Medical use: Scaffold-based biomedical application – parent material for bone implants	Silk fiber/ sheep wool fiber with PLA, composite films of PLA with microcrystalline cellulose extracted from jute fiber	[20, 21]

fiber composites are being researched and promoted for various commercial uses, some examples are listed in Table 10.1.

10.2 CONVENTIONAL MACHINING AND DRILLING OF SUSTAINABLE COMPOSITES

For their usage in semi-structural applications and functional component assembly, the machining of sustainable composites is frequently required. Drilling natural fiber composites is more challenging compared to metals, as the drill bit has to switch between matrix and reinforcement, each possessing distinct characteristics [22]. This process leads to various damage forms in the machined composite parts, including delamination, fiber pullouts, and inter-laminar crack propagation, leading to abundant difficulties [23, 24]. Delamination impacts dimensional precision and surface finish, resulting in a 60% rejection rate at the assembly stage. Drilling substantially deteriorates the surface and subsurface of biocomposites [25] and creep, fatigue, and wear mechanisms dominate in regions with elevated surface roughness and delamination. These damage forms undermine the load-carrying ability, effectiveness, and structural stability [26] of FRP composite components. Binaz et al. [27] demonstrated that the selected cutting method greatly affects the strength of the machined samples, with a strong dependence on the damage caused by the machining process. The common issues identified with the conventional machining of fiber-reinforced polymers are given in the Table 10.2. There is a pressing industrial need for advanced machining processes that produce damage-free components, which is achievable through the adoption of innovative non-conventional machining techniques.

TABLE 10.2

Issues Identified with Conventional Machining of Natural Fiber Reinforced Polymer Composites

Issues Identified with Conventional Machining	References
Machining induced damages Conventional machining processes often result in issues, such as, fiber pullout, delamination, subsurface damage, and extensive matrix removal. Damage caused by machining can compromise the composite structure's mechanical integrity and negatively impact its long-term performance.	[28, 29]
Tool wear The heterogeneous nature of composite materials exerts inconsistent force on the machining tool, resulting in accelerated tool wear, reduced machining efficiency, and subpar machining quality.	[30]
Tool failure Cutting tool encounters alternating cohesive regions, such as strong fibers embedded in a softer matrix. This simultaneous cutting of both fiber and matrix can lead to premature failure of the cutting tool.	[28]
Increased cutting forces and machining temperature Elevated tool wear contributes to a greater demand for cutting force, which in turn increases the machining temperature. This process can cause the composite surface to deteriorate and the matrix to degrade. Machining reinforced thermoplastic composites can be particularly difficult, as they are prone to melting due to the intense heat generated during cutting. If the molten material clogs the cutting blade, the cutting machinery's effectiveness will be significantly reduced.	[31]
Health and environmental hazards Generation of fine, powdery chips releasing substantial amount of dust into the atmosphere, leads to significant occupational health and environmental concerns. Furthermore, these harmful substances can infiltrate the operator's body through ingestion, inhalation, or direct skin contact.	[32]

10.3 ABRASIVE WATER JET MACHINING

10.3.1 Significance and Advantages of Abrasive Water Jet Machining

In order to overcome the shortcomings of traditional machining techniques, advanced approaches, such as, electric discharge machining (EDM), laser machining, ultrasonic machining, water jet and abrasive water jet machining (AWJM), have been employed. As natural fiber-reinforced polymer composites are non-conductive, the EDM process is not suitable for their machining. Laser drilling serves as a practical alternative to conventional drilling methods, as it eliminates issues related to tool wear, tool failure, and vibrations [33, 34]. Laser drilling, being a heat-driven technique, can cause composite ablation as a consequence of laser interactions while making holes. It is vital to minimize thermal damage and flaws in this process [35]. AWJM has captivated extensive attention in the fields of fiber reinforced polymer composites (FRPCs) to produce complex industrial components. AWJM is well suited for machining sustainable engineering materials and has become well known among researchers and production industries. Some of the advantages of using AWJM are given in Table 10.3.

TABLE 10.3

Advantages of Abrasive Water Jet Machining Process

Advantages of AWJM	References
Green machining process AWJM employs a water jet to cleanse the workpiece surface of eroded materials, considerably diminishing or even eradicating the risk of environmental pollution from fibrous substances. The small amount of residual scrap is devoid of chemical pollutants and can be repurposed, leading to conservation of resources, energy, and expenses for reusable materials. AWJM eliminates the need for coolants or lubricants, preventing the creation of chemically tainted waste. Furthermore, this technique does not produce harmful emissions and mandates only a minimal amount of preliminary processing.	[31, 32, 36, 37]
No heat-affected zones Avoids creation of heat-affected zones during the machining process by utilizing a water jet on the workpiece's surface. The water absorbs heat generated by the abrasive particles' impact on the material, thereby preventing thermal damage to the matrix, and hence, it is called a cold machining method.	[38]
Absence of high cutting forces AWJM exerts a relatively lower cutting force on the workpiece as compared to traditional methods, as the abrasive particles serve as the cutting edges, producing minimal or no tangential cutting forces. Additionally, an added benefit of AWJM is that securing the workpiece to the machine does not necessitate complicated tooling, which reduces preparation/setup time.	[39]
Capability of machining various engineering materials Regardless of the properties of a material, AWJM is a highly efficient technique for machining a wide range of engineering materials, including ceramics, composites, polymers, nonferrous metal alloys, stone, aluminium, brass, steel and glass.	[28]
Minimal tool wear and lower vibrations Since the cutting head and the workpiece do not make direct contact in AWJM, tool wear is significantly reduced compared to traditional cutting tools. This method also prevents issues related to deformation or vibrations. Moreover, AWJM achieves better dimensional precision as a result of minimal deformation. It is important to note that wear remains independent of the material being processed.	[40]
Faster process and higher productivity rate AWJM boasts of a machining speed up to ten times faster than conventional processes, resulting in a significant productivity boost over traditional methods. Additionally, AWJ technology demonstrates lower sensitivity to material types and offers exceptional versatility and adaptability in machining.	[30]

Dr. Norman C. Franz conceived the water jet machining concept in 1968, leading to the first commercial water jet in 1971. Later in 1980, Dr. Hashish introduced adding abrasives to water jets. An AWJ machine was then designed for cutting industrial materials [41]. Key AWJ system components include a pressure intensifier, nozzle positioning, water collection, abrasive supply, and mixing unit (Figure 10.1a). In the pressure intensifier unit, a motor pumps oil, and the intensifier serves as an amplifier, using low-pressure oil to generate a high-pressure water jet. The accumulator addresses water pulsations. After passing through an orifice and focusing the nozzle, high-pressure water becomes a high-velocity jet, entering the mixing chamber. Here, momentum transfers from the jet to abrasive particles, directed through the focusing

(a)

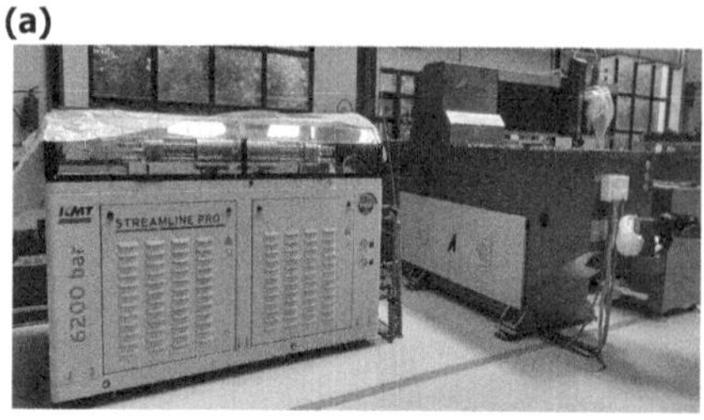

(b)

FIGURE 10.1   (a) Abrasive water jet machine, KMT Streamline Pro, with ultra-high pressure intensifier pumps of 6200 bar. (b) Hole-cutting operation on sustainable laminate using AWJ.

nozzle. The impact of accelerated abrasive particles with high kinetic energy causes material removal via erosion [42]. Abrasive substances like garnet generate erosive forces. High-resistant materials, such as diamond and tungsten carbide, are used for AWJ nozzles. A catcher tank under the workpiece collects waste and water jet. AWJ serves as a versatile machining hub, performing operations like linear cutting, hole drilling, and hole cutting. AWJ drilling efficiently produces holes in a workpiece. The AWJ hole-cutting technique, adjusting AWJ movement based on programmed diameter, enables various hole diameters unattainable through drilling alone. AWJ drilling of sustainable laminate is shown in Figure 10.1b.

10.3.2 Material Removal Mechanism

The core principle of abrasive water jet (AWJ) machining is the high-velocity collision of abrasive particles with workpiece material in a fluid medium, resulting in erosion wear [43]. This process consists of two primary erosion modes: cutting wear, which occurs at shallow impact angles and creates a smooth surface, and deformation wear, which happens at larger angles, producing a rough surface with striations. The balance between these modes depends on process variables [44]. Material removal in ductile erosion combines cutting and deformation wear mechanisms, with micro-cutting and micro-ploughing by particles. Conversely, brittle erosion involves crack growth and chipping due to stress from abrasive particle impact, contributing to material removal [45]. The removal mechanism in specific materials is affected by jet impact angles.

10.3.3 Influencing Factors of AWJM Process

In AWJM, the evaluation of the machined components relies on factors such as the material removal rate (MRR), delamination, kerf characteristics, and surface roughness (SR). Kerf tapering, a detrimental aspect influencing machining accuracy, transpires in conjunction with the penetration depth, presenting a broader jet entrance at the top as opposed to the jet exit. Delamination, an inter-layer occurrence, involves the debonding of layers due to the force of the water wedge. This phenomenon predominantly takes place at the entry and exit points as a result of shock wave initiation

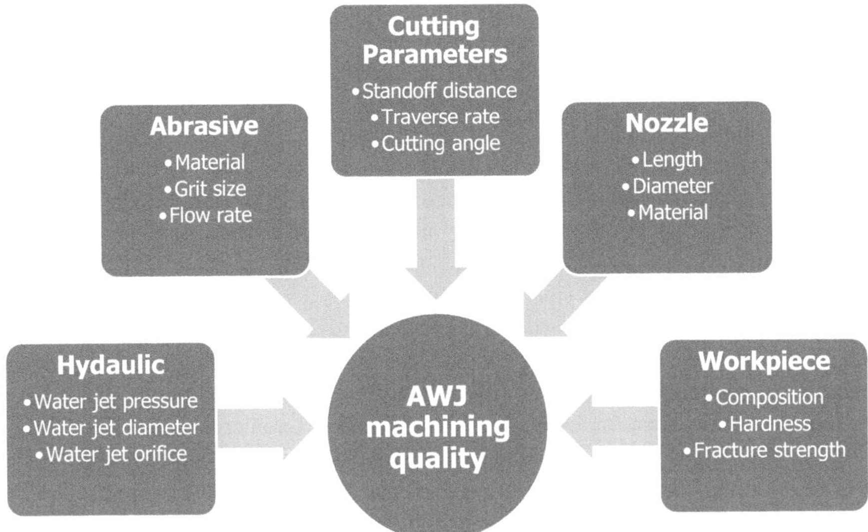

FIGURE 10.2 Process parameters affecting machined quality during AWJ cutting/ machining.

by the high-speed jet. The process parameters that affect the AWJ machined quality are given in Figure 10.2. A majority of existing research indicates that factors, such as jet pressure, stand-off distance (SOD), traverse rate, and abrasive material flow rate (AMFR) play a crucial role in determining machining performance [39].

Jet water pressure impacts the depth of penetration and material removal rate in abrasive water jet cutting, affecting water and abrasive particle dispersion in the stream. It is expressed in units like MPa, bar, or PSI. Traverse rate, measured in mm³/min, influences surface quality; slower rates increase contact time and allow more abrasive particles to impact the material. The AMFR, determined by the focusing nozzle's diameter and denoted in kg/min, affects cutting performance and surface finish. SOD is the gap between the workpiece and nozzle, maintained at an optimal level in millimeters, and significantly influences the kerf profile generated.

10.4   CASE STUDY: MAKING HOLES IN SUSTAINABLE COMPOSITES USING AWJM

The use of sustainable and biodegradable materials in product development has led to the exploration of natural fibers for reinforcing bio-composites. Identifying suitable fibers for eco-friendly and cost-effective composites is crucial. Mahajan et al. [46] found out that basalt fiber (BF) is an optimal choice for sustainable composites with desirable mechanical properties. BF also outperforms E-glass fibers in elasticity and strength while being more cost-effective than carbon fibers. BF has various applications, including automotive reinforcement and wind turbine blades. Its well-developed surface promotes matrix interaction, enhancing composite sustainability.

Consequently, BF is used to develop sustainable composites with poly(lactic acid) (PLA) and poly(butylene succinate) (PBS).

BF/PLA and BF/PBS laminates are primarily processed using compression molding technique. A KMT abrasive water jet cutter (45-kW Streamline Pro-III) with 6200-bar intensifier pump was used to cut laminates. Standard parameters were applied, including 1000-mm/min feed rate, 3-mm stand-off, and 250-mm/min traverse rate. The hole-cutting operation was done on these laminate samples and the machined samples were tested for their mechanical properties, particularly tensile strength and notched strength, using a universal testing machine (UTM; Instron: 5982, USA, ASTM D3039 standards, loading rate: 2 mm/min, gauge length: 100 mm). Figures 10.3a and 10.4a show the BF/PLA and BF/PBS samples that have been cut from laminates, respectively. Figures 10.3b and 10.4b show the holes made using the AWJM technique. For the purpose of identifying surface morphology and topographical patterns, the samples were studied under a stereomicroscope (Nikon, SMZ-745T; Figure 10.3c and 10.4c). Additionally, an analysis of the microstructure in the machined area was conducted employing a field emission scanning electron microscope (FESEM; Zeiss Ultra Plus, Sigma 500; Figures 10.3d and 10.4d).

The inspection of the cutting edge (utilizing both stereomicroscopy and FESEM) discloses wear patterns from erosion at the apex of the cutting area, as well as the emergence of matrix lumps at the terminal point of these tracks. This wear is caused by stray abrasive elements propelled along the outer limits of the cohesive jet.

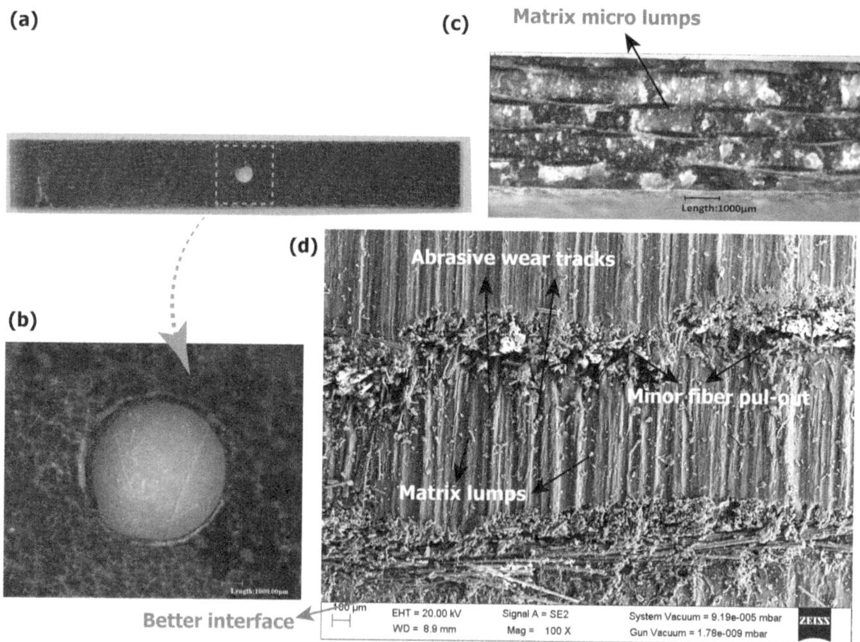

FIGURE 10.3   (a) BF/PLA specimen drilled using AWJ. (b) Stereomicroscope image of the hole produced. (c) Stereomicroscope image of BF/PLA specimen's side surface cut using AWJM. (d) FESEM of BF/PLA specimen cut using AWJM.

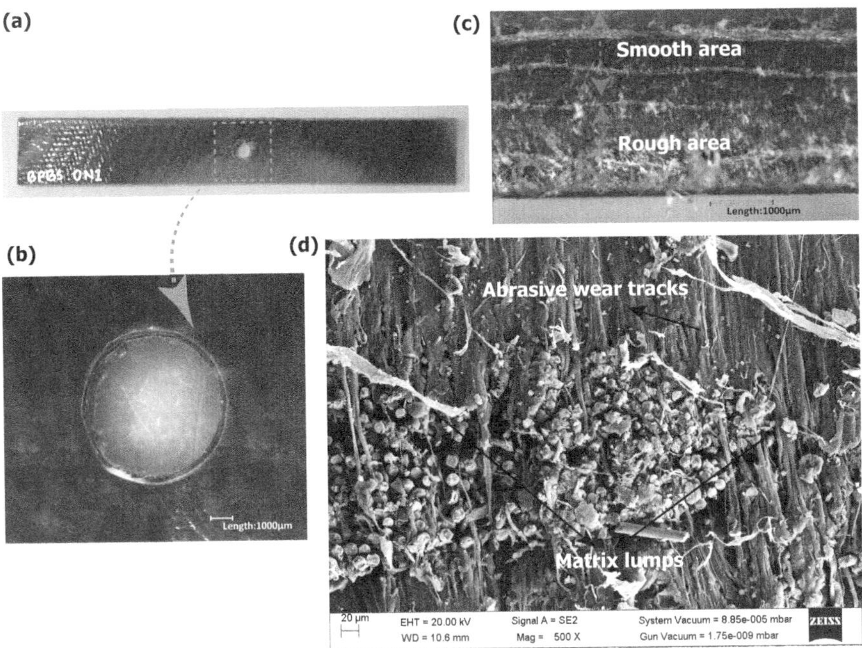

FIGURE 10.4 (a) BF/PBS specimen drilled using AWJ. (b) Stereomicroscope image of the hole produced. (c) Stereomicroscope image of BF/PBS specimen's side surface cut using AWJM. (d) FESEM of BF/PBS specimen cut using AWJM.

Furthermore, the top segment shows an area with a relatively smooth texture, while the lower section demonstrates a relatively coarser surface, a characteristic more pronounced in the BF/PBS sample (Figure 10.4c). Minimal incidents of fiber displacement are also detected, yet the extent of the inflicted damage is significantly less than with traditional machining methods. Thermal damage and matrix scorching are also virtually nonexistent. A unique attribute identified in samples drilled by the AWJ is the presence of a minor damaged zone or matrix crack at the upper layer boundaries, which is significantly more visible in the BF/PLA sample (Figure 10.3b). This might be a consequence of the high-energy abrasive particles' impact on the laminate. Nevertheless, AWJM consistently yields superior and evenly distributed surface profiles for sustainable composites, and the machined surface is fairly smooth, exhibiting negligible inconsistency in texture throughout the sample depth.

Water jet pressure significantly affects machining-induced damage, particularly in polymer composites. Hence holes are cut in BF/PLA and BF/PBS samples at 3500 (N1) and 4500 (N2) bar pressures to assess their effect on open-hole tensile strength. Figures 10.5a and 10.5c show the tensile failed specimens of BF/PLA and BF/PBS respectively. When subjected to a tensile test, the open-hole samples begin to exhibit crack development. Longitudinal cracks are visible in Figures 10.5a and 10.5c, more evident in the BF/PBS samples. These cracks initiate at the hole's edge and progress until they reach their maximum size, culminating in fracture. For the BF/PBS

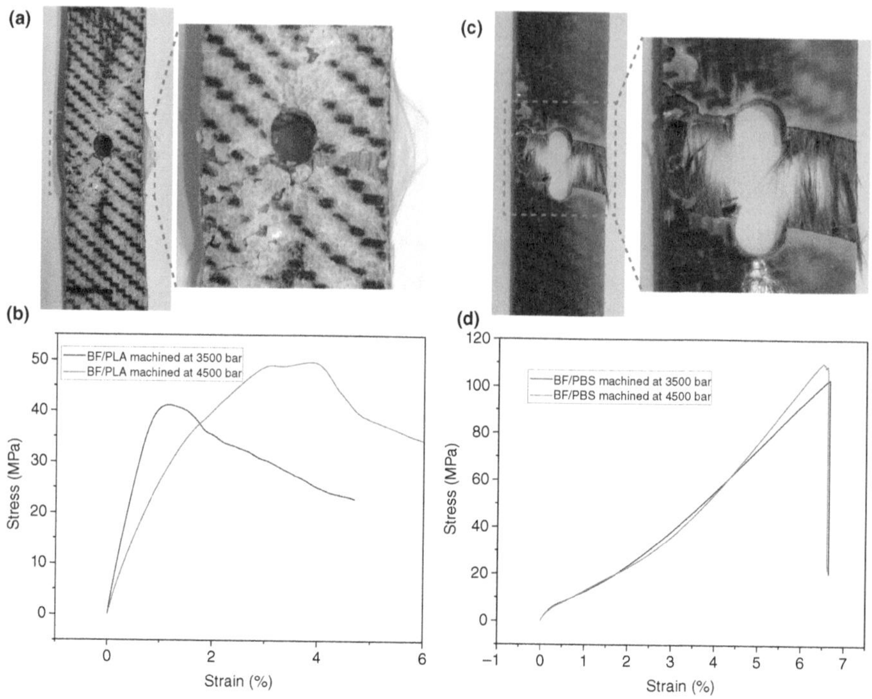

FIGURE 10.5　(a) Failed BF/PLA specimen under tensile loading. (b) Stress–strain curve for BF/PLA machined at water pressures of 3500 and 4500 bar. (c) Failed BF/PBS specimen under tensile loading. (d) Stress–strain curve for BF/PBS machined at water pressures of 3500 and 4500 bar.

samples, at the moment of fracture, the fibers and matrix particles debond completely, resulting in a total break.

Figures 10.5b and 10.5d present the stress–strain graphs for the laminates under two distinct water pressure conditions. It is apparent that the samples cut at elevated water pressure demonstrate superior tensile strength for both types of laminates, and this significantly impacts the stress–strain behavior. This improvement can be linked to the fact that, greater water pressure while cutting, results in less defect formation around the machined area, enhancing material integrity [47, 48]. This also leads to a more homogeneous microstructure, which consequently improves the mechanical properties. The BF/PBS samples display brittle characteristics, while the BF/PLA samples exhibit relatively better elasticity [49, 50]. A more thorough investigation of the failure zones in the composite samples is required to understand the impact of water jet pressure on failure modes in sustainable composites.

10.5　CONCLUSION

The significance of sustainable composites as greener substitutes for traditional materials is highlighted in this chapter. The case study using BF/PLA and BF/PBS laminates shows how AWJM may be successfully applied to achieve dependable and efficient hole-making procedures. The results highlight the improved surface profiles and minimal damage obtained with AWJM, together with barely perceptible texture variation across the sample depth. In addition to its non-contact, non-thermal character and lack of hazardous waste creation, the use of AWJM in sustainable composite machining offers a number of benefits that make it an eco-friendly and viable method for future applications. The study also emphasizes how water jet pressure affects open-hole samples' tensile strength, highlighting the significance of optimizing machining settings to improve material integrity and mechanical properties. Overall, this study highlights the possibility of AWJM as a feasible machining solution for sustainable composites.

10.6 FUTURE SCOPE

It would be advantageous to broaden the study's focus in subsequent studies by examining the effects of different variables, such as abrasive particle size and nozzle diameter, on the machining of sustainable composites. Additionally, investigating the machined components' long-term dependability and durability would offer insightful information for useful applications. In addition, assessing the AWJM technique's economic viability and scalability in large-scale industrial environments would be essential for its broad implementation.

REFERENCES

[1] Karim N, Zhang M, Afroj S, Koncherry V, Potluri P, Novoselov KS. Graphene-based surface heater for de-icing applications. *RSC Adv* 2018;8:16815–23. https://doi. org/10.1039/C8RA02567C

[2] Scribante A, Vallittu PK, Özcan M, Lassila LVJ, Gandini P, Sfondrini MF. Travel beyond clinical uses of fiber reinforced composites (FRCs) in dentistry: A review of past employments, present applications, and future perspectives. *Biomed Res Int* 2018;2018. https://doi.org/10.1155/2018/1498901

[3] Vajdová I, Jenčová E, Szabo S, Melníková L, Galanda J, Dobrowolska M, et al. Environmental impact of burning composite materials used in aircraft construction on the air. *Int J Environ Res Public Health* 2019;16:4008. https://doi.org/10.3390/IJERPH 16204008

[4] Maiti S, Islam MR, Uddin MA, Afroj S, Eichhorn SJ, Karim N. Sustainable fiber-reinforced composites: A review. *Adv Sustain Syst* 2022;6:2200258. https://doi.org/10. 1002/ADSU.202200258

[5] Kaushik D, Gairola S, Varikkadinmel B, Singh I. Static and dynamic mechanical behavior of intra-hybrid jute/sisal and flax/kenaf reinforced polypropylene composites. *Polym Compos* 2023;44:515–23. https://doi.org/10.1002/PC.27114

[6] Vinod A, Sanjay MR, Suchart S, Jyotishkumar P. Renewable and sustainable biobased materials: An assessment on biofibers, biofilms, biopolymers and biocomposites. *J Clean Prod* 2020;258:120978. https://doi.org/10.1016/J.JCLEPRO.2020.120978

[7] Imoisili PE, Ukoba K, Jen TC. Physical, mechanical and thermal properties of high frequency microwave treated plantain (Musa Paradisiaca) fibre/MWCNT hybrid epoxy nanocomposites. *J Mater Res Technol* 2020;9:4933–9. https://doi.org/10.1016/J. JMRT.2020.03.012

[8] Al-Oqla FM, Sapuan SM. Natural fiber reinforced polymer composites in industrial applications: feasibility of date palm fibers for sustainable automotive industry. *J Clean Prod* 2014;66:347–54. https://doi.org/10.1016/J.JCLEPRO.2013.10.050

[9] Cestari SP, Silva Freitas D de F, Rodrigues DC, Mendes LC. Recycling processes and issues in natural fiber-reinforced polymer composites. *Green Compos Auto Appl* 2019:285–99. https://doi.org/10.1016/B978-0-08-102177-4.00012-4

[10] Suriani MJ, Zainudin HA, Ilyas RA, Petrů M, Sapuan SM, Ruzaidi CM, et al. Kenaf fiber/pet yarn reinforced epoxy hybrid polymer composites: Morphological, tensile, and flammability properties. *Polymers* 2021;13:1532. https://doi.org/10.3390/POLYM13091532

[11] Fogorasi MS, Barbu I. The potential of natural fibres for automotive sector - review. *IOP Conf Ser Mater Sci Eng* 2017;252:012044. https://doi.org/10.1088/1757-899X/252/1/012044

[12] Peças P, Carvalho H, Salman H, Leite M. Natural fibre composites and their applications: A review. *J Compos Sci* 2018;2:66. https://doi.org/10.3390/JCS2040066

[13] Morales AP, Güemes A, Fernandez-Lopez A, Valero VC, de La Rosa LS. Bamboo–Polylactic Acid (PLA) composite material for structural applications. *Materials* 2017;10:1286. https://doi.org/10.3390/MA10111286

[14] Dahy H. Natural Fibre-Reinforced Polymer Composites (NFRP) Fabricated from Lignocellulosic fibres for future sustainable architectural applications, case studies: Segmented-shell construction, acoustic panels, and furniture. *Sensors* 2019;19:738. https://doi.org/10.3390/S19030738

[15] Ashraf MA, Zwawi M, Mehran MT, Kanthasamy R, Bahadar A. Jute based bio and hybrid composites and their applications. *Fibers* 2019;7:77. https://doi.org/10.3390/FIB7090077

[16] Nascimento LFC, Louro LHL, Monteiro SN, Gomes AV, Marçal RLSB, Lima ÉP, et al. Ballistic performance of mallow and jute natural fabrics reinforced epoxy composites in multilayered armor. *Mater Res* 2017;20:399–403. https://doi.org/10.1590/1980-5373-MR-2016-0927

[17] Umair M, Khan RMWU. Fibers for sports textiles. In: Ahmad, S, Rasheed, A, Nawab Y, Eds., *Fibers for Technical Textiles*, Topics in Mining, Metallurgy and Materials Engineering 2020, 93–115; Springer, Cham. https://doi.org/10.1007/978-3-030-49224-3_5/COVER

[18] Dicker MPM, Duckworth PF, Baker AB, Francois G, Hazzard MK, Weaver PM. Green composites: A review of material attributes and complementary applications. *Compos Part A Appl Sci Manuf* 2014;56:280–9. https://doi.org/10.1016/J.COMPOSITESA.2013.10.014

[19] Pil L, Bensadoun F, Pariset J, Verpoest I. Why are designers fascinated by flax and hemp fibre composites? *Compos Part A Appl Sci Manuf* 2016;83:193–205. https://doi.org/10.1016/J.COMPOSITESA.2015.11.004

[20] Rahman MM, Afrin S, Haque P, Islam MM, Islam MS, Gafur MA. Preparation and characterization of jute cellulose crystals-reinforced poly(l-lactic acid) biocomposite for biomedical applications. *Int J Chem Eng* 2014. https://doi.org/10.1155/2014/842147

[21] Mangat AS, Singh S, Gupta M, Sharma R. Experimental investigations on natural fiber embedded additive manufacturing-based biodegradable structures for biomedical applications. *Rapid Prototyp J* 2018;24:1221–34. https://doi.org/10.1108/RPJ-08-2017-0162/FULL/PDF

[22] Arputhabalan J, Prabhu S, Palanikumar K, Venkatesh S, Vijay K. Assay of machining attributes in drilling of natural hybrid fiber reinforced polymer composite. *Mater Today Proc* 2019;16:1097–105. https://doi.org/10.1016/J.MATPR.2019.05.201

[23] Feito N, Díaz-Álvarez J, Díaz-Álvarez A, Cantero JL, Miguélez MH. Experimental analysis of the influence of drill point angle and wear on the drilling of woven CFRPs. *Materials* 2014;7:4258–4271. https://doi.org/10.3390/MA7064258

[24] Giasin K, Ayvar-Soberanis S, Hodzic A. An experimental study on drilling of unidirectional GLARE fibre metal laminates. *Compos Struct* 2015;133:794–808. https://doi.org/10.1016/J.COMPSTRUCT.2015.08.007

[25] Vinayagamoorthy R, Rajmohan T. Machining and its challenges on bio-fibre reinforced plastics: A critical review. *J Reinf Plast Compos* 2018;37:1037–50. https://doi.org/10.1177/0731684418778356/ASSET/IMAGES/LARGE/10.1177_0731684418778356-FIG7.JPEG

[26] Heidary H, Karimi NZ, Minak G. Investigation on delamination and flexural properties in drilling of carbon nanotube/polymer composites. *Compos Struct* 2018;201:112–20. https://doi.org/10.1016/J.COMPSTRUCT.2018.06.041

[27] Binaz V, Deepak K, Singh I. Comparative assessment of cutting processes in the mechanical behavior of basalt fiber/poly(lactic acid) matrix composites. *Express Polym Lett* 2023;17:152–68. https://doi.org/10.3144/expresspolymlett.2023.11

[28] Jani SP, Kumar AS, Khan MA, Kumar MU. Machinablity of hybrid natural fiber composite with and without filler as reinforcement. *Mater Manuf Processes* 2016;31:1393–9. https://doi.org/10.1080/10426914.2015.1117633

[29] Kim G, Denos BR, Sterkenburg R. Influence of different piercing methods of abrasive waterjet on delamination of fiber reinforced composite laminate. *Compos Struct* 2020;240:13. https://doi.org/10.1016/j.compstruct.2020.112065

[30] Dhanawade A, Kumar S. Study on carbon epoxy composite surfaces machined by abrasive water jet machining. *J Compos Mater* 2019;53:2909–24. https://doi.org/10.1177/0021998318807278

[31] Ares PFM, Mata FG, Ponce MB, Gomez JS. Defect analysis and detection of cutting regions in CFRP machining using AWJM. *Materials* 2019;12:15. https://doi.org/10.3390/ma12244055

[32] Jagadish BS, Ray A. Prediction and optimization of process parameters of green composites in AWJM process using response surface methodology. *Int J Adv Manuf Technol* 2016;87:1359–70. https://doi.org/10.1007/s00170-015-8281-x

[33] Rao S, Sethi A, Das AK, Mandal N, Kiran P, Ghosh R, et al. Fiber laser cutting of CFRP composites and process optimization through response surface methodology. *Mater. Manuf. Process* 2017;32(14):1612–21. https://doi.org/10.1080/10426914.2017.1279296

[34] Hejjaji A, Singh D, Kubher S, Kalyanasundaram D, Gururaja S. Machining damage in FRPs: Laser versus conventional drilling. *Compos Part A Appl Sci Manuf* 2016;82:42–52. https://doi.org/10.1016/J.COMPOSITESA.2015.11.036

[35] Malik K, Ahmad F, Gunister E. Drilling performance of natural fiber reinforced polymer composites: A review. *J Nat Fibers* 2022;19:4761–79. https://doi.org/10.1080/15440478.2020.1870624

[36] Prabu VA, Kumaran ST, Uthayakumar M. Performance evaluation of abrasive water jet machining on banana fiber reinforced polyester composite. *J Nat Fibers* 2017;14:450–7. https://doi.org/10.1080/15440478.2016.1212768

[37] Kim G, Denos BR, Sterkenburg R. Influence of different piercing methods of abrasive waterjet on delamination of fiber reinforced composite laminate. *Compos Struct* 2020;240:13. https://doi.org/10.1016/j.compstruct.2020.112065

[38] Vigneshwaran K, Venkateshwaran N. Statistical analysis of mechanical properties of wood-PLA composites prepared via additive manufacturing. *Int J Polym Anal Charact* 2019;24:584–96. https://doi.org/10.1080/1023666x.2019.1630940

[39] Natarajan Y, Murugesan PK, Mohan M, Khan S. Abrasive water jet machining process: A state of art of review. *J Manuf Process* 2020;49:271–322. https://doi.org/10.1016/j.jmapro.2019.11.030

[40] Thakur RK, Singh KK. Abrasive waterjet machining of fiber-reinforced composites: A state-of-the-art review. *J Braz Soc Mechan Sci Eng* 2020;42:25. https://doi.org/10.1007/s40430-020-02463-7

[41] Pang KL, Nguyen T, Fan JM, Wang J. Modelling of the micro-channelling process on glasses using an abrasive slurry jet. *Int J Mach Tools Manuf* 2012;53:118–26. https://doi.org/10.1016/j.ijmachtools.2011.10.005

[42] Shanmugam DK, Masood SH. An investigation on kerf characteristics in abrasive water-jet cutting of layered composites. *J Mater Process Technol* 2009;209:3887–93. https://doi.org/10.1016/j.jmatprotec.2008.09.001

[43] Meng HC, Ludema KC. Wear models and predictive equations - their form and content. *Wear* 1995;181:443–57. https://doi.org/10.1016/0043-1648(95)90158-2

[44] Hashish M. Visualization of the abrasive-waterjet cutting process. *Exp Mech* 1988;28:159–69. https://doi.org/10.1007/bf02317567

[45] Wang YF, Yang ZG. Finite element model of erosive wear on ductile and brittle materials. *Wear* 2008;265:871–8. https://doi.org/10.1016/j.wear.2008.01.014

[46] Mahajan A, Binaz V, Singh I, Arora N. Selection of natural fiber for sustainable composites using hybrid multi criteria decision making techniques. *Compos Part C: Open Access* 2022;7:100224. https://doi.org/10.1016/J.JCOMC.2021.100224

[47] Doreswamy D, Shivamurthy B, Anjaiah D, Sharma NY. An investigation of abrasive water jet machining on graphite/glass/epoxy composite. *Int J Manuf Eng* 2015. https://doi.org/10.1155/2015/627218

[48] Varikkadinmel B, Singh I. Fracture behaviour analysis of sustainable basalt-reinforced polymer composites subjected to thermal cycling and open holes. *Eng Fail Anal*, 2024;163. https://doi.org/10.1016/j.engfailanal.2024.108481

[49] Madhu S, Balasubramanian M. Influence of nozzle design and process parameters on surface roughness of CFRP machined by abrasive jet. *Mater. Manuf. Process*, 2017;32(9). https://doi.org/10.1080/10426914.2016.1257132

[50] Varikkadinmel B, Kaushik D, Singh I. Effect of thermal cycling on open-hole tensile strength of sustainable composites: An experimental investigation. *Polym Compos*, 2024;45(4):3169–3183. https://doi.org/10.1002/pc.27981

11 Machine Learning Approaches for Drilling GFRP Laminate

A Study for Thrust Force and Torque-induced Damage

Hitesh Sharma, Pawan Kumar Rakesh, and Inderdeep Singh

11.1 INTRODUCTION

It has been demonstrated that glass fibre–reinforced plastics (GFRPs) are superior materials for structural applications and are already being utilised in these roles [1–3]. Due to the anisotropic structure of these materials, drilling them requires a whole different technique than drilling conventional materials like metals. This is because the tool being used must simultaneously cut through several different phases of the material. The result of drilling is contingent on a number of variables, including the characteristics of the reinforcement and the matrix, the fibre–volume fraction, the fibre orientation, and so on. When GFRP is drilled, it causes several different types of damage both inside the composite and on its surface [4]. Damage caused by drilling, such as delamination, fibre pull-out, hole ovality, deformation, and cracking can substantially negatively affect the structural integrity of composite parts. The drilled hole quality is primarily determined by a number of elements, the most important of which are the drilling process parameters tool material, cutting speed, feed rate, and tool shape. The tuning of these parameters, numerical modelling, and artificial intelligence (AI) are some of the primary study areas that have been performed to optimise the drilling operation in order to obtain damage-free holes. Various researchers have tried reducing delamination worldwide. In today's rapidly developing technological world, perceptive computing techniques such as fuzzy logic, neural networks (deep), and signal processing systems play an essential part in today's most advanced manufacturing systems. These techniques allow for more accurate sensor monitoring and control of machining procedures [5]. These are only a few of the significant research areas that have been undertaken. For the purpose of optimisation in a variety of domains, several artificial intelligence practices, such as adaptive network-based fuzzy inference systems (ANFISs), artificial neural networks (ANNs), genetic algorithms, and so on, can be utilised [6]. The field of artificial intelligence (AI) has developed a method known as machine learning, which trains computers to learn from the experiences they have had in the past. Instead of depending on an equation as a model, machine learning algorithms employ computer technology to "learn" information straight from the data. This is in contrast to traditional methods of data analysis, which rely on mathematical models. Because of this, it is possible for machine learning algorithms to function without the use of models. The algorithms are able to respond to an increase in the amount of samples that are available for learning by dynamically improving their performance. Deep learning is a subfield of machine learning that presents its own distinct set of challenges to researchers. The process of machine learning makes use of not one but two separate learning strategies: supervised learning and unsupervised learning. Supervised learning is the process of teaching a model with data that has known inputs and outputs so that the model can forecast future outcomes. Unsupervised learning, by comparison, looks for intrinsic structures or hidden patterns in input data in order to learn from them. An ANN is a paradigm for processing data that is conceptually similar to the way the brain and other biological nerve systems work. It is made up of many neurons (processing elements) that are tremendously well connected and work collectively to address specific issues. ANNs are able to learn from examples by modifying the synaptic connections between their neurons. There are many composites-related domains where ANNs have found use. To reduce the risk of graphite-epoxy laminate delamination during drilling, Stone and Krishnamurthy [7] created a thrust force controller. Kadi [8] employed an ANN to simulate the mechanical properties of polymer-fibre composites. The density and tensile properties of metal matrix composites that are particle-reinforced can be predicted using the ANN modelling approach were first presented by Altinkok and Koker [9]. The purpose of this study is to utilise a supervised machine learning approach to create a neural network-based damage prediction model for woven GFRP laminates that may be used prior to drilling operations.

11.2 EXPERIMENTAL

For the purpose of developing composite specimens, a woven E-glass fibre was utilised in this work. To create the GFRP laminates, epoxy resin AW 108 and hardener HV953 IN were utilised in the manufacturing process. In order to make the GFRP laminates, the hand lay-up process was utilised. After that, the specimens were chopped into the desired shape and dimensions in order to carry out the drilling operations. The newly created GFRP laminates were drilled on a drilling machine (radial) with carbide drills (JO, parabolic, and twist) Figure 11.1.

This allowed for a more precise and accurate drilling process. The operating variables used in the current study were cutting speed (90–2800 rpm), feed rate (0.03–0.30 mm/rev), drill diameter (4 mm and 8 mm), and drill geometry. The newly created GFRP laminates were drilled with solid carbide drills (JO [1], parabolic [2], and twist [3]) on a radial drilling machine. This allowed for a more precise and

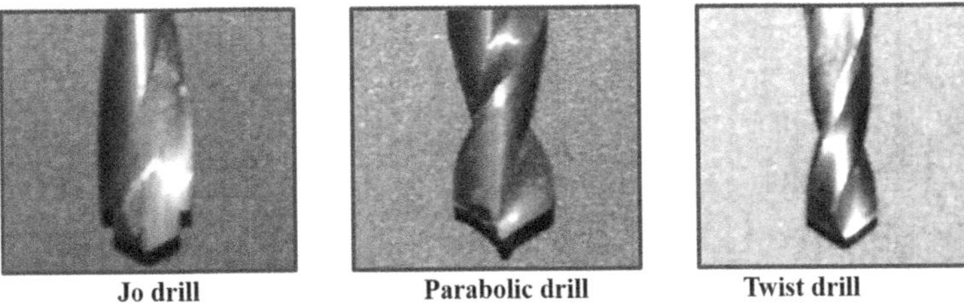

FIGURE 11.1 Various drill bit geometries.

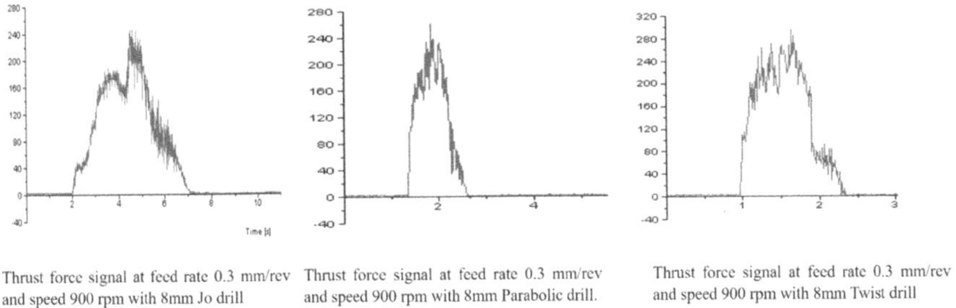

FIGURE 11.2 Thrust force signals with different 8 mm drills at 0.3mm/rev and speed 900 rpm.

accurate drilling process. For the purpose of recording the signals of torque and thrust force, a piezoelectric drill dynamometer with four components (Make: Kistler, Type 9272) was utilised. The dynamometer is connected to the charge metres (Make: Kistler, Type-5015), and the output from the charge metres is fed into the personal computer through a card (analogue/digital). The charge metres were manufactured by Kistler. After obtaining the signals, they were evaluated, and research was conducted to determine how the forces and torque were linked to the damage that was caused surrounding the hole, Figure 11.2.

When it comes to quantifying drilling-induced damages or delamination, there are a few different approaches to choose from. Dye-penetrating testing is, however, still the method that is utilised most frequently. Dye that had gotten into the damaged area about the hole presented a vibrant photograph of that area when it was viewed in the light. Using a digital camera, we were able to record digital photographs of the damage that was caused by the hole. Image J, a piece of open-source software, was utilised to determine the extent of the damage, as depicted in Figure 11.3.

11.3 PREDICTION OF DRILLING-INDUCED DAMAGE USING AN ANN MODEL

Developing an ANN model that is capable of predicting thrust forces and torques requires the identification of both input and output components, both of which are of utmost importance. drill diameter (d), cutting speed (V), drill geometry, and feed rate

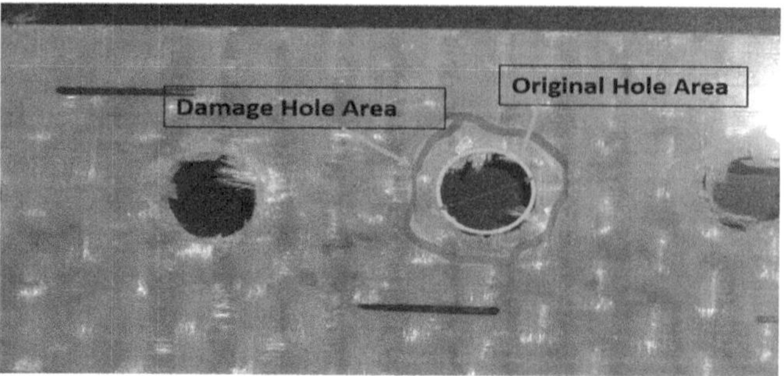

FIGURE 11.3   Damage around the hole in a GFRP laminate.

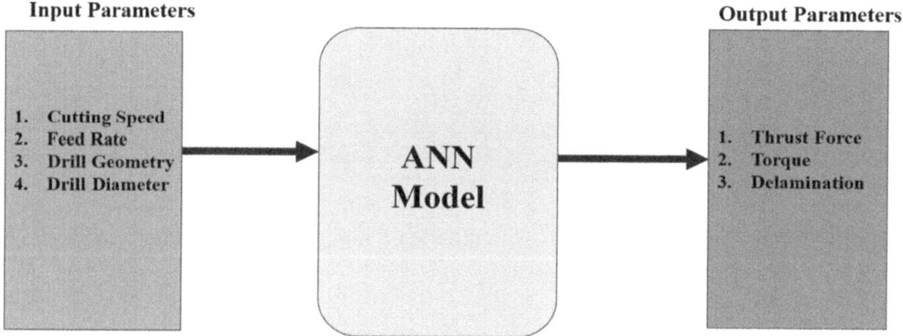

FIGURE 11.4   Graphic illustration of an ANN model for various input–output parameters.

(f) are the input parameters that are taken into consideration by the model that is being projected. In advance, these parameters can be chosen, which indicates that they are controlled in nature and may be chosen prior to carrying out an experiment. This also suggests that the parameters can be selected in advance. The following is an illustration of the input/output data sets of the exemplary ANN that are going to be explained in order to forecast torque, thrust force, and delamination. Figure 11.4 depicts the schematic diagram of an ANN model for various input–output parameters.

Processing the experimental data into patterns is one of the most critical activities that need to be finished before the ANN can be trained by delivering the data set to the network and the input–output mapping. This task must be finished before the ANN can be trained. Before the input–output data set is supplied to the network, training and testing pattern data points are created and evaluated. A total of 396 holes were drilled by utilising various combinations of the input values. In total, there were 396 data points used, 316 of these 396 holes were used for training, while 80 of these holes were used for testing. A neural network that is of the multilayer feed-forward back-propagation type has been taken into consideration in the model that has been developed. MATLAB software was used to create the neural network that was being used. In this particular model, the input layer is comprised of four neurons that correspond to a total of four drilling parameters, and the output layer is made up of three neurons that correlate to a total of three outputs. The ANN configuration is shown as 4–10–3 in this representation. There are ten neurons located in the hidden layer, four neurons located in the input layer, and three neurons located in the output layer. After creating and analysing a large number of networks, which might differ in a number of ways including their structure, transfer function, training methodology, and so on, the approach of trial and error is useful for estimating the number of neurons that should be placed in the hidden layer. There are 316 different training patterns that are taken into account for ANN modelling. After training, the weights are set in stone, and the produced model is validated by the use of eighty different testing points. The learning rate was retained at 0.5 throughout the training process, 1000 training epochs were determined to be the maximum number allowed, and the ANN was trained for a total of 1000 iterations. It is generally agreed that mean squares with lower values are superior. The mean square error steadily got better and finally reached a stable state at epoch 25 (Figure 11.5). The training sustained for more iterations before reaching a point where it became unproductive. The result shows that the error is less than 8%, which is within acceptable limits. Four axes are used to represent the overall data, testing data, training data, and validation data respectively. The line that represents the target is shown as a dashed line along each axis (Figure 11.6). This line is generated by deducting the outputs from the ideal results. The solid line

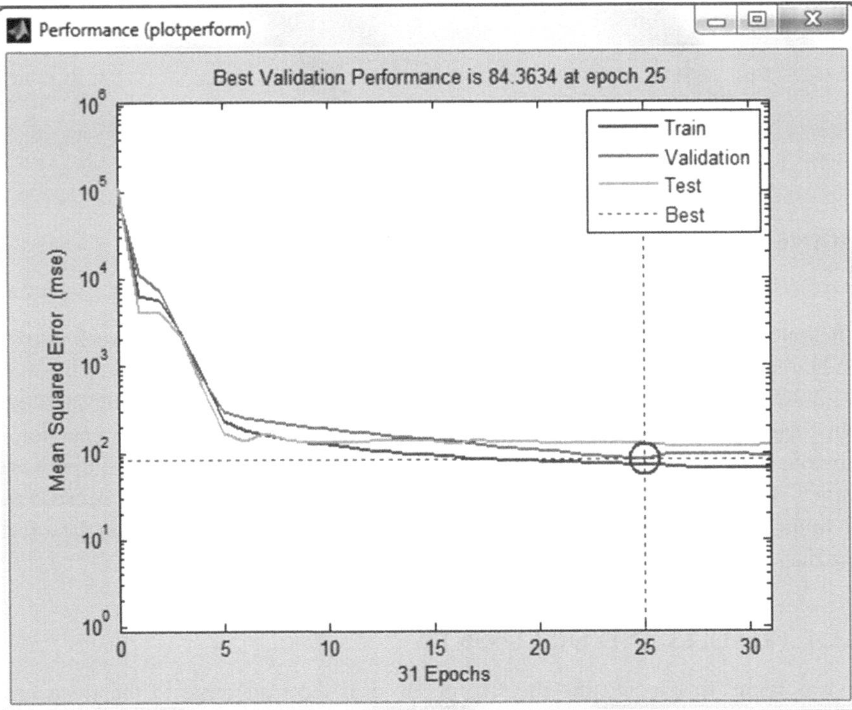

FIGURE 11.5 Performance plot.

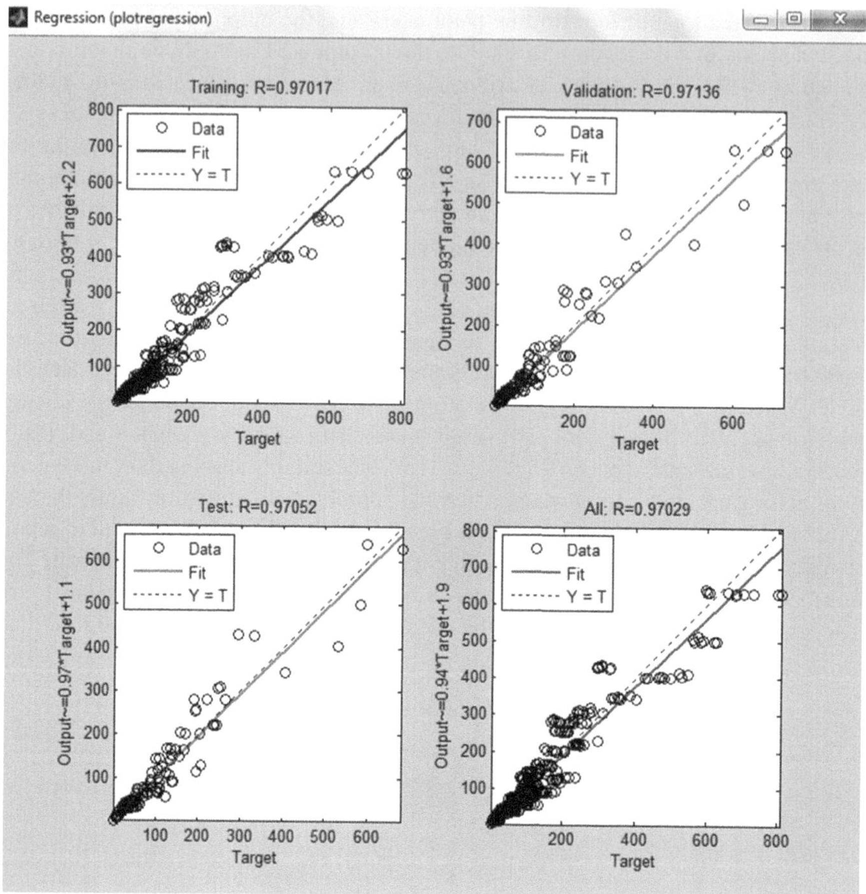

FIGURE 11.6   Regression plots.

illustrates the best-fitting linear regression line, which can be found between outputs and targets. This solid line illustrates the linear regression line that provides the most accurate fit. A value of one for R indicates a close link between the targets and outputs being measured, whereas a value of zero for R indicates that the connection is random. Regression is a metric that assesses the connection between outputs and targets. We may state that there is a close link between aims and outputs because the R-value is nearing 0.97 for all the data, including training data, validation data, testing data, and the overall data.

11.4   RESULTS AND DISCUSSION

It was found that increasing the size of the drill led to a surge in the shear area amount, which further led to an intensification of the thrust force and torque signals. Because of this, it is safe to say that the dimensions of the drill had some effect on the thrust forces and torque. It has come to everyone's attention that the thrust forces are the primary cause of damage to GFRPs when drilling. When the feed rate is slowed, the thrust forces are reduced, which can then help in reducing the amount of induced delamination [10].

Based on the results of the experiment, we are able to establish particular ranges of force and torque, which enables us to determine particular values of torque and thrust force, the particular inputs like drill type, drill diameter, feed rate, and cutting speed. The thrust force ranged anywhere from 24.28 N to 806.77 N, the torque ranged anywhere from 6.97 N-cm to 117.27 N-cm, and the delamination ranged anywhere from 1 to 2.45 accordingly. Graphs were drawn to visually compare the anticipated values for various outputs like thrust force, torque, and delamination (Figures 11.7–11.9) to the various experimental values that matched them.

The results demonstrate conclusively that the estimated accuracy is high and settles well with the experimental data. Thrust, torque, and delamination can all be estimated with the use of the prediction model. Drilling-induced damages should be reduced, and process efficiency should be raised, through process optimisation. The operator receives an array of values for the three output parameters and based on these parameters can choose whether to proceed with the drilling procedures for the supplied input drilling parameters. The output parameters give the user with the following outcomes:

- It was discovered that increasing the drill diameter increases the thrust force and torque values due to an increase in shear area, implying that drill dimensions have an effect on thrust force and torque.
- The results show that true and precise holes are obtained at a restrained spindle speed, a low feed rate, and the minutest drill diameter.

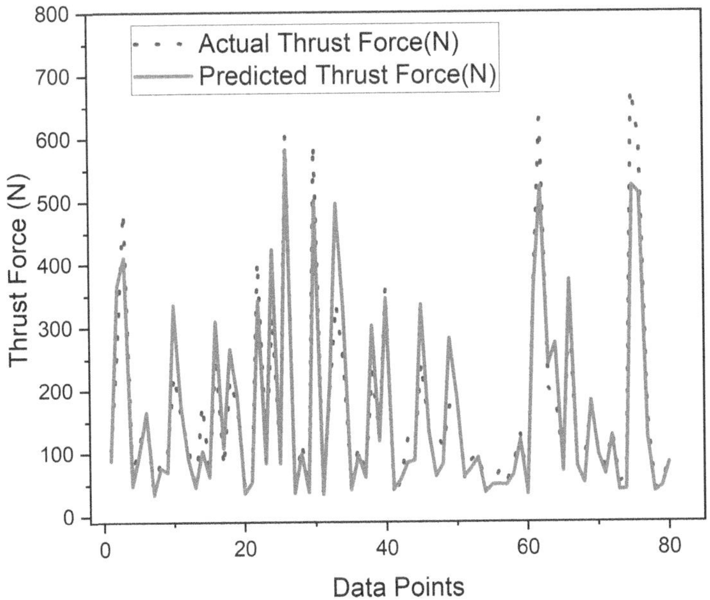

FIGURE 11.7 ANN anticipated thrust force vs. experimental thrust force.

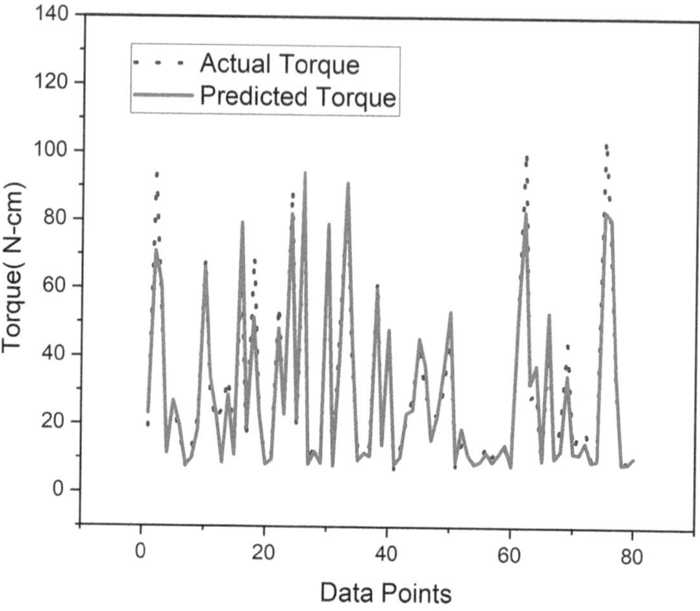

FIGURE 11.8 ANN anticipated torque vs. experimental torque.

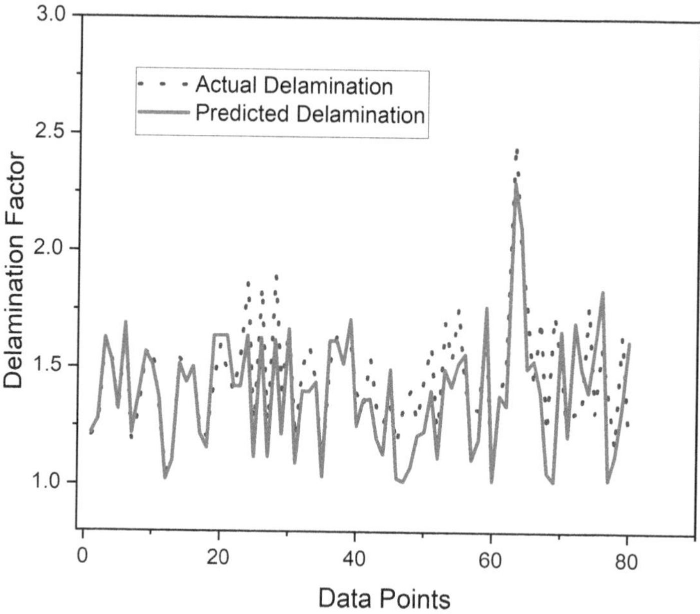

FIGURE 11.9 ANN anticipated damage vs. experimental damage.

11.5 CONCLUSION

Experiments were carried out on GFRP laminates that were constructed using the hand lay-up technique. The numerous input factors such as drill diameter, spindle speed, feed rate, and drill geometry were taken into consideration during the experimental process. For the purpose of forecasting thrust force, torque, and delamination, ANN-based AI techniques were utilised. A total of 396 holes were drilled by utilising various combinations of the input parameters. The current research effort opens the door to the following potential outcomes:

- The drilling parameters have an effect on the drilling forces, which in turn has an effect on the damage caused by the drilling.
- The thrust force and torque rise as the drill diameter increases.
- Intermediate spindle speed, moderate feed rate, and the drill with the smallest diameter allowed for the most precise holes to be drilled.
- The delamination factor was determined to be the lowest with the JO drill and the highest with the twist drill.

Of late, various advancements have been reported by various researchers worldwide. For instance, for the purpose of drilling fiber-reinforced polymer composites, a rotary-mode ultrasonic drilling technology has been conceived and developed by Debnath et al. (2015) [11]. It was reported that drilling performance in terms of the MRR of glass-epoxy laminates can be greatly enhanced by making extensive modifications to the typical ultrasonic drilling method [11]. To examine the drilling behaviour of sisal/epoxy and nettle/epoxy laminates, Debnath et al. (2016) [12] conceived of, planned, developed, and implemented a new tool geometry. The effectiveness of the new instrument was related to that of the conventional twist drill. Forces and the resulting drilling-induced damage were used to evaluate the levels of performance. The proposed tool shape was shown to be more effective than the conventional twist drill at reducing drilling forces and minimising material loss [12]. The use of unconventional energy sources like lasers and microwaves to investigate the drilling characteristics has also been introduced in recent years [13–15].

11.6 FUTURE SCOPE OF WORK

The recommendations that follow have been made in order to further increase the scope of the application of the machine learning approach in the field of drilling of GFRP laminates:

1. Because ANNs are adaptive systems, adding more training data will help these systems become even more accurate in their output.
2. The data used in this study only cover a short range of values due to technical limitations. More data that range over a wide range of values could be gathered in order to increase the applications of machine learning.
3. To make future predictions more adaptable, various increased input data, such as specimen specifications, the material of the drill, cutting fluid use, and others can be included to further improve the predictions.

REFERENCES

[1] Palanikumar, K. "Experimental investigation and optimisation in drilling of GFRP composites." *Measurement* 44, no. 10 (2011): 2138–2148.

[2] Kilickap, E. "Optimization of cutting parameters on delamination based on Taguchi method during drilling of GFRP composite." *Expert Systems with Applications* 37, no. 8 (2010): 6116–6122.

[3] Kumar, S., S. R. Chauhan, P. K. Rakesh, Inderdeep Singh, and J. P. Davim. "Drilling of glass fiber/vinyl ester composites with fillers." *Materials and Manufacturing Processes* 27, no. 3 (2012): 314–319.

[4] Singh, Inderdeep, Pawan Kumar Rakesh, and Jagannath Malik. "Predicting forces and damage in drilling of polymer composites: soft computing techniques." In *Mechatronics and Manufacturing Engineering*, pp. 227–258. Woodhead Publishing, 2012.

[5] Panchagnula, Kishore Kumar, Naga Vamsi Krishna Jasti, and Jayaprakash Sharma Panchagnula. "Prediction of drilling induced delamination and circularity deviation in GFRP nanocomposites using deep neural network." *Materials Today: Proceedings* 62 (2022): 7118–7123.

[6] Mishra, R., J. Malik, and Inderdeep Singh. "Prediction of drilling-induced damage in unidirectional glass-fibre-reinforced plastic laminates using an artificial neural network." *Proceedings of the Institution of Mechanical Engineers, Part B: Journal of Engineering Manufacture* 224, no. 5 (2010): 733–738.

[7] Stone, R., and K. Krishnamurthy. "A neural network thrust force controller to minimize delamination during drilling of graphite-epoxy laminates." *International Journal of Machine Tools and Manufacture* 36, no. 9 (1996): 985–1003.

[8] El Kadi, Hany. "Modeling the mechanical behavior of fiber-reinforced polymeric composite materials using artificial neural networks—A review." *Composite Structures* 73, no. 1 (2006): 1–23.

[9] Altinkok, Necat, and Rasit Koker. "Modelling of the prediction of tensile and density properties in particle reinforced metal matrix composites by using neural networks." *Materials & Design* 27, no. 8 (2006): 625–631.

[10] Rakesh, P. K., Inderdeep Singh, and D. Kumar. "Failure prediction in glass fiber reinforced plastics laminates with drilled hole under uni-axial loading." *Materials & Design* 31, no. 6 (2010): 3002–3007.

[11] Debnath, Kishore, Inderdeep Singh, and Akshay Dvivedi. "Rotary mode ultrasonic drilling of glass fiber-reinforced epoxy laminates." *Journal of Composite Materials* 49, no. 8 (2015): 949–963.

[12] Debnath, Kishore, Mohit Sisodia, Ashutosh Kumar, and Inderdeep Singh. "Damage-free hole making in fiber-reinforced composites: an innovative tool design approach." *Materials and Manufacturing Processes* 31, no. 10 (2016): 1400–1408.

[13] Tewari, Renu, Manoj Kumar Singh, Sunny Zafar, and Satvasheel Powar. "Parametric optimization of laser drilling of microwave-processed kenaf/HDPE composite." *Polymers and Polymer Composites* 29, no. 3 (2021): 176–187.

[14] Kumar, Gaurav, Sanjay Mavinkere Rangappa, Suchart Siengchin, and Sunny Zafar. "A review of recent advancements in drilling of fiber-reinforced polymer composites." *Composites Part C: Open Access* 9 (2022): 100312.

[15] Lautre, Nitin Kumar, Apurbba Kumar Sharma, Pradeep Kumar, and Shantanu Das. "Characterization of drilled hole in low melting point material during low power microwave drilling process." *Materials Research Express* 6, no. 9 (2019): 095329.

12 Process Optimization of Aluminum-Reinforced Epoxy Composites for Aerospace Application
A Case Study for Mechanical and Microstructural Behavior

Adefemi Adeodu, Rendani Maladzhi,
Mukondeleli Grace Kana-Kana Katumba,
Ilesanmi Daniyan, and Tobilola Sholanke

12.1 INTRODUCTION

According to Giacomo [1], composites are composed of two or more mechanically and metallurgically combined materials that are heterogeneous. These materials possess properties that are distinct from those of their individual components. Composites are widely used in various applications because they offer tailored mechanical behavior, such as stiffness, strength, weight, and high temperature, which cannot be achieved by a single material. Examples of composites that have been used for their special qualities include steel reinforced by concrete and cellulose reinforced by lignin (wood). To get the desired engineering features, manufacturers can customize composite production. Polymer matrix composites (PMCs), metal matrix composites (MMCs), ceramic matrix composites (CMCs), and carbon matrix composites (CAMCs) are the four main categories into which composites can be categorized [2]. Reinforced plastics and advanced composites make up PMCs. Low-stiffness glass fibers (E-glass) are often used in reinforced plastics since they are reasonably inexpensive. For the past 30–40 years, they have been utilized in a variety of products, including pipe, corrugated sheet, automotive panels, sporting goods, and boat hulls. The aircraft sector has been using advanced composites, which are relatively expensive, for about 15 years. They have a fiber and matrix mix for improved strength and stiffness. High-performance continuous fibers like high-stiffness glass

DOI: 10.1201/9781032665375-12

"""

(S-glass), graphite, aramid, or other organic fibers are typically present in them in significant amounts. This assessment is primarily concerned with the market potential for advanced composites.

According to Giacomo [1], polymer–matrix composites have been widely used due to their high mechanical qualities and low specific weight in many different sectors, with aerospace and military industries being among the first to adopt them. The cost implications of manufacturing these composites, especially for larger and more complex parts, have been a challenge. In recent years, the focus of research has shifted toward improving production processes and materials for industrial sectors, such as automotive, electrical/electronic, and high-level sports. Adeodu et al. [2] suggest that electromagnetic (microwave) curing may be a viable alternative to expensive autoclave processes. Transfer molding has been used since the 1960s for the manufacture of propeller blades and lightweight airframes, replacing aluminum alloys, as mentioned by Giacomo [1]. If more effective production procedures are used, high-quality components can be produced at a pace of up to 10,000 parts per year, such as microwave-assisted molding [3]. These processes have been used in the production of structural parts for cars such as the DB7 Aston Martin and the BMW Z-1. Marsh [4] notes that in the aerospace industry, the percentage application of polymer-matrix composites due to weight has reached 50% with the Boeing 787 and 53% with the Airbus 350.

The microwave-assisted production process is intricate, and a random approach to process design and modeling can be risky. The modeling stage is crucial for making choices that align with desired specifications and production requirements. Manufacturing issues in composite parts can stem from resin degradation, dry spots, and porosity caused by air entrapment or incomplete filling. Simulation and optimization are essential to anticipate the impact of decision variables on the final product, determine process parameters that maximize quality, minimize defects, increase production rate, and reduce process costs. Microwave-assisted heating has drawbacks in composite material processing. Process simulations employ physical-based models to predict cure time and other properties in the final part.

Additionally, during the curing process of polymer–matrix composites, some may undergo condensation chemical reactions that produce volatile products. Traditional heating techniques often result in faster curing at the surface, leading to the formation of impermeable skin that can trap these volatiles, causing voids and delamination that render the finished part useless. The electromagnetic heating of composites via microwave also presents negative effects such as low penetration of the energy, difficulty in heating materials with low dielectric loss, and the potential for hot spots or thermal runaway. The slow conduction of heat in an autoclave and the exothermic nature of the matrix curing reaction also impose limitations on the manufacturing process, leading to prolonged processing times and potential thermal degradation.

Process simulations for composite curing use physical-based models to forecast cure time and final properties in order to address these problems. Meeting quality standards and cutting processing time require effective production strategies.

The goal of the study is to determine the ideal process parameters for manufacturing while examining the effects of autoclave curing procedures on the mechanical and microstructural behavior of aluminum–epoxy composites. The study's goals are to investigate the effects of curing techniques on mechanical properties and microstructure; examine production process parameters like cure temperature, actual cure cycle, and degree of cure; and identify the ideal production process parameters that minimize processing time by analyzing the ideal temperature evolution and degree of cure.

12.2 OVERVIEW OF PROCESS OPTIMIZATION OF POLYMER–MATRIX COMPOSITES

In the field of composite process optimization, researchers have focused on three main areas: cure, impregnation, and drape optimization [1]. Within each area, single and multiple objective optimizations have been explored. Cure optimization has been studied for two main processes: pultrusion and batch (lay-up) [1]. For single-objective optimization, techniques based on gradients have been used to optimize residual stress and process time. Multiple objective optimizations for pultrusion have considered full speed and energy consumption. In batch processes, three different combinations of objectives have been examined: process time and degree of cure, process time and maximal exotherm, and process cost and process time [1]. Genetic algorithms (GAs) have been utilized to solve multiple objective cure problems. This study focuses on single-objective optimization in the batch process, considering process time, degree of cure, and maximum exotherm.

12.2.1 SINGLE-OBJECTIVE OPTIMIZATION OF CURE CYCLE PROBLEMS

Finding the least point of an objective function is a necessary step in solving an optimization problem. Subject to appropriate equality and inequality criteria, the same formal technique can be used with a decreasing function of the purpose if finding a maximum point is necessary [5]. The format of a single optimization task is as follows:

$$Minimise: f(x) \quad gi(x_1, x_2\ldots\ldots\ldots x_n) \geq 0, \text{ With } i = 1,2,3\ldots r \tag{12.1}$$
$$Subject:$$

$$hj(x_1, x_2\ldots\ldots\ldots x_n) = 0, \text{ With } j = 1,2,3\ldots p, \tag{12.2}$$

where the third inequality constraint is gi, the fourth equality constraint is hj, and the nth parameter is x_n. The parameters are counted as n, the inequality constraints as r, and the equality constraints as p.

It is possible to find sites in the design space where each of the set values for the optimization problem satisfies the constraints of the problem. If there are no other feasible points in the design space where all the objectives are equal or better than it, then that feasible point is the Pareto optimum one. The following is a mathematical definition of such a point for a minimization problem:

$$\left(x_1, x_2, x_3 \ldots x_n\right) \text{ is the Pareto optimal, iff} \tag{12.3}$$

$$\text{There is no}\left(x_1', x_2', x_3' \ldots x_n'\right) \text{such that} \tag{12.4}$$

$$V \in \{1,2,3\ldots m\}, f_i\left(x_1', x_2', x_3' \ldots x_n'\right) \le f_i\left(x_1, x_2, x_3 \ldots x_n\right). \tag{12.5}$$

12.2.1.1 Gradient-Based Technique

By just using data from the objective function and its gradient, the gradient-based technique seeks to minimize the value of the objective function at each step [1]. Only when the problem's landscape is well known and devoid of local minima or when the terrain has a convex space can the gradient base technique be used with confidence. A position that is near the local minimal will prevent the process from reaching the global minimal; otherwise, the outcome of this algorithm depends significantly on the beginning point selection. This kind of algorithm starts by choosing an initial design point x^k and setting the iteration counter $k = 0$, then choosing a search direction d_k that must be specified in the design space. The next choice is a positive step size of a_k in the direction of d_k. The following definition for the iteration's next step is

$$x^{k+1} - x^k + a_k d_k. \tag{12.6}$$

In a minimization problem, a desirable direction is the one able to reduce the current value of the objective function as

$$f\left(x^{k+1}\right) \angle f\left(x^k\right). \tag{12.7}$$

Substituting Equation 12.6 into Equation 12.7 and applying Taylor's expansion, the following equation is obtained:

$$f\left(x^k\right) + a_k\left(c_k d_k\right) \angle f\left(x^k\right) \tag{12.8}$$

$$c_k = \nabla f\left(x^k\right) \tag{12.9}$$

The condition to satisfy is therefore

$$c_k.d_k \angle 0. \tag{12.10}$$

Because the function's gradient is known, a desirable direction can be defined as any d_k satisfying Equation 12.10. Cauchy's oldest first-order approach is the steepest descent method [6]. The strategy is based on the basic property of a function's gradient to indicate in direction of maximal increase [5]. As a result, the maximum reduction is in the other direction. The steepest decline is then indicated by the negative gradient vector.

$$d = -c \tag{12.11}$$

12.2.1.2 Algorithm-Based Approach

Random searches of all potential solutions serve as the foundation of stochastic algorithms. The best solution will emerge as a result of the random search after a huge number of solutions are generated at random and evaluated. A stochastic search strategy called simulated annealing (SA) is based on statistical mechanics, most especially the Boltzmann distribution [1]. In a large search space with numerous local minima, it can offer a solution that is close to the global optimum. Finding the system's lowest energy state is the basis of the technique. The SA makes an effort to mimic the physical annealing procedure, which enables atoms to achieve a minimal energy configuration after heating up to a high energy state. A high temperature (control parameter) is used at the beginning of the process to accommodate a wide range of input variability. The temperature decreases and the range of input fluctuation decreases as the program runs. This behavior leads to better solutions [7]. The main principle of genetic algorithms, which are founded on the ideas of natural selection, is the survival of the fittest. A population of individuals (strings of bits) that each represent a solution make up a generation. At each generation, the goodness of each solution is given a fitness score, and the current generation evolves into the following one by using certain mathematical operators (reproduction, crossover, mutation) that try to bias the survival of the fittest solution from the previous generation in a favorable direction [8].

12.2.2 Batch Optimization of the Cure Cycle to Minimize Process Time

Determining the appropriate cure cycle involves finding the temperature profile that will produce a part with a tolerable level of cure, state of stress, and void content [1]. To achieve these objectives, it is necessary to take into account the following factors:

1. In the shortest amount of time, the degree of cure must achieve a minimum defined value (e.g., 0.8).
2. The internal temperature of the component must be kept below the permitted maximum.
3. There must be a minimal amount of residual stress.
4. The vacancy rate needs to be respectable.
5. Extra resin must be taken out to guarantee equal consolidation [1, 2, 9].

By improving the temperature cycle, the first three issues can be addressed, and the last two can be improved by improving the pressure profile. The work does not address pressure profile optimization.

The application of some constraints can help batch curing optimization shorten the process duration. In this kind of optimization, typical limitations include the highest exothermic temperature permitted inside the part to prevent resin degradation and uneven curing [1], the lowest temperature permitted to start curing [10], and the minimum degree of cure at the beginning of the reaction [10]. The problem is computationally difficult due to its large dimensions. A thick polyester–carbon fiber composite laminate's cure time was shortened by using the look-ahead method [11]. In order to find a close-to-ideal solution, this approach aims to strike a compromise between two techniques: a straightforward gradient-based descent method and a heuristic search. The key idea is that the look-ahead approach is used to approximate the original issue with one that includes local optimization criteria; as a result, the method is also known as local criterion optimization (LCO). A heuristic function h is assessed for each potential decision for each time step, and the one with the lowest value is chosen for that step. It takes a long time to compute this problem. This issue could be resolved using parallel processing [1]. The cure time was lowered from 300 minutes to 170 minutes (43%), there was less exotherm, and there was less temperature gradient inside the part when compared to the recommended cure profile. Selecting the ideal dwell temperature is a challenge. The quality of the final product is not significantly impacted by the number of dwells for thin parts but is impacted by the number of dwells for thick parts. The comparison of a thin and thick laminate composite with epoxy resin cure profile optimization was done using an sequential quadratic programming (SQP) algorithm based on thermo-chemical analysis [12]. The optimal cure time for a thin part (2 mm) was discovered to be 588 seconds with one dwell and 582 seconds with two dwells, according to a study by Li and Tucker [13]. By comparison, the ideal cure profile required 5024 seconds with a single dwell, 3702 seconds with two dwells, and 1544 seconds with three dwells for a thick section (40 mm). The investigation came to the conclusion that heat dissipation concerns make it important to choose the dwell time, cooling rate, and heating rate in thick sections. The cure time can be greatly decreased by carefully choosing the cure cycle. Li and Tucker discovered in a subsequent investigation that failing to take into account consolidation while determining the cure profile results in partially thick parts. But taking consolidation into account results in an ideal cure cycle that yields fully cured and consolidated parts with a 48% to 63% reduction in cure time.

The SQP method was employed in a different study by Udaykumar and Shyy [14] to improve the temperature profile of a cure cycle with five stages among heating and cooling zones for a polyester glass fiber system. According to the findings, an optimized cure profile performed 56% better than a heuristic strategy and 80% better than a conventional profile.

Using GAs, Chang et al. [15] determined the ideal cure profile for thick carbon fiber–epoxy resin prepreg composite laminates of 5 cm and 10 cm while taking the maximum temperature and maximum heating rate into consideration as restrictions. The outcomes demonstrated how effective and adaptable GAs algorithms were in determining the ideal cure profile. According to the study, the cure time was slashed by 62% for laminates measuring 5 cm and 67% for components measuring 10 cm. To shorten cure times and raise the caliber of the finished product, researchers have created a variety of algorithms and techniques to optimize the cure profile of composite materials. For instance, Yang and Lee [16] reduced the cure time of thick carbon fiber–epoxy resin prepreg laminates by 36% to 80% compared to the conventional profile by optimizing the cure profile using an iterative gradient-based technique. With the help of a GA, Skordos and Partridge [17] were able to reduce the cure time of a carbon fiber–epoxy resin system by 26% as compared to the normal cure profile. While Dufour et al. [19] used the Levenberg Marquardt algorithm to optimize the cure profile of a thick composite laminate while taking into account residual stresses and the maximum temperature difference between center and surface, Rai and Pitchumani [18] used a SQP algorithm to find the optimal cure and pressure profiles for a thick graphite–epoxy laminate. For both glass fiber–epoxy and glass fiber–polyester systems, Mawardi and Pitchumani [20] significantly reduced cure times and improved cure uniformity by combining the SA and simplex methods with resistive heating components. Recent work by Carlone [21] demonstrated the enhanced capabilities of the Nelder–Mead approach when paired with SA for the simulation-based optimization of the cure profile of a thick carbon fiber–epoxy resin composite laminate.

12.3 MATERIALS AND METHOD

The study utilized aluminum powder mixed with epoxy resin as the base matrix, along with graphene nanoplatelets with a purity of 99.9%. Particles of grade D, ranging from >150 μm to <45 μm, were added to the matrix. Diglycidyl ether of bisphenol A (DGEBA) was the specific epoxy resin utilized, and Huntsman Advanced Materials also provided the hardener (XB 3473), which contained diethyltoluenediamine (80–92% concentration) and 1,2-diaminocyclohexane (4–10% concentration). All samples were created using polyvinyl chloride (PVC) as a mold release agent and a stoichiometric ratio of DGEBA to diamino diphenyl methane (DDM).

12.3.1 LAMINATION OF THE NANOCOMPOSITES AND CURING

The composites were laminated by hand lay-up after the constructed molds had been adequately cleaned and prepped. To achieve homogeneity, the necessary quantities of fillers, matrix, and other additives were combined. Polymer matrix composites of 200 grams with varied loadings of 10%, 20%, and 30% weighted aluminum were created using the mixture. Each sample was laminated into a square mold measuring 30 × 30 × 30 mm, rubbed with a mold-releasing agent, and then dried for 24 hours before being post-cured for 6 hours in an autoclave oven at a maximum temperature of 86 °C. The mixing ratios for the composites and post-curing profiles are shown in Tables 12.1, 12.2 and 12.3, respectively.

12.3.2 CHARACTERIZATION OF THE SAMPLES

The cured samples were cut to size using a machine in compliance with the necessary analytical and American Society for Testing and Materials (ASTM) requirements.

12.3.2.1 Tensile Strength Analysis

Using a versatile Instron 3369 tensile machine, three epoxy/aluminum samples (designated 1, 2, and 3) were machined and evaluated for tensile properties in accordance with ASTM E8/E8M-13a. Before each specimen was secured to the machine's grips,

TABLE 12.1

Mixing Ratio of the Composite Samples

Sample	Percentage Fraction of Filler	Weight Fraction of Matrix (g)	Weight Fraction of Filler (g)	Weight Fraction of Additive (g)	Mass of Composite (g)
1	10	135	20	45	200
2	20	120	40	40	200
3	30	105	60	35	200

TABLE 12.2

Degree of Cure Profile for Aluminium–Epoxy Composites

30% Aluminum-Filled Epoxy		20% Aluminum-Filled Epoxy		10% Aluminum-Filled Epoxy	
Time (min)	Degree of Cure (%)	Time (min)	Degree of Cure (%)	Time (min)	Degree of Cure (%)
5	0.2	5	0.25	5	0.33
10	0.40	10	0.40	10	0.42
20	0.50	20	0.48	20	0.45
30	0.55	30	0.50	30	0.50
50	0.65	50	0.62	50	0.60
110	0.75	110	0.70	110	0.69
170	0.85	170	0.80	170	0.80
230	0.96	230	0.92	230	0.92
290	0.96	290	0.92	290	0.92

TABLE 12.3

Temperature Profile for Aluminum–Epoxy Composite Samples

30% Aluminum-Filled Epoxy		20% Aluminum-Filled Epoxy		10% Aluminum-Filled Epoxy	
Time (min)	Temp (°C)	Time (min)	Temp (°C)	Time (min)	Temp (°C)
5	32	5	33	5	32
10	39	10	39	10	40
20	44	20	45	20	42
30	52	30	54	30	50
50	58	50	58	50	58
110	62	110	62	110	62
170	70	170	68	170	68
230	80	230	78	230	80
290	86	290	90	290	88
350	86	350	90	350	88

its dimensions were measured using a strain-measuring sensor and monitor. The specimens were put under 1 N of strain and 0.5 mm/min of crosshead speed until they broke. On the monitor, the resulting tensile strength was displayed automatically.

12.3.2.2 Fracture Toughness Test Analysis

With a focus on strain energy release rate (G) and tensile load application utilizing a double cantilever beam in accordance with ASTM D5528, the fracture toughness of a rectangular, uniform thickness, and unidirectional laminated composite was characterized using the interlaminar (delamination) method. Loading blocks were positioned on the top and bottom surfaces of the specimen arms during fabrication, and a non-adhesive Teflon film was applied to serve as a delamination initiator. With a steady crosshead speed of 1–5 mm per minute, the specimen's delaminated end was subjected to quasi-static loading. The interlaminal fracture toughness critical strain energy rate (Gc) is calculated using Equation 12.12 [22]:

$$G_c = \frac{3P\delta}{2b(a+\Delta)},\tag{12.12}$$

where P represents the load, δ represents the displacement, b represents specimen width and a represents delamination length. The Δ may be determined experimentally by generating a least squares plot of the cube root of compliance, $C^{1/3}$ as a function of delamination length (Figure 12.1).

12.3.2.3 Thermal Conductivity Analysis

Thermal conductivity is a measure of a material's responsiveness to heat. The LaserComp Fox 50 equipment was used in this investigation to test the thermal conductivity of the samples in line with the ASTM C518 and ISO 8301 standards. The samples were placed in a test stack between two isothermal plates, and a temperature gradient was applied across their thickness. The upper plate was fixed while the lower plate was modified to maintain good contact with the sample and reduce interfacial resistance [24]. Equation 12.13 [23] expresses the one-dimensional Fourier–Biot law, which is used by the Fox instrument.

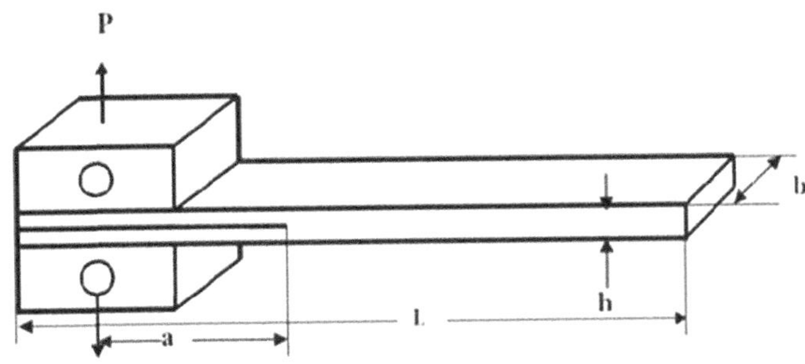

FIGURE 12.1 Double cantilever beam specimen with load blocks used for Mode I testing.

$$Q = -\lambda\left(dt/dx\right), \tag{12.13}$$

where q is the heat flux (W/m^2) flowing through the samples, λ is the thermal conductivity of the sample (W/m/k), and dt/dx is the temperature gradient (k/m) on the isothermal plates.

The temperature gradient within the sample equals the ratio of the temperature difference between the surfaces and the thickness of the sample (Δx) is expressed by Equation 12.14.

$$\Delta T = \left(T\_hot - T\_cold\right)/\Delta x = \Delta T / \Delta x \tag{12.14}$$

The thermal resistance of the sample Rs is equal to the ratio of the sample thickness to the thermal conductivity as expressed by Equation 12.15:

$$R\_s = \Delta x / \lambda. \tag{12.15}$$

12.3.2.4 Impact Analysis

The Charpy impact tester was used to evaluate the impact strength and energy needed to deform or fracture the samples under a rapid load. Four samples were created as Charpy specimens with a central V-notch in accordance with ASTM E23 specifications. The pendulum was released using the prescribed mechanism after the specimens were placed in the tester's vise with the notch facing the hammer. For each sample, this procedure was carried out three times, starting with an immediate measurement of the impact energy on the scale. The figures for impact strength were derived from the average of the measured impact energy.

12.3.2.5 Hardness Analysis

According to the ASTM E384 standard, the hardness of the samples was assessed using Brinell's hardness testing apparatus with a 1.5-mm diameter ball and a 100-kg load. As soon as the indenter made contact with the specimens, they were taken off the testing table. The diameter of the resulting depression was measured with a micrometer and under a microscope. To determine the Brinell hardness number, use Equation 12.16:

$$BHN = \frac{2P}{\pi D\left(D\sqrt{D^2 - d^2}\right)}, \tag{12.16}$$

where D is the diameter of the indenter in millimeters, d is the indent diameter in millimeters, and P is the load applied in (N).

12.3.2.6 Scanning Electron Microscopy

Scanning electron microscopy (SEM) was utilized to examine filler dispersion and the interfaces between the fillers and the matrix in polymer nanocomposites.

TABLE 12.4
Summary of the Numerical Experimentation

Notation	Independent Variables	Levels		
		−1	0	1
A	Aluminum-filled epoxy (%)	10	20	30
B	Curing time (min)	3	177.5	350
C	Curing temperature (°C)	30	60	90

The JEOL-JSM 5800 electron microscope was used for the SEM examination, with the necessary sample size fixed tightly on a specimen holder and a 15-kV accelerating voltage used for the desired magnification. Samples were prepared by sonicating a tiny amount of Al in acetone, then transferring several drops of the liquid to a gold-coated silicon substrate and drying in a vacuum oven [25, 26].

12.3.3 OPTIMIZATION OF THE CURING PROCESS

The settings for the curing process were optimized using the response surface methodology (RSM). RSM is a design of experiment (DoE) that discovers workable combinations of process parameters and the relationships between their cross-effects. RSM was used because it can produce a mathematical model that predicts how the specified experiment would respond in relation to the process factors [27–29]. According to curing time, temperature, and the percentage of aluminum-filled epoxy, this model is able to forecast the degree of cure under identical circumstances. For the application of RSM, Design Expert software (version 8) was employed, and 20 experimental trials based on workable combinations of process parameters were carried out. The results were recorded and presented in Table 12.4, with the degree of curing (%) as the response of the designed experiment. The process parameters considered in this study include percent aluminum-filled epoxy (10–30%), curing time (5–350 min), and curing temperature (30–90 °C), informed by similar studies carried out by [30, 31].

To confirm the accuracy of the numerical experiments, physical experiments were conducted, followed by an analysis of variance (ANOVA) and statistical analysis of the results obtained from both the physical and numerical experiments. A valid model is indicated by a "p-value Prob > F" less than 0.050, with a statistically insignificant "lack of fit" in comparison to pure error, and Adeq. Precision greater than 4 [32, 33]. Additionally, a correlation coefficient such as R-squared, predicted R-squared, and adjusted R-squared with a value near 1 indicates a statistically significant model [32, 33].

12.4 RESULTS AND DISCUSSION

In this section, the effects of aluminum on the mechanical characteristics of polymer composites are examined.

12.4.1 RESULTS OF TESTING

12.4.1.1 Test of Tensile Strength Results

The relationship between the tensile strength of the polymer composites and the percentage weight fractions of aluminum employed in them is shown in Figure 12.2. The data from the tensile characterization of the polymer composites show a gradual rise in tensile strength for 10%, 20%, and 30% weight fractions of aluminum, respectively, from 94.20, 98.4, to 105.5 MPa. The presence of aluminum particles as reinforcement improves the interfacial bonding, thereby enhancing the tensile property of the composites [24]. Furthermore, the particle sizes of aluminum promote their easy dispersion into the polymer phase to form a homogeneous mixture, which consequently enhances the tensile strength of the composite [23, 34].

12.4.1.2 Fracture Toughness Test Result

Figure 12.3 depicts the impact of aluminum percentage weight fractions on the fracture toughness of polymer composites. The fracture toughness characterization of polymer composites reveals decreasing trends of 13.92, 12.14, and 6.24 MPaM$^{1/2}$ for 10%, 20%, and 30% weight fractions of aluminum in the polymer composite, respectively. The declining trend was caused by the composite's ductility property at 10% aluminum deteriorating to brittleness as the weight fraction of the reinforcement increased; that is, less energy is required at higher brittleness states [35]. This demonstrates that the matrix and reinforcement are debonded [36].

12.4.1.3 Thermal Conductivity Result

Figure 12.4 depicts the influence of reinforcement dispersion on the thermal conductivity of polymer composites. Figure 12.4 shows that the thermal conductivity property increases as the percentage weight fractions of aluminum increase. When

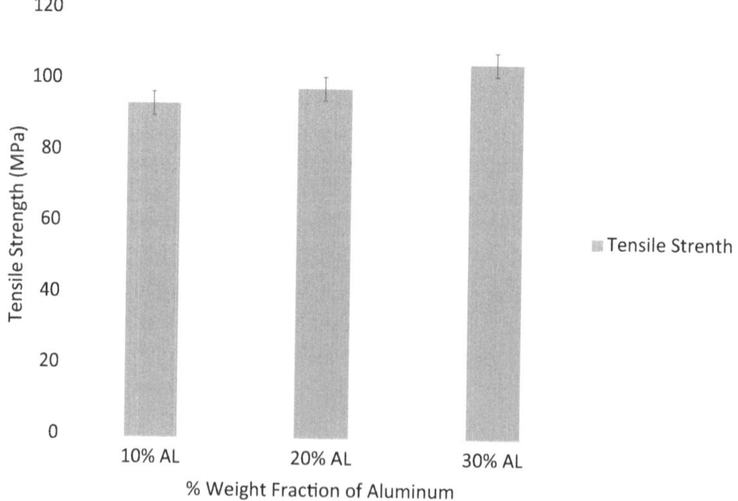

FIGURE 12.2 Tensile strength against percentage weight fraction of aluminum.

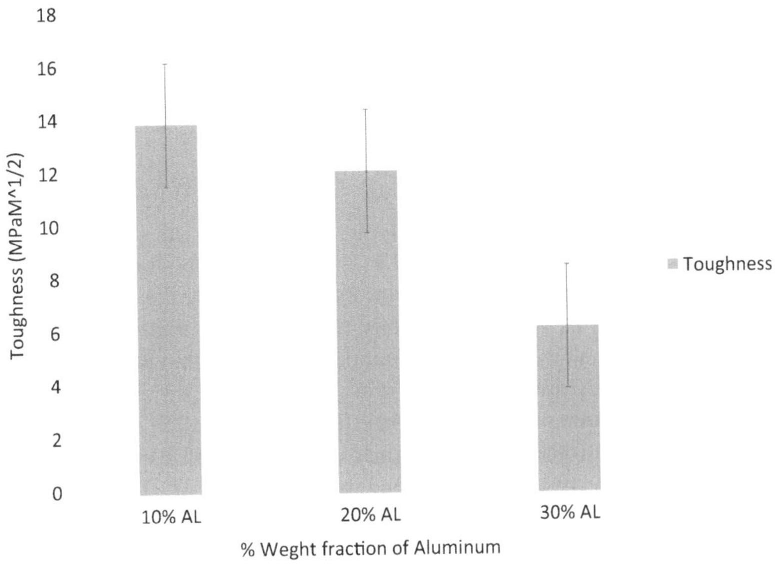

FIGURE 12.3   Fracture toughness against percentage weight fraction of aluminum.

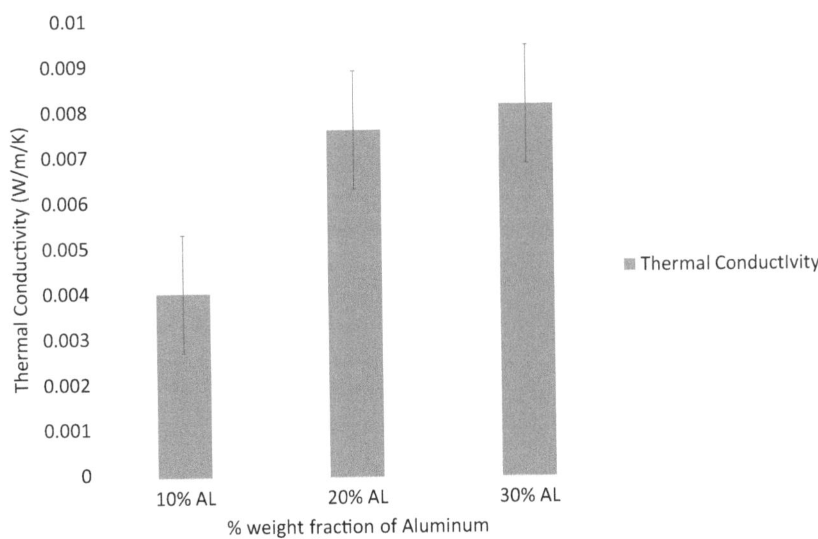

FIGURE 12.4   Thermal conductivity against percentage weight fraction of aluminum.

compared to other samples, the sample with a 10% weight fraction of reinforcement has exceptionally low thermal conductivity (0.00405 W/m/k). Thermal energy is transferred between contacting interfaces by the interaction of free electrons and lattice vibration, as described by Kapitza resistance [37]. As a result, poor coupling at polymer surfaces has had a major impact on heat resistance [23, 25]. According to the research, numerous aspects such as filler nature, orientation, aspect ratio, and processing style may be to blame [26, 38–41]. Furthermore, as the percentage weight fractions of aluminum grow, so does the concentration of the reinforcement in the composite, resulting in networks of aluminum that lower the mean free path of the heat carrier.

12.4.1.4   Impact Test Result

The variation in energy absorbed by polymer composites with various percentage weight fractions of aluminum powder (10%, 20%, and 30%) is shown in Figure 12.5. As the percentage weight fractions of reinforcements climb, the amount of impact energy absorbed rises as well. Additionally, the composite with a weight fraction of 30% showed a notable increase. Aluminum polymer composites' ductility and deformation capabilities are enhanced by reinforcements [42, 43]. The reinforcement encourages the dislocating motion necessary for the deformation of composite plastic [43]. The cohesion force between the matrix and the reinforcements is also used to explain how composite materials respond to impacts [44].

12.4.1.5   Hardness Test Result

Hardness testing using the common Brinell hardness measurement was used to quantify the degree of brittleness of composites made of aluminum and polymer reinforcement. Figure 12.6 shows how composite materials' Brinell hardness values fluctuate as their percentage weight fractions of reinforcements increase. The hardness number rises as the percentage weight fractions of the reinforcements do. The increase is ascribed to the reinforcements' considerable homogeneous dispersion

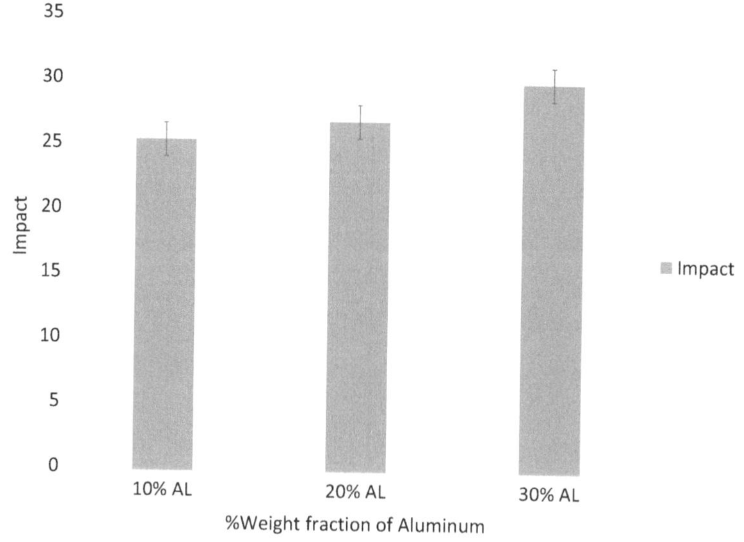

FIGURE 12.5   Impact strength against percentage weight fraction of aluminum.

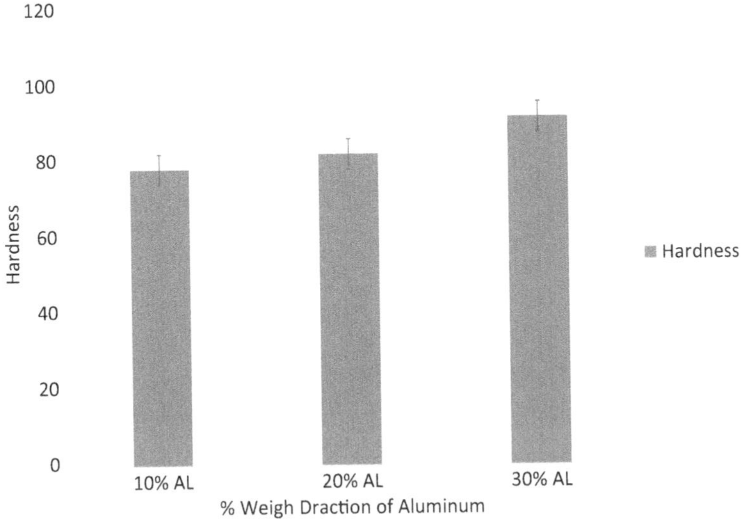

FIGURE 12.6 Hardness against percentage weight fraction of aluminum.

within the matrix [45], which serves as a physical impediment to dislocation motion [46]. Composites' hardness is influenced by their reinforcement and matrix [42].

12.4.1.6 Microstructural Analysis

Figures 12.7a–c show SEM images of changing dispersion of aluminum particles in epoxy matrix weight fractions of 10%, 20%, and 30%. At higher magnification [47, 48], the micrographs exhibit matrix-particle interactions. These micrographs show the rather uniform distributions of reinforced aluminum particles in the matrix [49]. A weak-linked dendritic colony was detected within the polymer composite in Figure 12.7a due to the matrix's failure to evenly mix with the aluminum particles [50]. The presence of weakly bound particles within the matrix causes discontinuities in the composite flow during load transfer [50]. In Figures 12.7b and 12.7c, the weakly bonded dendritic colonies strengthen as the proportion of aluminum increases. The matrix, the weight fraction/arrangement of the reinforcement, and the interface binding all influence the properties of the polymer composite [49]. The micrographs demonstrating good wettability between the reinforcement and matrix [51, 52] show no holes. As the weight percentage of the reinforcement increases, the average grain size of the epoxy matrix decreases [49]. It is also suggested that mechanical characteristics increase when interfacial bonding between the reinforcement and matrix improves [49].

12.4.2 Results of the Process Optimization

The outcomes of numerical and physical studies that established the process parameters leading to a workable combination and degree of cure are shown in Table 12.5. The percentage of epoxy that is filled with aluminum, the curing time, and the curing

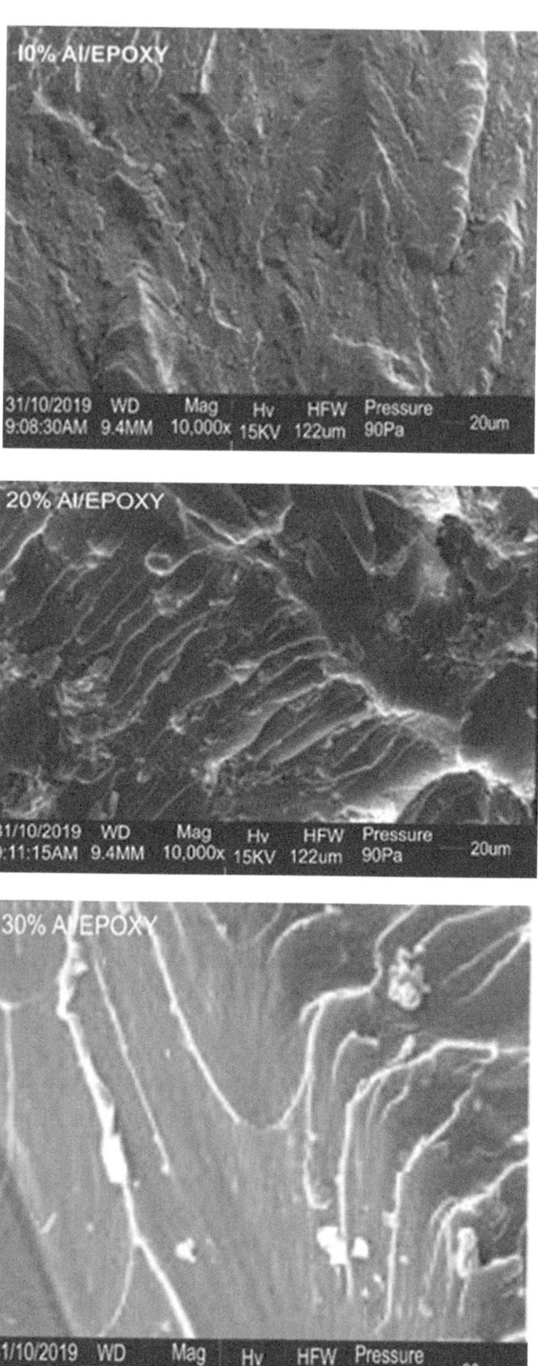

FIGURE 12.7   Microstructural analysis of aluminum epoxy composites for 10%, 20%, and 30% weight fraction of the reinforcement.

TABLE 12.5

The Results Obtained from the Numerical and Physical Experimentations

Trials	Factor A: Aluminum-Filled Epoxy	Factor B: Curing Time (min)	Factor C: Curing Temperature (°C)	Actual Degree of Cure (%)	Predicted Degree of Cure (%)
1.	30.00	5.00	32.00	0.20	0.1771
2.	10.00	10.00	39.00	0.40	0.3821
3.	30.00	20.00	44.00	0.50	0.4280
4.	10.00	30.00	52.00	0.55	0.5840
5.	20.00	50.00	58.00	0.65	0.7203
6.	10.00	110.00	62.00	0.75	0.6500
7.	20.00	170.00	70.00	0.85	0.9370
8.	30.00	230.00	80.00	0.96	0.9020
9.	30.00	290.00	86.00	0.50	0.5205
10.	20.00	350.00	86.00	0.62	0.6945
11.	10.00	177.50	80.00	0.70	0.7596
12.	30.00	177.50	68.00	0.80	0.7563
13.	20.00	112.61	78.00	0.92	0.8629
14.	10.00	5.00	90.00	0.60	0.5377
15.	30.00	177.50	90.00	0.69	0.7509
16.	20.00	177.50	60.00	0.80	0.7811
17.	30.00	177.50	60.00	0.92	0.9061
18.	20.00	177.50	62.00	0.92	0.9530
19.	10.00	350.00	68.00	0.45	0.5823
20.	20.00	350.00	90.00	0.50	0.5670

temperature are the process variables that produce the best degree of cure. To forecast the degree of cure under comparable circumstances, a predictive model derived from the RSM was also applied. Significant agreement between the actual and anticipated values of the degree of cure in Figure 12.8 suggests that the proposed model is appropriate for prediction purposes.

Table 12.6 shows the statistical analysis of the quadratic model that was established to predict the degree of cure as a function of the independent process parameters, which are the percentage of aluminum-filled epoxy, the curing time, and the curing temperature time, while Table 12.7 shows the ANOVA of the model that was developed.

The statistical model's F-value of 7.56, which denotes that there is a small likelihood that the high F-value is caused by noise, demonstrates the model's significance. The model terms are statistically significant because the p-value Prob > F was .0020, which is significantly lower than .050. The adjusted R-squared value (.7564) and the R-squared value (.8718) are both close to 1 and have a difference of less than 0.2, proving that the model is suitable for making predictions. A significant model is typically indicated by a value of appropriate precision (9.0140) that is larger than 4, which is another indicator.

The RSM was used to statistically assess the outcomes of the numerical and physical experiments in order to create a prediction model. Equation 12.17 plots the

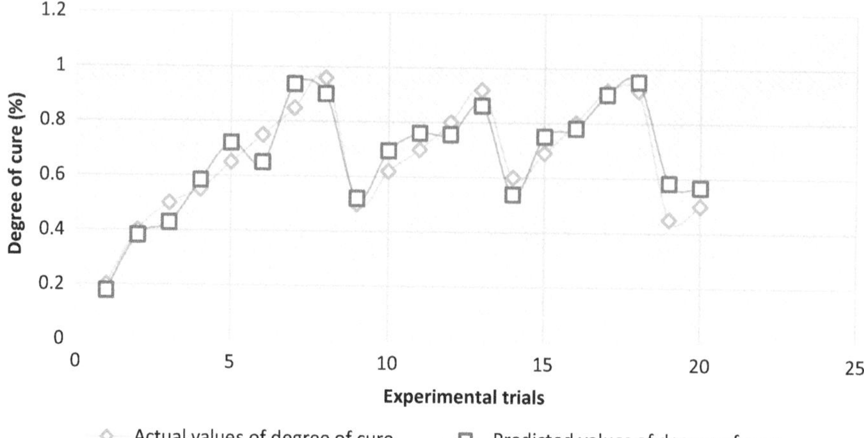

FIGURE 12.8   The plots of the actual and predicted values of the degree of cure.

TABLE 12.6
The Statistical Analysis of the Quadratic Developed Model

Statistical Parameters	Sum of Squares	df	Mean Square	F-Value	p-Value Prob > F	Remarks
Model	0.69	9	0.077	7.56	.0020	Significant
A—Percent aluminum-filled polyester	0.019	1	0.019	1.89	.1993	
B—Curing time	1.532E-004	1	1.532E-004	0.015	.9094	
C—Curing temperature	4.983E-004	1	4.983E-004	0.49	.5006	
AB	0.011	1	0.011	1.03	.3334	
AC	2.182E-003	1	2.182E-003	0.21	.6536	
BC	4.022E-003	1	4.022E-003	0.39	.5441	
A^2	0.024	1	0.024	2.35	.1561	
B^2	0.054	1	0.054	5.26	.0448	
C^2	0.021	1	0.021	2.11	.1774	
Residual	0.10	10	0.010			
Cor. total	0.80	19				

TABLE 12.7
The Analysis of Variance for the Developed Quadratic Model

Parameter	Value	Remarks
R-Squared	.8718	Weighty
Adjusted R-Squared	.7564	Weighty
Adequate Precision	9.0140	Weighty

degree of cure (dependent variable) against the percentage of aluminum-filled epoxy, curing time, and curing temperature (independent process factors):

$$\text{Degree of cure} = \begin{aligned} &+ 0.87 + 0.058A \\ &+ 0.013B + 0.10C \\ &+ 0.073AB - 0.033AC \\ &- 0.083BC - 0.083A^2 \\ &- 0.20B^2 - 0.21C^2, \end{aligned} \tag{12.17}$$

where A is the aluminum-filled epoxy (%), B is the curing time (min), and C is the curing temperature (°C).

Table 12.8 displays the quadratic model's regression analysis, where a smaller standard error indicates that the observations are closer to the fitted line. The variance inflation factor (VIF) measures the variance of the estimated regression coefficient that is inflated by the correlation among predictor variables. A VIF value of 1 indicates no correlation between predictor variables, while a value exceeding 5 may indicate moderated correction, and a value exceeding 10 may indicate multicollinearity. In Table 12.8, the VIF values do not exceed 5, indicating that the model is substantially accurate.

Figure 12.9 illustrates the normal plot of residual, analyzing the quadratic model's accuracy in predicting the degree of cure. The plot shows a normal distribution of data, with the diagonal line representing average values. The slight deviation of data points from the diagonal line, within ±10% of the average line, indicates the suitability of the quadratic model for prediction. Furthermore, the close alignment between

---

TABLE 12.8

The Regression Analysis of the Quadratic Developed Model

Factor	Coefficient Estimate	df	Standard Error	95% CI (Low)	95% CI (High)	Variance Inflation Factor
Intercept	0.87	1	0.061	0.74	1.01	
A—Percent aluminum-filled polyester	0.058	1	0.042	−0.036	0.15	2.23
B—Curing time	0.013	1	0.10	−0.22	0.24	8.86
C—Curing temperature	0.10	1	0.15	−0.23	0.43	3.74
AB	0.073	1	0.072	−0.088	0.23	3.12
AC	−0.033	1	0.071	−0.19	0.13	2.76
BC	−0.083	1	0.13	−0.38	0.21	6.91
A^2	−0.083	1	0.054	−0.20	0.037	1.30
B^2	−0.20	1	0.088	−0.40	−5.649E-003	2.81
C^2	−0.21	1	0.15	−0.53	0.11	5.73

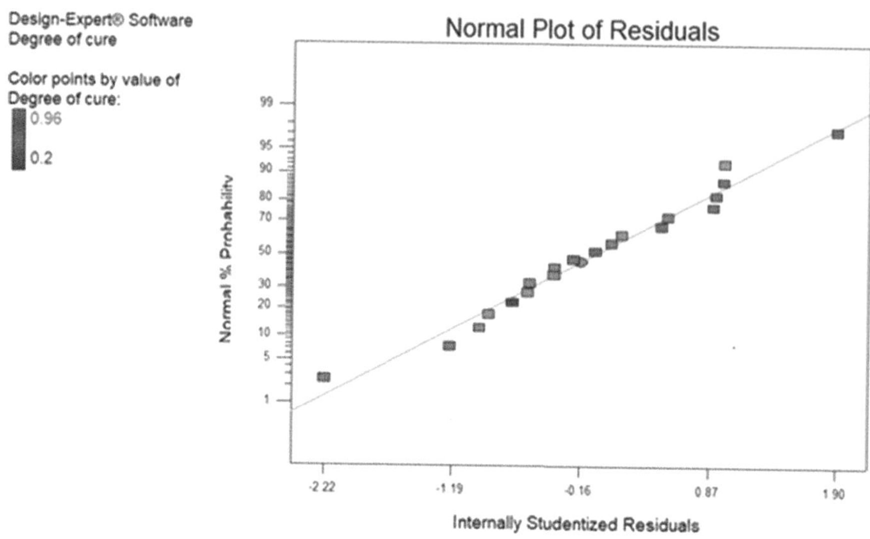

FIGURE 12.9 The Normal plot of residual.

the actual and the predicted values of the degree of cure is evident in Figure 12.9, with data points exhibiting minimal deviation from the diagonal line.

The contour and three-dimensional (3D) plots of the influence of percentage aluminum filled epoxy and time on degree of cure are shown in Figures 12.10 and 12.11. The degree of cure increased from 0.677656 at about 50 °C and 20% aluminum-filled epoxy to an optimal value of 0.815815 at approximately 177.5 °C and 15% aluminum-filled epoxy. Increases in the magnitude of the temperature and the percentage of aluminum-filled epoxy resulted in a reduction in the degree of cure. This is due to the composite achieving full crosslinking, often known as the ultimate cure [30].

The contour and 3D graphs of the influence of percentage aluminum-filled epoxy and the temperature on degree of cure are shown in Figures 12.12 and 12.13. The degree of cure increased from 0.499983 at about 35 °C and 10% aluminum-filled epoxy to an optimal value of 0.814475 at about 72 °C and 10% aluminum-filled epoxy. Increases in the magnitude of the temperature and the percentage of aluminum-filled epoxy resulted in a reduction in the degree of cure. This is due to the composite achieving full crosslinking, often known as the ultimate cure [30].

Figures 12.14 and 12.15 depict the relationship between time, temperature, and the degree of cure. Initially, as time and temperature increase, the degree of cure also increases until reaching an optimal value of 0.788184. However, further increases in time and temperature lead to a decrease in the degree of cure due to reaching the point of full crosslinking, known as the ultimate cure [30].

According to Table 12.5, the ideal values of the process parameters that achieved the highest degree of cure (0.9530) through physical experiments are 30% aluminum-filled epoxy, 177.50 minutes curing time, and 62 °C curing temperature. This physical experimentation result verifies the numerical experimentation results, which show that low process parameter values resulted in lower degrees of cure. However, as the

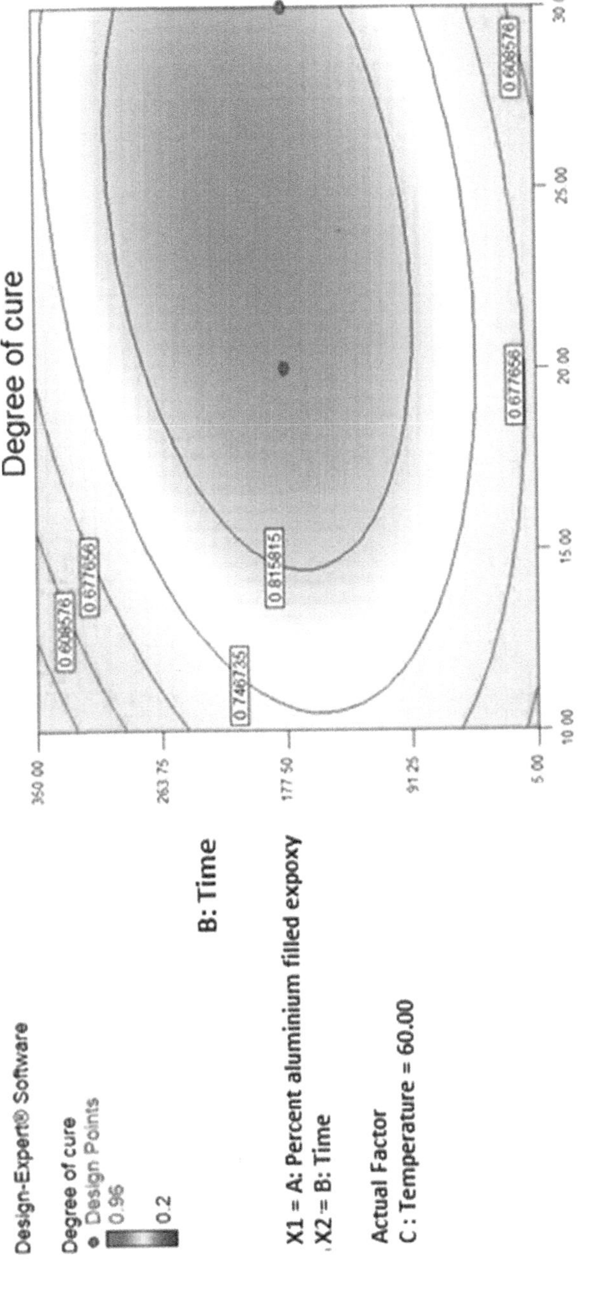

FIGURE 12.10 The contour plot of the effect of the curing time and percent aluminum-filled epoxy on the degree of cure.

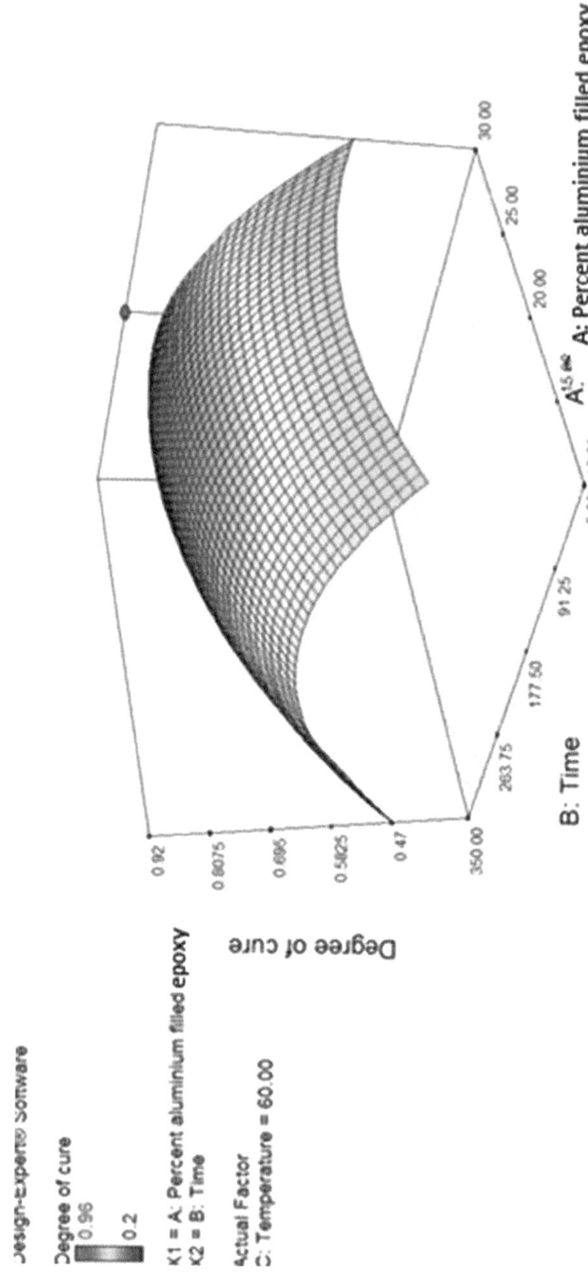

FIGURE 12.11   The 3D plot of the effect of curing time and percent aluminum-filled epoxy on the degree of cure.

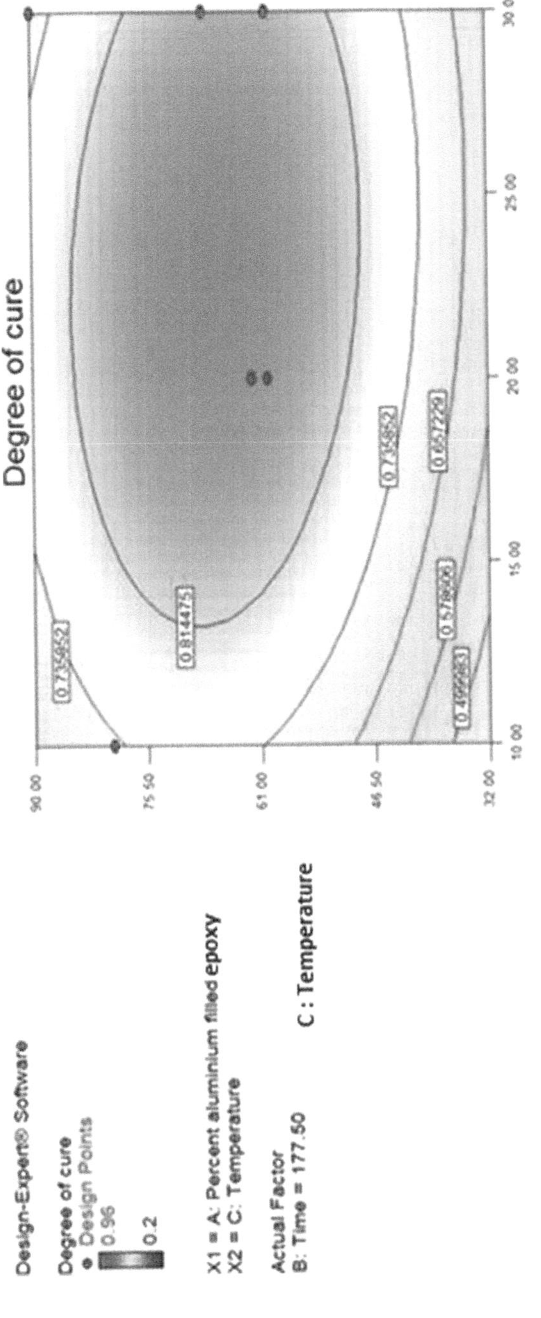

FIGURE 12.12 The contour plot of the effect of percentage aluminum-filled epoxy and the temperature on the degree of cure.

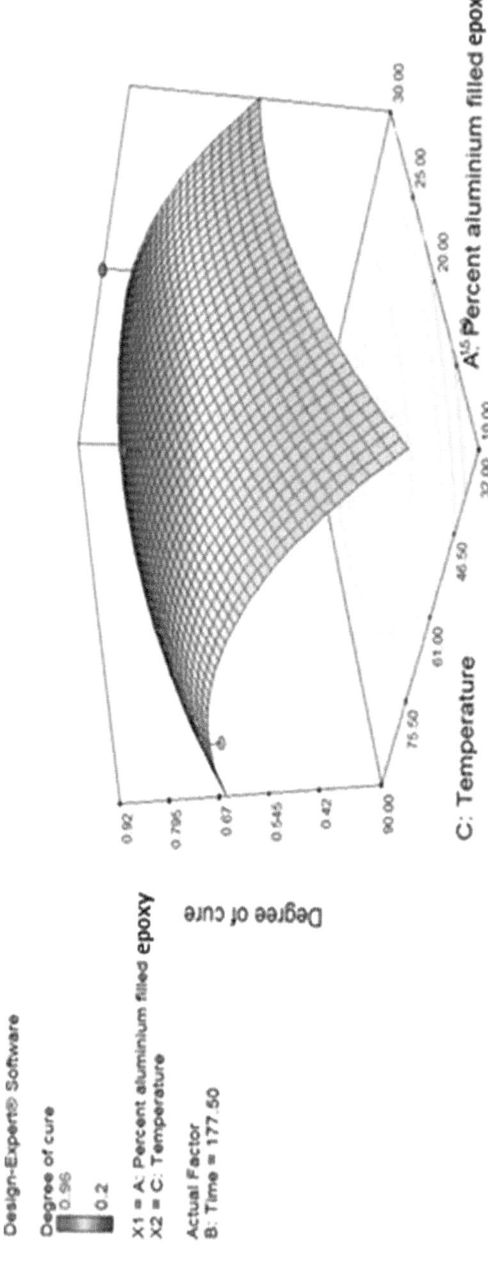

FIGURE 12.13 The 3D plot of the effect of percentage aluminum-filled epoxy and the temperature on the degree of cure.

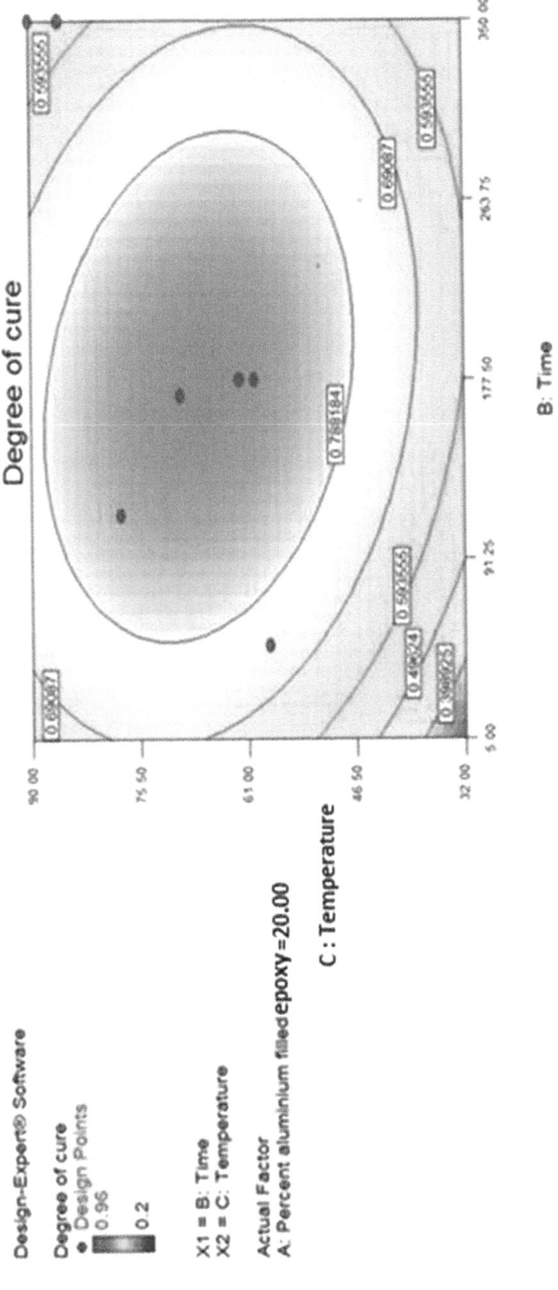

FIGURE 12.14 The contour plot of the effect of time and temperature on the degree of cure.

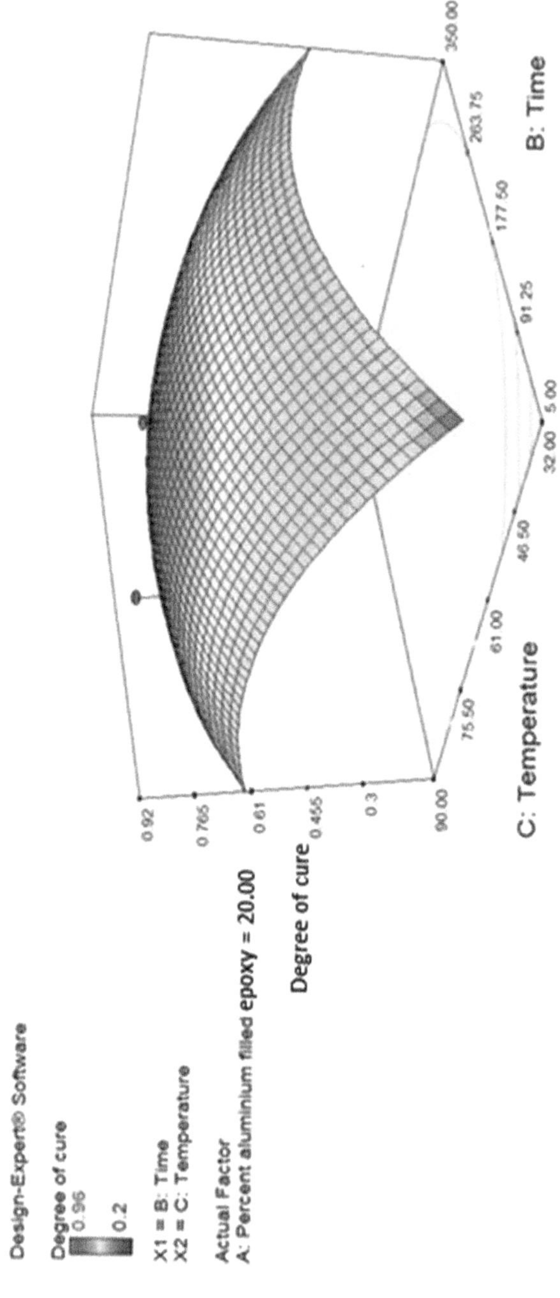

FIGURE 12.15 The 3D plot of the effect of time and temperature on the degree of cure.

process parameter values are increased to the optimum, the degrees of cure improve noticeably but drop when the process parameter values exceed the optimum.

12.5 CONCLUSION

Hand lay-up processing was used to fabricate aluminum powdered–reinforced epoxy polymer composites, which were subsequently post-cured in a conventional auto-clave oven. The post-curing process had a significant positive impact on the mechanical and microstructural properties of the composites. Tensile strength, hardness, impact resistance, and thermal conductivity increased by 12.12%, 17.4%, 17.4%, and 100%, respectively. However, the energy required to fracture the composites decreased by 55.2% due to a shift from ductile to a more brittle nature. SEM images revealed a homogeneous dispersion of aluminum particles in the matrix after post-curing, contributing to the enhanced mechanical properties of the composites.

Using the Design Expert software, a single-objective batch optimization process was conducted. The optimal process parameters that resulted in the highest degree of cure (0.884206) were identified as 20% aluminum-filled epoxy, a curing time of 170 minutes, and a curing temperature of 70 °C. Comparing these optimal values to those obtained through physical experimentation, it can be concluded that the process parameters' range established from the numerical optimization and physical experimentation are similar.

REFERENCES

[1] Giacomo, S., Optimisation of the vartm process. *A Ph. D. Thesis, School of Aerospace, Transport and Manufacturing.* Cranfield University, United Kingdom, 2014.

[2] Adeodu, A.O., Anyaeche, C.O. Oluwole, O.O., Omohimoria, C.U. Effect of Microwave and Conventional Heating on the Cure Cycles of Particulate Reinforced Polymer Matrix Composites. *International Journal of Materials Science and Applications*, 4(4), 229–240, 2015. doi: 10.11648/j.ijmsa.20150404.12

[3] Adeodu, A.O., Anyaeche, C.O., Oluwole O.O., Alo, O. Effect of Microwave and Conventional Autoclave Curing on the Mechanical and Microstructural Properties of Particulate Reinforced Polymer Matrix Composites. *Adv. Mater.*, 4(5), 85–94, 2015. doi: 10.11648/j.am.20150405.11

[4] Marsh, G. Airbus 350-XWB Update. *Reinforced Plas.*, 54, 6, 2010.

[5] Arora, J. S. *Introduction to Optimum Design.* 2nd ed. Boston: Elsevier, 2004.

[6] Yuan, Y. A new step-size for the steepest descend method. *J. Comput. Math.*, 24(2), 149–156, 2006.

[7] Kirkpatrick, S., Gelatt, C.D., and Vecchi, M.P. Optimization by simulated annealing. *Science*, 220(4598), 671–680, 1983.

[8] Goldberg, D.E. *Genetic Algorithms in Search, Optimization and Machine Learning*, Addison-Wesley, Boston, 1989.

[9] Thostenson, E.T., and Chou, T.W. *Compos. A: Appl. Sci. Manuf.*, 30, 1055, 1999.

[10] Min Li, Qi Zhu, Philippe, H., Geubelle and Charles, L. Optimal curing for thermoset matrix composites: Thermochemical considerations. *Polym. Compos.*, 22(1), 118–131, 2001.

[11] Pillai, V., Beris, A.N., and Dhurjati, P. Heuristics guided optimization of a batch autoclave curing process. *Comput. Chem. Eng.*, 20(3), 275–294, 1996.

[12] Li, M., Zhu, Q., Geubelle, P.H., and Tucker III, C.L. Optimal curing for thermoset matrix composites: Thermochemical considerations. *Polym. Compos.*, 22(1), 18–131, 2001.

[13] Li, M. and Tucker III, C.L. Optimal curing for thermoset matrix composites: Thermochemical considerations. *Polym. Compos.*, 23(5), 739–757, 2002.

[14] Udaykumar, H.S., and Shyy, W. Optimal cure cycles for thermoset composites manufacture. *29th National Heat Transfer Conference*, 241, 23, 1993.

[15] Chang, M.H., Chen, C.L., and Young, W.B. Optimal design of the cure cycle for consolidation of thick composite laminates. *Polym. Compos.*, 17(5), 743–750, 1996.

[16] Yang, Z.L., and Lee, S. Optimized curing of thick section composite laminates. *Mater. Manuf. Processes*, 16(4), 541–560, 2001.

[17] Skordos, A.A., and Partridge, I.K. Inverse heat transfer for optimization and on-line thermal properties estimation in composites curing. *Inverse Prob. Sci. Eng.*, 12(2), 157–172, 2004.

[18] Rai, N., and Pitchumani, R. Optimal cure cycles for the fabrication of thermosetting-matrix composites. *Polym. Compos.*, 18(4), 566–581, 1997.

[19] Dufour, P., Michaud, D. J., Touré, Y. and Dhurjati, P. S. A partial differential equation model predictive control strategy: application to autoclave composite processing. *Comput. Chem. Eng.*, 28(4), 545–556, 2004.

[20] Mawardi, A., and Pitchumani, R. Optimal temperature and current cycles for curing of composites using embedded resistive heating elements. *J. Heat Transf.*, 125(1), 126–136, 2003.

[21] Carlone, P. P. A Simulation based metaheuristic optimization of the thermal cure cycle of carbon epoxy composite laminates. *The 14th International ESAFORM Conference on Material Forming: ESAFORM 2011*. Belfast, UK, pp. 5–10. 2011.

[22] ASTM D5528 Standard test method for Mode I inter-laminar fracture toughness of unidirectional fiber reinforced polymer matrix composites. 1994.

[23] Cao, G. Multi-Functional Epoxy/Graphene Nanoplatelets Composites. Doctoral Thesis, University of Manchester, 2016.

[24] Pitchayyapillai, G., P. Seenikannan, P. Balasundar, P. Narayanasamy. Effect of Nano-Silver on Microstructure, Mechanical and Tribological Properties of Cast 6061 Aluminium Alloy. *Trans. Nonferrous Met. Soc. Chin.*, 27, 2137–2145, 2017.

[25] Hossain, M.E., Hossain, M.K., Hosur, M.V., & Jeelani, S. Study of Mechanical Responses and Thermal Expansion of CNF-modified Polyester Nanocomposites Processed by Different Mixing Systems. *Cambridge J Online*, 2011. https://doi.org/10.1557/opl.2011.111

[26] Hossain, M.K., Hossain, M.E., Hosur, M.V., & Jeelani, S. Flexural and Compression Response of Woven E-Glass/Polyester–CNF Nanophased Composites. *Comp. Part A*, 42(11), 1774–1782, 2011.

[27] Daniyan, I. A., Tlhabadira, I., Mpofu, K. and Adeodu, A. O. Development of numerical models for the prediction of temperature and surface roughness during the machining operation of titanium alloy (Ti6Al14V). *Acta Polytech. J.*, 60(5), 369–390, 2020. https://doi.org/10.14311/AP.2020.60.0369

[28] Daniyan, I. A., Fameso, F. Ale, F., Bello, K. and Tlhabadira, I. Modelling, simulation and experimental validation of the milling operation of titanium alloy (Ti6Al4V). *Int. J. Adv. Manuf. Technol.*, 109(7), 1853–1866, 2020. https://doi.org/10.1007/s00170-020-05714-y

[29] Daniyan, I. A., Mpofu, K. and Adeodu, A. O. Optimization of welding parameters using Taguchi and response surface methodology for rail car bracket assembly. *Int. J. Adv. Manuf. Technol.*, 100, 2221–2228, 2019. https://doi.org/10.1007/s00170-018-2878-9

[30] Adeodu, A. O., Anyaeche, C. O., and Oluwole, O. Optimum cure cycles of autoclave process for polymer matrix composites using Genetic Algorithm. *2nd International Conference on Sustainable Materials Processing and Manufacturing*. University of Johannesburg, South Africa. March 8–10, 2019. Procedia Manufacturing 35, pp. 16–26.

[31] Adeodu, A.O., Kanakana-Katumba, M.G., Maladzhi, R.W. and Daniyan, I.A. Optimization of injection process parameters of plastic reinforced composites using response surface methodology and central composite design. *36th Technical Conference of America Society for Composites. USA*. September 19–22, 2021. ISBN: 978-1-60595-665-7

[32] Daniyan, I. A., Tlhabadira, I., Phokobye, S. N., Siviwe, M. and Mpofu K. Modelling and optimization of the cutting forces during Ti6Al4V milling process using the response surface methodology and dynamometer. *MM Sci. J.*, 128, 3353–3363, 2019. https://doi.org/10.17973/MMSJ.2019_11_2019093

[33] Daniyan, I.A., Mpofu, K., Adeodu, A. O. and Momoh, O. Development and automation of a 12 kW-Capacity Gasifier for energy generation 2020 Southern African Universities Power Engineering Conference/Robotics and Mechatronics/Pattern Recognition Association of South Africa (SAUPEC/RobMech/PRASA). *IEEE Xplore*, 2020, 68–73. https://doi.org/10.1109/SAUPEC/RobMech/PRASA48453.2020.9040963

[34] Adeodu, A.O., Daniyan, I.A., Akokhia, A. and Adelowo, O. Effect of graphene nanoplatelets dispersion on the enhancement of tensile strength, thermal and electrical conductivity of polymer nanocomposites. *35th Annual Technical Conference and ASTM D-30 Committee meeting of American Society for Composites*. New York University, USA. September 14–17, 2020. ISBN: 978-1-60595-665-7

[35] Sham Prasad, M.S., Venkatesha, C.S. and Jayaraju, T. Experimental methods of determining fracture toughness of fiber einforced polymer composites under various loading conditions. *J. Miner. Mater. Charact. Eng.*, 10(13), 1263–1275, 2011.

[36] Dharan, C.K.H. Fracture mechanics of composite materials. *J. Eng. Mater. Technol.*, 100: 223–247, 1978.

[37] Wei, J., Vo, T. and Inam, F. Epoxy/graphene nano-composites processing and properties: A review. *RSC Adv.*, 5, 73510–73524, 2015.

[38] Veca, L.M., Meziani, M. J., Wang, W., Wang, X., Lu, F., Zhang, P., Lin, R., Fee, Y., Connell, J.W. and Sun, Y.P. *Adv. Mater.*, 21, 2088–2092, 2009.

[39] Shiu, S.C., Tsai, J.L. Characterizing thermal and mechanical properties of graphene/epoxy nanocomposites. *Compos. Part B*, 56, 691–697, 2014.

[40] Nan, C.W., Birringer, R., Clarke, D.R., Gleiter, H. Effective thermal conductivity of particulate composites with interfacial thermal resistance. *J. Appl. Phys.*, 81, 6692–6699, 1997.

[41] Hu, L., Desai, T., Keblinski, P. Thermal transport in graphene-based nanocomposites. *J. Appl. Phys.*, 110, 1–6, 2011.

[42] Nishida, Y. *Introduction to Metal Matrix Composites: Fabrication and Recycling*. Springer Science & Business Media: Berlin, Germany, 2013.

[43] Razzaq, A.M., Majid, D.L., Ishak, M.R. and Basheer, U.M. Effect of fly ash addition on the physical and mechanical properties of AA6063 alloy reinforcement. *Metallomics* 7, 477, 2017.

[44] Srivastava, A.K. and Das, K. Microstructural and mechanical characterization of in situ TiC and (Ti, W) C-reinforced high manganese austenitic steel matrix composites. *Mater. Sci. Eng. A*, 516, 1–6, 2009.

[45] Flore-Compos, R., Deaquino-Lara, R., Ponce, A., Estrada-Guel, I., Miki-Yoshida, M.J., Herrrera-Ramirez, M. and Martinez-Sanchez, R. Composite of Aluminium 7075 Alloy with Silver Nano particles prepared by Mechanical Milling. *3rd International Conference on Multidisciplinary Research*. Saltillo, Coach. Mexico. May 22–24. 2009.

[46] Ravindran, P., Manisekar, K., Narayanasamy, P., Selvakumar, N. and Narayanasamy, R. Application of factorial techniques to study the wear behaviour of Al hybrid composites with graphite addition. *Mater. Design*, 39, 42–54, 2012.

[47] Liu, J., Sun, K., Zeng, L., Wang, J., Xiao, X., Liu, J., Guo, C. and Ding, Y. Microstructure and properties of copper–graphite composites fabricated by spark plasma sintering based on two-step mixing. *Metals*, 10, 1506, 2020.

[48] Malaki, M., Xu, W., Kasar, A.K., Menezes, P.L., Dieringa, H., Varma, R.S. and Gupta, M. Advanced metal matrix nanocomposites. *Metals*, 9, 330, 2019.

[49] Juliyana, S.J., Prakash, J.U., Salunkhe, S., Hussein, H.M.A. and Gawade, S.R. Mechanical characterization and microstructural analysis of hybrid composites (LM5/ZrO$_2$/Gr). *Crystals*, 12, 1207, 2022. https://doi.org/10.3390/cryst12091207

[50] Bello, S.A. Fracture toughness of reinforced epoxy aluminum composite. *Compos. Commun.*, https://doi.org/10.1016/j.coco.2019.11.006

[51] Singh, M., Garg, H.K., Maharana, S., Yadav, A., Singh, R., Maharana, P., Nguyen, T.V., Yadav, S. and Loganathan, M.K. An experimental investigation on the materil removal rate and surface roughness of a hybrid aluminum metal matrix composite Al6061/sic/gr. *Metals*, 11, 1449, 2021.

[52] Khan, A., Abdelrazeq, M.W., Mattli, M.R., Yusuf, M.M., Alashraf, A., Matli, P.R. and Shakoor, R.A. Structural and mechanical properties of Al-SiC-ZrO$_2$ nanocomposites fabricated by microwave sintering technique. *Crystals*, 10, 904, 2020.

13 Artificial Intelligence-Driven Modeling and Optimization for Machining of Engineering Composites

Gobinda Chandra Behera, Suman Saha, and Sankha Deb

13.1 INTRODUCTION

The standard of living of human society is intimately linked to a great extent to the materials and technologies available to its members at the concerned period. Historically, from the very existence of humans, the temporal advancement of civilization was predominantly fueled by the competency of the concerned members to produce and manipulate materials for different requirements. Early humans possessed only a narrow range of naturally available materials such as stone, wood, clay, bone, skin, and feathers [1]. Mastering the art of controlled utilization of the fire opened up a new horizon for the development of promising materials superior to the naturally available ones. As named by archaeological anthropologists based on the wide availability of the materials at the concerned epochs, ancient periods gradually evolved through the Stone Age, the Copper Age, the Bronze Age, and the Iron Age [2]. Human conscience about materials continued in the Common Era, leading to the emergence of steel, aluminum, plastics, semiconductors, and superconductors. Even in the twenty-first century, within the realm of the Fourth Industrial Revolution (Industry 4.0), the development of new materials and tailoring the material properties to meet the customized demands are challenging areas for metallurgists and manufacturers.

In general, solid engineering materials are conveniently classified as metals, polymers, and ceramics based on their atomic structures and chemo-mechanical properties. When two or three of the previously mentioned types of solid materials are combined, the resulting group is called composites. Although a wide variety of composites occurs and exists in nature, composite materials can also be synthesized through an amalgamation of two or more mutually non-soluble constituent materials having distinct physical and chemical properties in order to obtain multiple properties simultaneously that cannot be fetched by its individual constituent. In a composite material, the phase that remains continuous is known as the matrix, whereas the phase that remains dispersed is known as the reinforcement. The matrix phase usually offers load-bearing capacity and structural integrity, whereas the reinforcing phase contributes to enhancing the mechanical and tribological properties. Depending on the matrix material, a composite can be metal, ceramic, or polymer matrix composite. Such artificial composites can be fabricated through different processes ranging from powder metallurgy to additive manufacturing.

As manufactured, composite components often require machining and finishing to impart the required shape, incorporate the intended feature, improve the dimensional tolerance, and enhance the surface quality. Machining is one subtractive manufacturing process that uses a sharp-edged, hard cutting tool to gradually eradicate work material in the form of a chip to impart the required shape, size, and finish [3]. However, machining composite material is challenging owing to its inherent inhomogeneity, anisotropy, reinforcement-abrasiveness, and non-ductile behavior. Thus, composites are usually regarded as the "difficult-to-machine" materials [4]. Despite the challenges, a slew of investigations were carried out in the literature to explore the feasibility of machining and finishing composite materials. Nowadays, information technology is also increasingly adapted when looking to digitally transform such machining processes to embrace Industry 4.0.

In general, Industry 4.0 drives the traditional "high-touch" (manpower-intensive) sectors for digital transformation toward "high-tech" (technology-intensive) sectors through strategic adaptation of the internet and information technology. Several digital tools such as data engineering, artificial intelligence (AI), the Internet of Things, big data, cloud computing, wireless sensors, digital twins, and augmented reality [5] are empowering digitalization of the core manufacturing factories for computer-aided "smart" production. The fundamental principle of AI involves advanced algorithms and computational capabilities to scrutinize extensive datasets, identify patterns, and make intelligent decisions or forecasts [6]. The field of AI involves various methodologies, including machine learning, deep learning, and natural language processing, which facilitate the acquisition of knowledge from data and the ability to adjust to dynamic circumstances. This "decision" can be of different types, ranging from path planning to product quality and life cycle. In fact, a digital twin is built focusing on a specific objective, and thus, the twin may not present information other than its intended purpose [7].

This chapter aims to present descriptive details regarding the process of implementation of artificial neural networks (ANNs), fuzzy logic (FL), adaptive neuro-fuzzy inference systems (ANFISs), and other machine learning approaches to discuss and focus on improving the process performance and output prediction when machining composite materials.

13.2 BASIC CONCEPTS OF MACHINING COMPOSITES

13.2.1 Engineering Composites—Common Types, Properties, and Applications

By virtue of its synthesis principle, engineering composites can be of different types based on the material of the matrix phase (Figure 13.1). A metal matrix composite (MMC) is characterized by the combination of a metallic matrix with ceramic, polymer, or organic fiber reinforcement. MMCs possess desirable mechanical and thermal properties rendering them appropriate for implementation in the aerospace, automobile, and electronics industries. A ceramic matrix composite (CMC) contains a ceramic matrix with another ceramic (usually) as reinforcement. Due to their admirable resistance toward thermal and chemical, CMCs are preferred for high-temperature applications like jet engines, gas turbines, and heat exchangers. A polymer matrix composite (PMC) contains reinforcements like carbon or glass fibers that are woven into a polymer matrix. PMCs are highly desirable in the industries of aeronautical, automobile, and sports equipment due to their low weight, high corrosion resistance, and long lifespan. In addition, a carbon–carbon composite (CCC) is often considered a separate group that is composed of carbon fibers in a matrix phase (such as graphite, coke, and sintered carbon). In addition to these basic groups, a few common composite types have widespread applications as discussed next.

Fiber-reinforced composites (FRCs) are produced by integrating a matrix substance with fiber reinforcement, such as carbon or glass fibers. The matrix material consists of a variety of substances including polymers, metals, or ceramics. FRCs are recognized for possessing a great strength-to-weight ratio, stiffness, and resistance to corrosion, rendering them highly appropriate for applications in the aeronautical, automobile, and sports equipment industries. Sandwich composites are produced through the process of adhering two slender external layers of high-strength substances, such as composites reinforced with fibers, to a core material of low density, such as foam or honeycomb. Sandwich composites exhibit excellent mechanical properties, including high stiffness

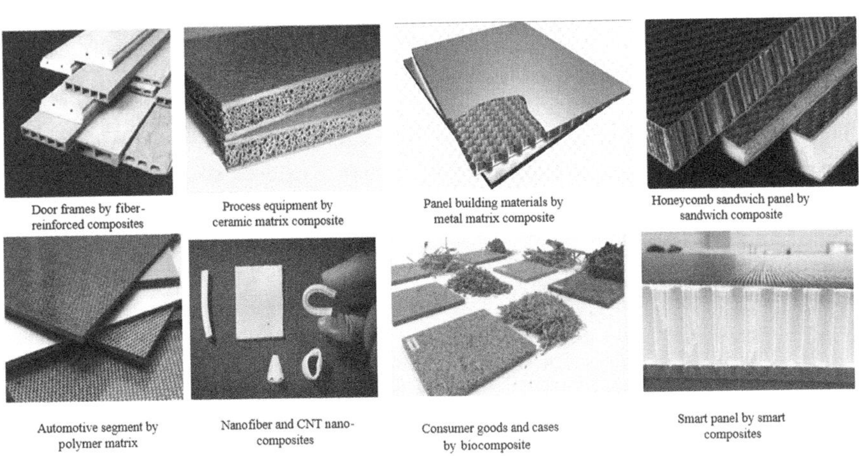

FIGURE 13.1 Different types of composites.

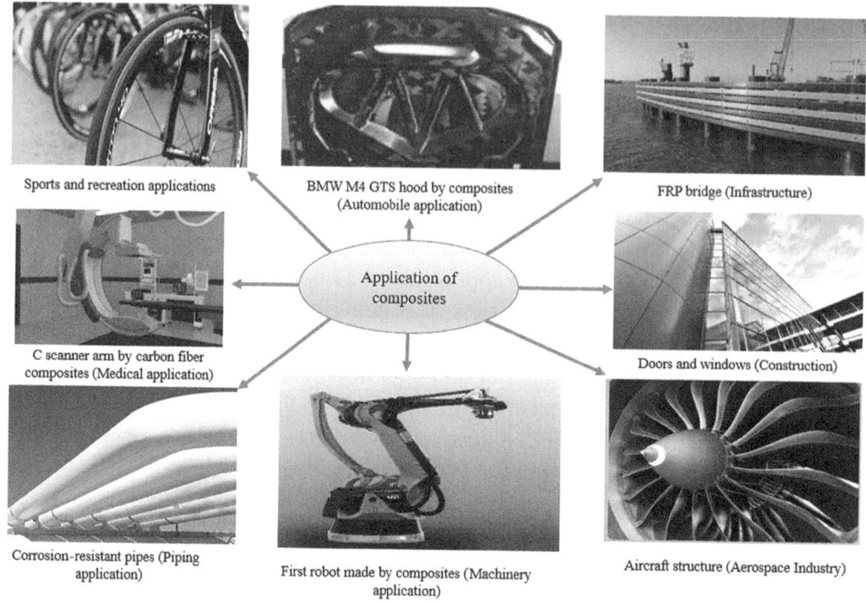

FIGURE 13.2   Different areas of application of the composites.

and strength, coupled with a low weight profile, rendering them highly suitable for deployment in lightweight structures, such as aircraft and boats. Nanocomposites involve the incorporation of nanoparticles, such as carbon nanotubes or graphene, into a matrix material, such as polymer or metal. Nanocomposites exhibit distinctive characteristics, such as enhanced mechanical robustness, rigidity, and thermal and electrical conductivity, rendering them highly suitable for a diverse range of uses, encompassing electronics, biomedical, and energy sectors. Biocomposites are produced by blending a matrix material, like a biodegradable polymer, with natural fibers like flax or hemp. Biocomposites are preferred for usage in the building and automotive industries as well as in packaging and consumer goods since they are lightweight, ecologically friendly, and have good mechanical qualities. Smart composites are engineered to exhibit a responsive behavior toward various external stimuli, including alterations in temperature, pressure, or humidity. Smart composites possess the ability to modify their characteristics, such as rigidity or configuration, in reaction to external stimuli, rendering them highly suitable for employment in the fields of aerospace, robotics, and medicine.

To summarize, engineering composites possess a diverse set of distinctive characteristics, rendering them well-suited for a multitude of uses across diverse sectors. Composites are crucial in the advancement of novel materials with enhanced characteristics across various fields, including aerospace, sports equipment, and biomedical applications (Figure 13.2). In general, different composites possess different properties:

- Strength-to-weight ratio—Carbon fiber–reinforced polymers (CFRPs), aramid fiber–reinforced polymers (AFRPs), and glass fiber–reinforced polymers (GFRPs) exhibit a high strength-to-weight ratio. Conversely, natural fibers and low-strength synthetic fibers demonstrate a low strength-to-weight ratio.

- Stiffness—CFRPs, AFRPs, and GFRPs are composite materials with high stiffness. However, polymer composites usually exhibit low strength.
- Corrosion resistance—Thermoset polymers, including epoxies, phenolic, and polyesters, as well as ceramic-based composites such as silicon carbide, alumina, and zirconia, exhibit notable resistance to a wide range of corrosive environments. Conversely, thermoplastic polymers exhibit comparatively lower levels of corrosion resistance.
- Thermal resistance—CCCs and ceramic matrix composites exhibit high thermal resistance, while thermoplastics and certain low-temperature thermoset polymers, such as polyesters and vinyl esters, display lower thermal resistance.
- Chemical inertness—Composites composed of thermoset polymers as well as ceramics exhibit notable chemical resistance. Conversely, composites comprised of thermoplastics and MMCs demonstrate lower chemical resistance.
- Mechanical durability—Composite materials that incorporate carbon fiber, aramid fiber, and glass fiber exhibit superior durability compared to natural fibers or synthetic fibers with low strength, which have lower durability.
- Lightweight—The category of high-lightweight composites comprises CFRPs, AFRPs, and foam-core sandwich structures. However, low-lightweight composites are composed of metal matrix materials or high-density polymers).

13.2.2   NEED ANALYSIS FOR MACHINING AND FINISHING OF ENGINEERING COMPOSITES

Machining is considered a secondary manufacturing process that helps give the intended form and improve the surface quality and dimensional tolerance. Machining processes utilize a special-shaped cutting tool to eradicate excess materials from the work material following the subtractive principle. The shape of the cutting tool can vary from one machining process to another. Similarly, the type of operation and output product quality also vary from one process to another. Some widely used machining processes include turning (for generating cylindrical surface), drilling (for originating a hole), milling (for flat surface generation), and grinding (for smooth, flat surface generation). Machining aid in the manipulation of composite materials in order to conform to predetermined criteria, ultimately improving their functionality and ability to integrate with other constituent parts. The necessity of machining and finishing processes for engineering composites can be attributed to several factors:

Attaining the intended configuration and dimensions: Composites possess the capability to be fabricated into diverse geometries and magnitudes, however, they may not invariably conform to the specific requisites of a given application. The process of machining is crucial in achieving the desired performance characteristics and fitting the composite into the intended application by cutting and shaping the material to the required size and shape.

Enhancing the surface quality: Improving surface quality is a crucial aspect of composites as they usually possess rough and uneven surfaces after the fabrication process. Implementing finishing procedures, including sanding and polishing, can facilitate the attainment of a smoother and more uniform surface, thereby enhancing the aesthetic appeal of the material and mitigating the likelihood of defects.

Enhancing functional performance and suitability: Performance can be improved through the utilization of machining and finishing techniques, which can modify the properties of composites to enhance their performance characteristics. For example, the process of perforating holes in a composite material can potentially yield benefits such as decreased weight and heightened strength-to-weight ratio, making them suitable for different applications.

Achieving compatibility with other components: Machining and finishing techniques can be employed to ensure that composites are harmoniously integrated with other constituent parts within a larger assembly. Utilizing precise machining techniques can ensure the accurate fitting of composite parts with other metal or plastic components within a multifaceted system, ultimately ensuring the dependability and efficacy of the end product.

13.2.3   MACHINING PROCESSES COMMONLY USED FOR COMPOSITE PROCESSING

Machining composites can be difficult due to their anisotropic nature, heterogeneous composition, and a high degree of abrasiveness. Despite such challenges, a few traditional and modern subtractive manufacturing processes were employed for mechanically processing composite materials, catering to the varying degrees of challenges, as indicated in Figure 13.3 and summarized in Table 13.1.

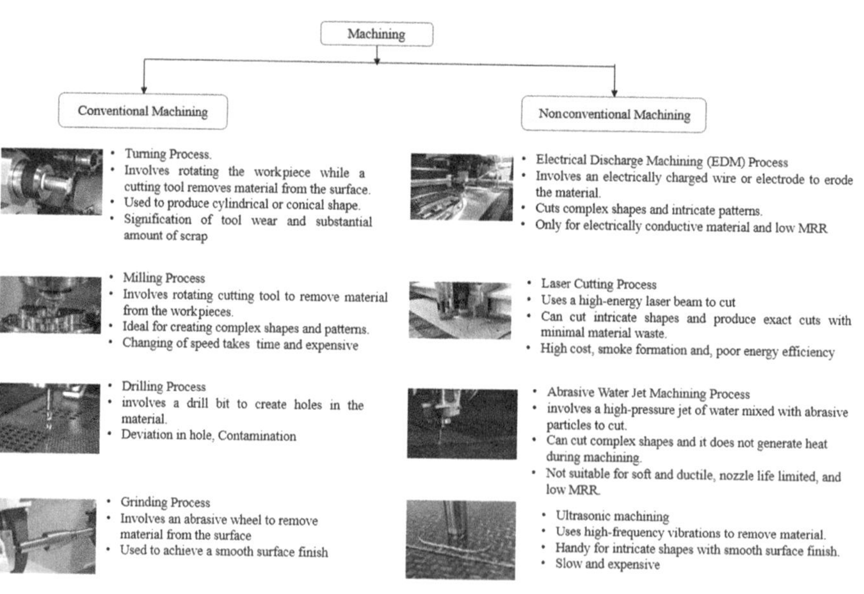

FIGURE 13.3   Common machining processes used for processing composite materials.

TABLE 13.1

Various Modeling Methodologies Employed for Composite Machining Processes

Authors	Machining Process	Parameters	Materials	Modelling Methods	Output Responses
Zhou et al. [8]	Milling	Cutting speed (V), cutting depth (d), feed rate (f), and fraction of SiC	Al/SiC metal matrix	Feed-forward ML-ANN	Surface roughness (Ra)
Chandrasekaran and Devarasiddappa [9]	Milling	Spindle speed, feed, depth, and SiC %	Composites	Fuzzy logic	Ra
Daniel et al. [10]	Milling	SiC (%), particle size, speed, feed, and cutting depth	Aluminum hybrid metal matrix	ANN and regression analysis	Ra, temperature, Material removal rate (MRR), feed, radial, and tangential force
Azmi [11]	End milling	Speed, feed, and depth	GFRP composites	ANFIS	Tool wear
Boga and Koroglu [12]	Milling	Cutting tools, spindle speed, and feed	Carbon fiber composites	ANN and GA	Ra
Latha and Senthilkumar [13]	Drilling	Spindle speed, feed, and drill diameter	GFRP composites	Fuzzy logic	Ra
Palanikumar et al. [14]	Drilling	Spindle speed, feed, and diameters of drill bit	Composites	ANN	Ra
Pawan Kumar [15]	Drilling	Feed, speed, diameter of drill bit, and point angle	GFRP composites	ANN	MRR
Bagci and Işık [16]	Turning	Speed, cutting depth, and feed	Unidirectional GFRP composites	ANN and RSM	Ra
Karim et al. [17]	Turning	Speed, feed, and cutting depth	SiC-Al alloy composites	ANN and ANFIS	Ra
Rajasekaran et al. [18]	Turning	Speed, feed, and depth	GFRP composites	Fuzzy logic	Ra
Chandrasekaran and Devarasiddappa [19]	Cylindrical grinding	Cutting speed of the grinding wheel, cutting speed of the workpiece, feed, and cutting depth	Al-SiC MMC	ANN and ANOVA analysis	Ra
Gopan et al. [20]	Grinding machining	Cutting speed, feed, and depth	Composites	Integrated ANN-GA	Ra
Aurtherson et al. [21]	Grinding machining	Number of passes, work speed, cutting depth, current duty ratio, and voltages.	AlSiC composites	ANFIS-based	Ra

(*Continued*)

TABLE 13.1 (Continued)

Authors	Machining Process	Parameters	Materials	Modelling Methods	Output Responses
Bhuyan et al. [22]	EDM	Peak current, pulse on time, and flushing pressure	Al-SiC MMC	RSM and fuzzy logic	Ra
Dewangan et al. [23]	EDM	Discharge current, pulse on time, tool work, and lift time.	Composites	Combined Grey regression analysis (GRA) and fuzzy logic	White layer thickness, surface crack density, and roughness
Hourmand et al. [24]	EDM	Voltage, current, pulse on time, and duty factors	Al-Mg$_2$Si composites.	RSM and ANFIS	Changes in microstructure and MRR
Chaki et al. [25]	Pulsed laser cutting	Speed, pulse energy, and pulse width	Composites	ANN and NSGA-II	Ra and MRR
Patel et al. [26]	Laser cutting	Laser power, speed, and gas pressure	Glass fiber composites	ANN	Ra
Najjar et al. [27]	Laser Cutting	Speed, air pressure, pulse frequency, pulse width, and lamp current	FRP composites	ANN and Chimp optimization model	Kerf quality
Kumaran et al. [28]	Abrasive water jet machining (AWJM)	Jet pressure (JP), traverse speed (TS), and standoff distance (SOD)	CFRP composites	ANFIS	Ra
Bhowmik and Ray [29]	AWJM	Grain size of abrasive, standoff distance, pumping system pressure, abrasive mass flow rate, and nozzle speed	Composites	Fuzzy logic	Ra
Shukla and Tambe [30]	AWJM	Water jet pressure, abrasive flow rate, and quality	Kevlar composite	ANN	Ra and kerf width
Popli et al. [31]	Rotary ultrasonic machining	Spindle speed, feed rate, power rating, abrasive size, and tool profiles	Composites	ANN	MRR
Banerjee et al. [32]	Ultrasonic machining	Pressure, amplitude, and thickness	GERF composites	ANN	MRR

13.2.3.1 Milling Ceramics

Milling involves a rotating cylindrical cutting tool to remove material from the workpieces. It allows precise control making it ideal for creating complex shapes and patterns. In order to examine the correlation between cutting variables, such as cutting speeds, cutting depth, feed, and fraction of SiC, with respect to surface roughness during the machining of Al/SiC MMC, a feed-forward multilayer ANN surface irregularity estimate model was put forth [8]. A multi-objective predictive model was introduced for the machining of aluminum-hybrid composites, utilizing ANN and regression techniques. The study generated several projected outputs, namely, average surface roughness (Ra), temperature, MRR, feed, radial, and tangential force. These outputs were derived from the input process variables including the fraction of SiC, the size of the particle, the cutting depth, the speed, and the feed as reported in a previous study [10]. A model for monitoring the condition of tool wear based on ANFIS was created to predict the wear of tools in relation to feed force during the end milling of GFRP composites. This technique optimized the speed, feed, and cutting depth parameters [11]. A predictive model for surface unevenness during milling machining of CFC was developed using a hybrid intelligence approach that combined ANN and GA. The aforementioned study employed a model to forecast surface roughness by considering various input process variables such as tools used during cutting, speed of spindle, and feed [12].

13.2.3.2 Drilling Composites

Drilling is a standard material removal process for composites in various industries. The process involves using a drill bit to create holes in the material. Drilling is often used to manufacture aircraft components such as wing pairs and fuselage frames. The authors presented a model based on FL to estimate the surface irregularity in the process of drilling GFRP composites. According to the literature [13], the input variables of the model comprise the speed of the spindle, the feed, and the diameter of the drill bit. The authors proposed an ANN based technique for predicting surface unevenness. The model utilized the speed of the spindle, the feed, and the diameter of the drill bit as input variables [14]. An analytical model was developed for the estimation of MRR from the perspective of drilling operations involving GFRP composites. The employed model incorporated input machining variables, including feed, speed, the diameter of drill bits, and the angle of point, and was founded on ANN technology [15].

13.2.3.3 Turning Composites

This machining process is used to produce rotational or conical shapes in composites. It involves rotating the workpiece while a cutting tool removes material from the surface. Turning is commonly used to manufacture composite shafts, bearings, and gears. Based on an ANN and an RSM, a surface irregularity estimate model was developed with input variables speed, cutting depth, and feed for turning of unidirectional GFRP composites [16]. The authors proposed a forecast model for surface texture during the machining of SiC–Al alloy composites, utilizing ANN and ANFIS. The input constraints of the model were identified as speed, feed, and depth, as reported [17]. The study investigated the impact of various machining factors, namely, speed, cutting depth, and feed on surface unevenness, during the turning of GFRP composites using a cubic boron nitride (CBN) tool. The objective was to introduce a fuzzy logic model for estimating surface roughness based on the observed responses [18].

13.2.3.4 Grinding Ceramics

Grinding is a process that involves the use of an abrasive wheel to remove material from the surface of the composite material. It is commonly used to achieve a smooth surface finish and eradicate extra material from the work material. In a recent study, a prototype was established for forecasting primary texture in machining Al-SiC MMC. The technique was based on the analysis of ANN and ANOVA analysis. The input parameters for the model included the cutting speed of the grinding wheel and workpiece, the feed, and the depth [19]. An integrated ANN–GA-based surface roughness forecast model was developed using speed, feed, and cutting depth as process variables during grinding machining [20]. A study was conducted to develop an ANFIS-based model for forecasting surface texture when machining AlSiC composites. The model incorporated inputs such as the number of passes, work material speed, depth, current duty ratio, and voltages [21].

The process of machining composites necessitates the utilization of specialized techniques that deviate from those employed for conventional materials such as metals. Composites exhibit anisotropic behavior, whereby their properties vary in different directions, and incorporate fibers that may induce abrasive wear on tools. The following are additional machining procedures that are frequently employed for composites.

13.2.3.5 Electrical Discharge Machining Composites

Electrical discharge machining (EDM) is a nontraditional material removal method that cuts intricate shapes and intricate patterns in composites. The process involves using an electrically charged wire or electrode to erode the material. EDM is often used in the manufacturing of composite molds and dies. In the context of EDM of Al-SiC MMCs, a proposed technique for estimating surface irregularity was based on RSM and fuzzy logic. The technique applied peak current, pulse on time, and flushing pressure as input process variables [22]. A multi-response predictive model was developed utilizing a combination of grey regression analysis (GRA) and FL for the purpose of forecasting the white-layer thickness, surface crack density, and surface roughness in EDM. The parameters employed in the study were discharge current, pulse duration, tool operation duration, and tool elevation duration, as reported by Dewangan et al. [23]. The study examined the impact of nanopowder-mixed EDM machining factors, including voltage, current, pulse on time, and duty factors, on alterations in microstructure and MRR during the EDM of Al-Mg$_2$Si composites, using RSM and ANFIS projected models [24].

13.2.3.6 Laser Cutting Composites

Laser cutting uses a high-energy laser beam to cut through the composite. This process is favorable for cutting intricate profiles and can produce exact cuts with minimal material waste. In a recent study, a novel integrated model of ANN and non-dominated sorting genetic algorithm II (NSGA-II) was proposed for the optimization of input process variables, including speed, pulse energy, and pulse width, during pulsed Nd: YAG laser cutting. The aim of this model was to estimate surface roughness and MRR, as reported in the literature [25]. A predictive model for surface roughness during laser cutting of GF composites was developed using ANN. The input parameters for this model were identified as laser power, speed, and gas pressure, as reported by Patel et al. [26]. In a previous study, a predictive model was developed for the quality of kerf during laser machining of FRP composites. The model developed an ANN and chimp optimization technique, with input parameters consisting of surface speed, air pressure, pulse frequency, pulse width, and lamp current [27].

13.2.3.7 Abrasive Water Jet Machining Composites

Abrasive water jet cutting involves using a high-pressure jet of water mixed with abrasive particles to cut through the composite. This process is crucial for cutting intricate forms and is non-thermal, so it does not generate heat that can damage the composite. A model based on ANFIS was proposed to predict surface roughness during abrasive water jet machining (AWJM) of CFRP composites. The model utilized jet pressure, traverse speed, and standoff distance (SOD) as inputs. The study exploited the grain size of the abrasive material, SOD, the pressure of the pumping system, abrasive mass flow rate, and the speed of the nozzle to construct a surface roughness prediction tool based on FL [29]. An ANN was utilized to develop a prognostic model for the determination of surface unevenness and kerf width during the machining process of Kevlar composites. The input parameters of the process comprise the pressure of the water jet, abrasive flow rate, and the level of quality, as stated by Shukla and Tambe [30].

13.2.3.8 Ultrasonic Machining Composites

It uses high-frequency vibrations to remove material from the composite. This process is handy for intricate shapes and can produce a smooth surface finish, but it can be slow and expensive compared to other methods. In a previous study, a model based on an ANN was investigated for the purpose of predicting MRR in rotary ultrasonic machining. The model was designed to take into account various input factors, including spindle speed, feed, power rating, diamond abrasive size, and tool profiles [31]. The involvement of ANN in forecasting MRR in the machining of GERF composites is proposed. The input parameters considered for this study are pressure, amplitude, and thickness, as reported by Banerjee et al. [32].

13.2.4 CHALLENGES ASSOCIATED WITH MACHINING COMPOSITES

The choice of material removal process depends on the specific use and the properties of the material being machined. Machining of composites is associated with a number of distinct challenges that are dissimilar to those encountered when machining

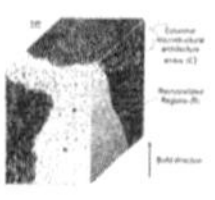

Material properties
- Possess distinctive mechanical and physical characteristics, such as anisotropy and heterogeneity.
- Challenging to fabricate.
- potential to induce complications as delamination, tool wear, and surface damage.

Kok et al. (2018)

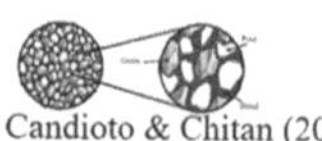

Abrasiveness
- The abrasiveness of composites can result in notable tool wear and reduced tool lifespan

Candioto & Chitan (2022)

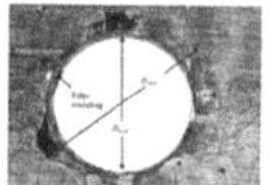

Delamination
- In delamination layers of material get separated.
- Potentially to compromise the structural integrity of the material and lead to a reduction in its overall strength.

Ahmad and Shinde (2016)

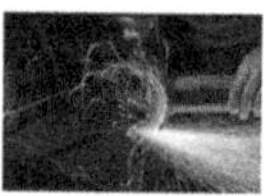

Heat generation
- Generation of heat is a common occurrence during the machining process.
- Excess generation will inflict harm upon the material, compromise dimensional precision, and diminish the longevity of the tool.

https://www.mecholic.com/

Dust and debris
- Can pose a potential threat to human health
- Sometimes cause thermal oxidation.

Kehren et al. (2019)

Surface finish
- The attainment of a smooth surface finish will a challenge during the machining of composites, owing to their propensity to chip and tear.

Karataş et al. (2021)

Cutting Tool selection
- The process of tool selection for machining composites is of utmost importance, as an inappropriate tool can result in material damage and prolonged machining duration.

Sheikh and Davim (2012)

FIGURE 13.4   Challenges in machining of composite materials.

conventional materials such as metals. Several significant challenges (Figure 13.4) arise during the process of machining composite materials:

- **Material properties**: Composite materials possess distinctive mechanical and physical characteristics, such as anisotropy and heterogeneity, which render them challenging to fabricate. The aforementioned properties have the potential to give rise to complications such as delamination, tool wear, and surface damage.
- **Abrasiveness**: The abrasiveness of composites can result in notable tool wear and reduced tool lifespan [33–35].
- **Heat generation**: Thermal energy production is a common occurrence during the machining process, which has the potential to inflict harm upon the material, compromise dimensional precision, and diminish the longevity of the tool [36].

- **Delamination**: Delamination is a common issue in composites, whereby the layers of the material become separated during machining. This phenomenon can compromise the structural integrity of the material and lead to a reduction in its overall strength [37–38].
- **Dust and debris**: The generation of dust and debris during the machining process of composites can be hazardous for human health, necessitating the implementation of appropriate protective equipment and ventilation measures [37–38].
- **Surface finish**: Surface finishing can be challenging during the machining of composites owing to their propensity to chip and tear. It is imperative to exercise caution in order to prevent any surface damage and attain the desired level of finish [39].
- **Tool selection**: The process of tool selection for machining composites is of utmost importance, as an inappropriate tool can result in material damage and prolonged machining duration. Carbide and diamond cutting tools are frequently employed in the machining process [40].
- **Machining parameters**: The selection of parameters such as surface speed, feed, and cutting depth is critical to prevent material damage and attain the desired tolerances.

13.2.5 ANALYTICAL MODELING AND SIMULATION OF COMPOSITE MACHINING PROCESSES

Various modeling and simulation techniques are employed to model and analyze composite machining performance, as illustrated in Figure 13.5. Table 13.2 exhibits the comparison between different analytical modeling techniques like finite element method, multiscale modeling, and Taguchi and research methodology applied for composite machining.

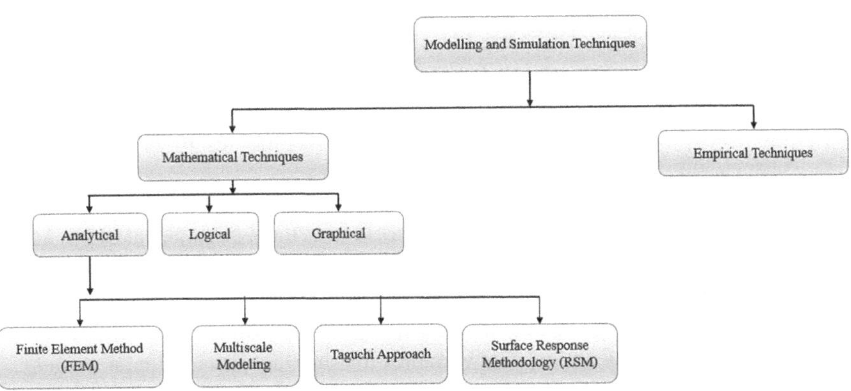

FIGURE 13.5 Classification of various methods used to model and analyze the process performance during machining of composites.

TABLE 13.2

Comparison among Different Modeling Techniques Applied for Composite Machining [4–52]

Approach	Crucial Attributes
Finite Element Method (FEM)	• Numerical technique • FEM can handle complex geometries and loading conditions. • FEM requires significant computational resources and time.
Multiscale modeling	• Computational approach • It can provide insights into the behavior of physical complex systems. • Multiscale modeling can be computationally intensive.
Analytical Modeling	• Mathematical approach • Analytical models are useful for simple systems and geometries. • They require minimal computational resources and time.
Taguchi Approach	• Statistical approach • It can optimize multiple parameters simultaneously and assume a linear relationship between factors and response. • It may require a large number of experiments to optimize the design, so it may lead to significant time.
Response Surface Methodology (RSM)	• Statistical approach • It can optimize multiple factors simultaneously and can model for nonlinear relationship between factors and response. • It may require a large number of experiments to optimize the design, so it may lead to significant time but less time compared to the Taguchi method.

13.2.6 Necessity of AI/Machine Learning Integration for Intelligence-based Modeling

The incorporation of AI and machine learning (ML) methodologies is becoming progressively indispensable for the intelligence-driven modeling of intricate systems, such as composite machining processes. Various advantages of AI–ML integration in the machining process modeling are discussed (Figure 13.6).

- Increased accuracy: Utilizing AI/ML algorithms can enhance precision by enabling the identification of patterns and relationships that may not be discernible through traditional analytical modeling. This phenomenon has the potential to result in predictions and models that are more precise and dependable.
- Improved efficiency: The system exhibits the ability to expeditiously scrutinize vast quantities of data and discern the most pivotal variables and correlations, thereby facilitating accelerated optimization of machining procedures.
- Enhancing comprehension of intricate systems: Composite machining is intricate and entails numerous variables that interact in a nonlinear manner. The identification of such interactions can offer valuable insights into the behavior of the system.
- Adaptive control: It has the potential to facilitate real-time parameter adjustments in machining processes based on system behavior. This approach can

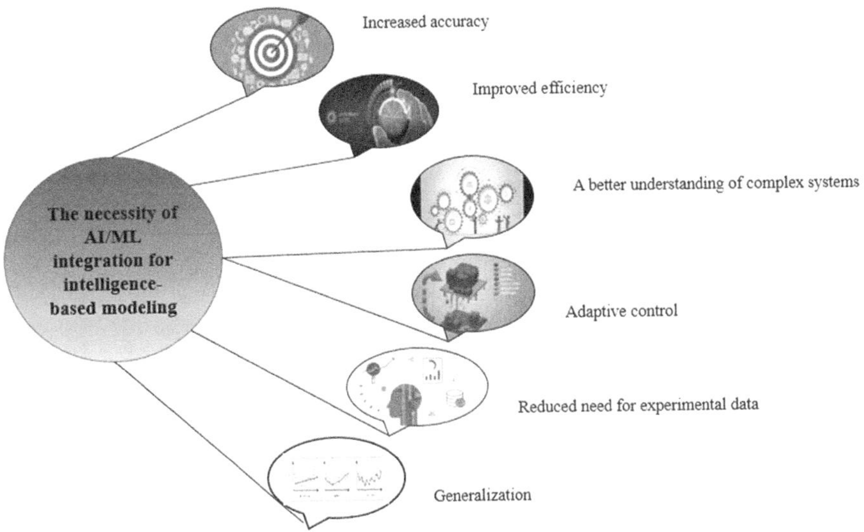

FIGURE 13.6 Various beneficial attributes of the AI–machine learning incorporation in machining process modeling.

effectively minimize the expenses and duration associated with experimentation.

- Reduced need for experimental data: Exploitation of existing data enables the system to generate predictions and models, thereby reducing the necessity for extensive empirical data.
- Generalization: The processes of generalization involve utilizing available data to make predictions about the behavior of novel/new materials and geometries.

13.2.7 Various Tools of AI and Their Functionalities

The field of AI is expanding at a swift pace, encompassing the development of software and systems capable of executing tasks that would conventionally necessitate human intelligence, such as learning, problem-solving, and reasoning. Various tools and techniques are employed in the field of artificial intelligence, each possessing unique functionalities (Figure 13.7):

ML: ML is a subfield of AI that focuses on the improvement of algorithms proficient in learning from data and making forecasts or decisions without the need for explicit programming. ML is a widely employed technique for performing tasks such as image recognition, speech recognition, and predictive modeling.

ANN: ANNs are a class of computational models that draw inspiration from the architecture and function of the human brain and are commonly employed in the field of machine learning. ANNs are composed of interconnected nodes, also known as neurons, which possess the ability to learn from data and subsequently make predictions or decisions. ANNs have the capability

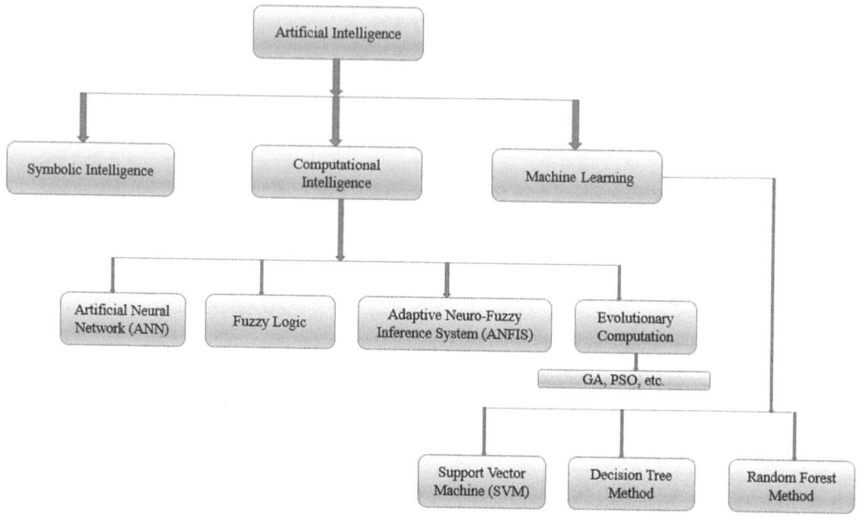

FIGURE 13.7 A general classification of AI showing common methodologies applied in machining.

to perform a multitude of tasks, such as pattern recognition, classification, regression, and control.

FL: FL is a mathematical paradigm that facilitates the process of reasoning and decision-making in scenarios that involve ambiguity and vagueness. FL is founded on the concept of fuzzy sets, which permit degrees of membership within a set, and fuzzy rules, which facilitate reasoning through the application of FL. This is a versatile tool that can be used in a variety of applications, such as regulating parameters, enhancing optimization, facilitating decision-making, and making predictions. Both ANNs and FL are applied in diverse domains such as control systems, image and speech recognition, predictive modeling, and decision-making. ANNs are highly advantageous in situations in which intricate patterns or relationships exist within the data, whereas fuzzy logic is a valuable tool for tasks that entail imprecision or uncertainty in the data. A comparative analysis is drawn in Figure 13.8. Integrating both tools in hybrid systems can lead to an enhancement in performance by leveraging their respective strengths.

ANFIS: ANFIS is a computational model that utilizes a hierarchical architecture consisting of a FIS and a neural network. The FIS enables the representation and reasoning of data that is uncertain or imprecise. On the contrary, the neural network offers the capacity to acquire knowledge from data and adapt its parameters to enhance its performance. The architecture of ANFIS comprises five layers:

- Input layer: This layer is responsible for receiving the input data and transmitting it to the subsequent layer.
- Fuzzy layer: The fuzzy layer applies fuzzy logic to process the input data and produce fuzzy sets that accurately represent the input data.

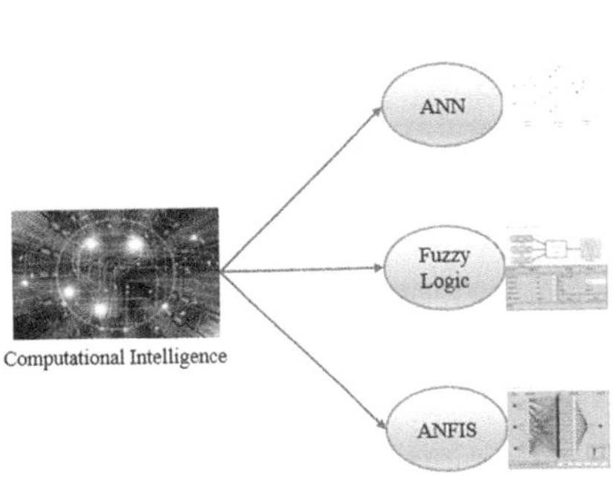

FIGURE 13.8   Crucial attributes of the ANN, FL, and ANFIS techniques.

- Normalization layer: The normalization layer is responsible for adjusting the membership values of fuzzy sets to accommodate any overlapping membership functions.
- Rule layer: This layer combines the fuzzy sets generated in the earlier layers to form fuzzy rules that represent the connection between input and output data.
- Output layer: The final output is generated by the combination of fuzzy rules in the output layer.

The ANFIS model is trained through a hybrid learning algorithm that integrates the gradient descent technique with the least squares method. Throughout the training process, the model modifies the parameters of both the FIS and neural network to minimize the discrepancy between the estimated output and the real output. ANFIS has been employed in diverse domains, such as control engineering, forecasting, and categorization. Utilizing data in situations in which uncertainty or imprecision is present and where learning from the data is necessary to enhance performance can be considered advantageous. The superiority of ANFIS over conventional fuzzy logic systems has been demonstrated in numerous applications, while simultaneously offering the interpretability and transparency characteristic of FL. Basic principles, advantages, disadvantages, and applications for previously discussed techniques as shown in Figure 13.8.

13.3   AI-DRIVEN SMART CONTROL AND MODELING OF COMPOSITE MACHINING PROCESSES

13.3.1   AI-BASED MODELING FOR SURFACE INTEGRITY PREDICTION IN COMPOSITE MACHINING

13.3.1.1   Implementation of ANN for Surface Integrity Prediction

Anticipating the surface integrity of a machined component is crucial in order to optimize the machining procedure and guarantee that the component conforms to the intended specifications. ANNs have the capability to forecast the surface integrity of a machined component by utilizing input parameters such as cutting parameters, material characteristics, and tool geometry. Neural networks have the potential to serve as a robust mechanism for forecasting surface integrity in machining operations due to their ability to comprehend intricate associations between the input and surface integrity metrics. The accuracy of forecasts is contingent upon the caliber and quantity of the input data, as well as the network architecture and training algorithms employed. The architecture of an ANN comprises three vital layers, namely, the input layer, the hidden layer(s), and the output layer. Each layer of an ANN is characterized by an activation function and a specific number of neurons. An ANN employs various activation functions such as transig, logsig, and linear activation functions. Occasionally, ReLU (rectified linear unit) and softmax activation functions are also utilized.

For linear transfer function $y = x$, log sigmoid $y = \dfrac{1}{1+e^{-ax}}$, tan sigmoid $y = \dfrac{e^{ax} - e^{-ax}}{e^{ax} + e^{-ax}}$, and a, is a coefficient that is going to decide what should be the slope of this particular curve. O is the output such that $O = (w \times x) + b$, v and w are connecting weights, b is the bias value. Error is calculated by subtracting the output value of the neural network value from the target value. The application of an ANN (Figure 13.9) involves the following steps:

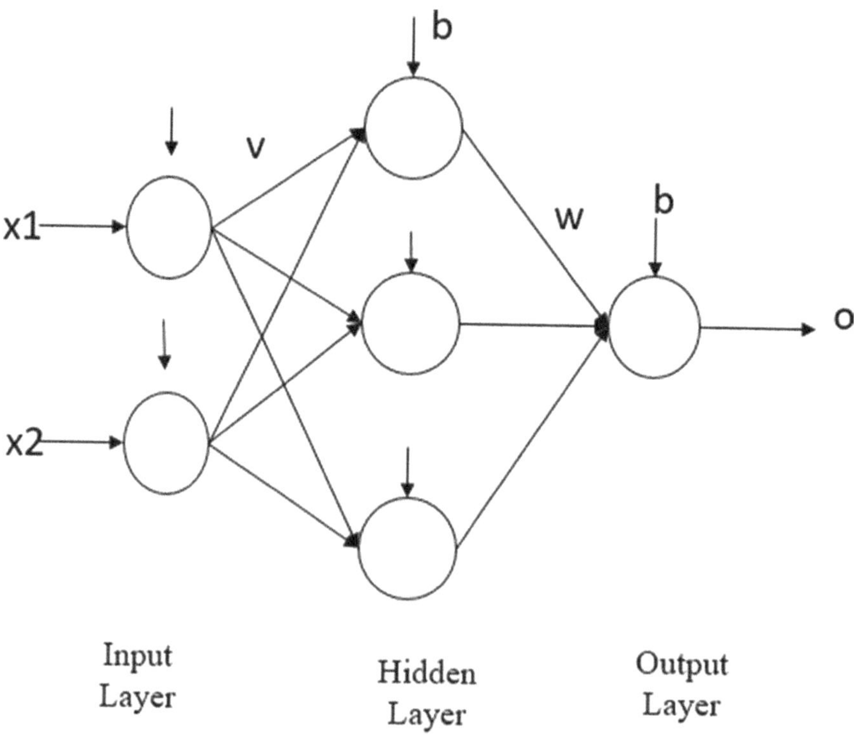

FIGURE 13.9 Basic architecture of neural network (here x_1 and x_2 are input for the neural network).

- Data collection: Collect data on the input variables and the corresponding surface integrity measurements. The data can be collected through experimental testing or simulations.
- Data preprocessing: Clean and preprocess the collected data by removing any outliers or errors and normalizing the data to a standard range.
- Network architecture: Choose an appropriate network architecture for the prediction problem. The number of hidden layers and nodes in each layer can be determined using trial and error or optimization techniques.
- Training the network: Train the network using preprocessed data, which involves feeding the input data to the network and adjusting the weights and biases of the network using Levenberg–Marquardt (LM) backpropagation, and gradient descent algorithms to minimize the prediction error.
- Testing and validation: Test the trained network on a separate data set to evaluate its accuracy and generalization ability. Validation techniques have been applied to avoid overfitting training data, such as k-fold cross-validation.
- Deployment: Once the ANN has been trained and validated, it is deployed for predicting new surface integrity input data.

The surface integrity of composites in conventional machining methods is primarily affected by the surface speed, depth of cut, feed rate, and occasionally by the workpiece speed. ANNs were employed to forecast the surface texture during the machining process of composite materials. The LM backpropagation algorithm is a commonly employed optimization technique for training ANN systems, owing to its superior convergence rate when dealing with small data sets. The process of generating training samples involves the consideration of various levels of process parameters, and the experimental results obtained from this data are assumed to be the target values. The basic working principle of an ANN for predicting machining performances is shown in Figure 13.10.

To further comprehend the concepts of ANN, FL, and ANFIS, a case study is demonstrated using the data sets retrieved from experimental research on turning GFRP composite material conducted by Bagci and Işık [16]. The data set comprises independent input parameters, including cutting speed (V), feed rate (f), and depth of cut (d), while the predicted parameter is the surface roughness, as presented in Figure 13.11.

Network architecture and training: The ANN model was constructed with different layers. The proposed model consists of an input layer with three process parameters,

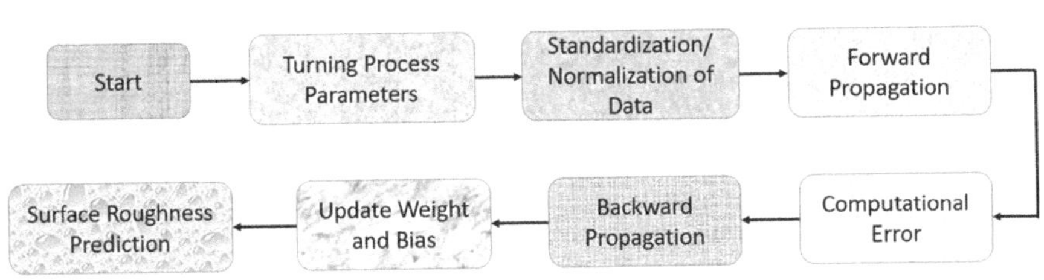

FIGURE 13.10 Flow chart of working principle of neural network.

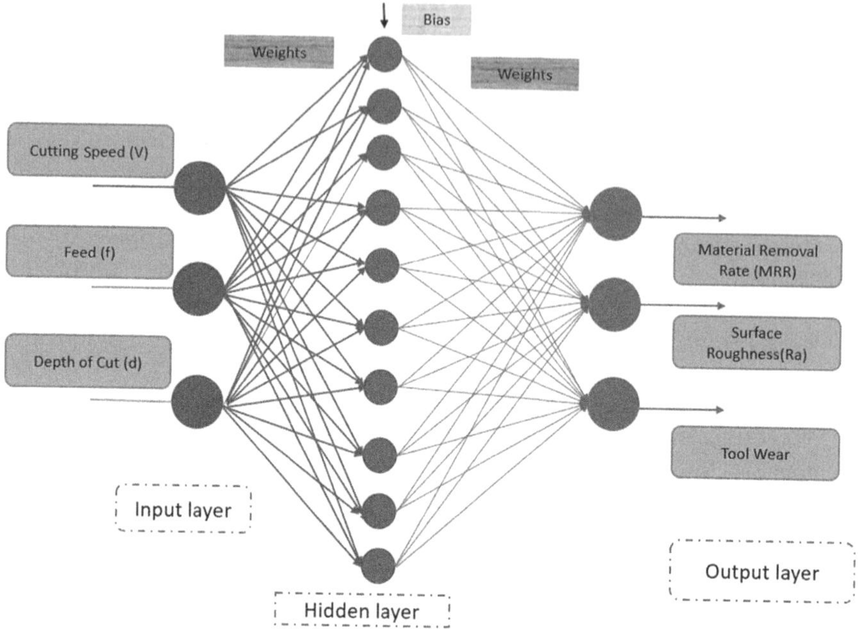

FIGURE 13.11　The architecture for a neural network to predict surface roughness, MRR, and tool wear.

namely, speed, feed, and depth. The hidden layer comprises 10 neurons, and the output layer is characterized by a linear transfer function. The optimal characteristics of a network are determined by the neurons located in the hidden layer. A neural network may not have properly learned the input–output pattern if it has a lower number of neurons, leading to higher error rates. Conversely, a higher number of hidden layers can improve accuracy in prediction and provide greater flexibility to the ANN. The neural network architecture was modeled employing MATLAB 2021A. The input layer is connected in a fully connected manner to the hidden layer, which, in turn, is linked to the output layer.

Figure 13.12 exhibits variation of experimentally observed roughness data with that obtained by three models. It has been noticed that the difference between the experimental and the predicted in the case of FL is more when compared to the other two models. The comparison revealed that the ANFIS model outperformed the ANN and FL models in terms of predicting the surface roughness of composites during machining. The ANFIS model is a hybrid approach that integrates the strengths of both FL and neural network techniques.

13.3.1.2　FL-based Surface Integrity Assessment

FL is also widely employed for surface integrity prediction during composite machining. FL is a mathematical construct that addresses the handling of imprecise and uncertain data. It can be utilized in intricate model systems in which conventional rule-based systems may prove inadequate. In the process of material removal composites,

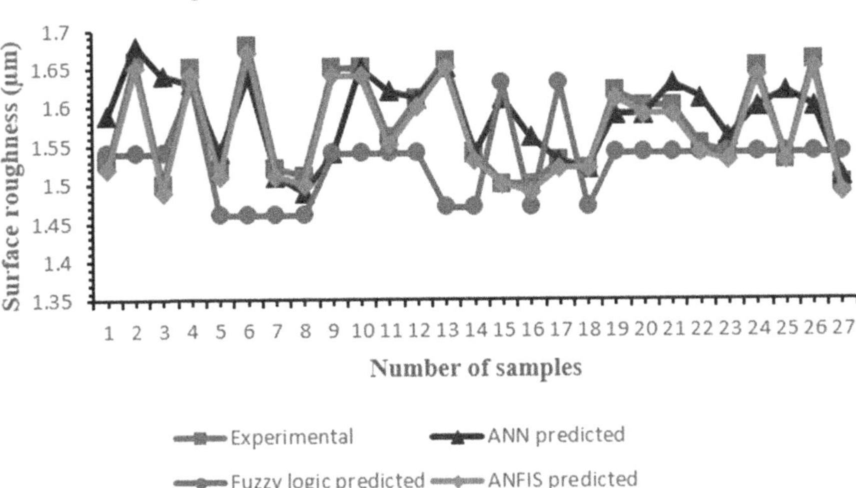

FIGURE 13.12 Comparison between experimentally observed roughness data with that obtained data by ANN, FL, and ANFIS.

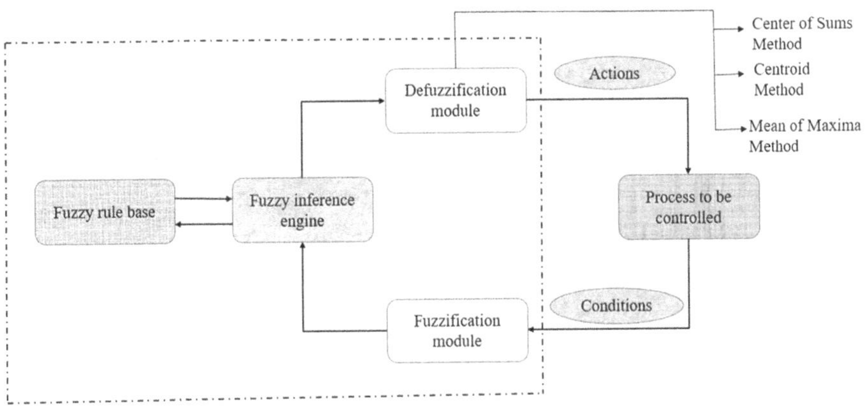

FIGURE 13.13 Basic principle of an FLC.

various factors including cutting speed, feed rate, tool wear, and tool geometry can have an impact on the surface texture of the machined components. The application of fuzzy logic is essential in the modeling of the correlation between the aforementioned factors and the surface texture of machined components. Figure 13.13 exhibits the basic principles of FL controller (FLC) imposed for predictions.

The initial stage involves the establishment of input and output variables for the purpose of constructing a surface integrity prediction system based on FL. The independent variables in this study may include surface speed, feed rate, wear in tool, and tool geometry, while the dependent variables may consist of surface roughness, delamination, and fiber pullout. Subsequently, it is imperative to establish the

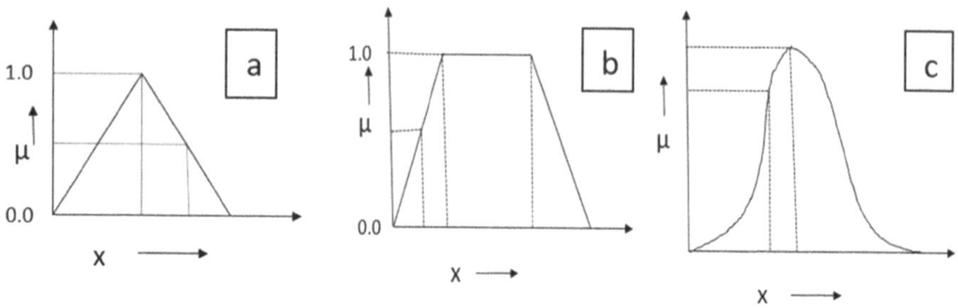

FIGURE 13.14 Types of membership implemented in FL.

imprecise sets for every individual variable. Fuzzy sets provide a means of representing information that is uncertain or imprecise. The cutting speed can be categorized into three levels, namely, low, medium, and high, with different membership functions (Figure 13.14). Subsequent to the establishment of the fuzzy sets, the subsequent stage involves the formulation of a collection of fuzzy rules that facilitate the relationship between the input and output variables. The fuzzy rules establish the link between the input and the output responses. In instances where the speed is elevated and the feed is reduced, it is probable that the surface roughness will be heightened. After the establishment of the fuzzy rules, the subsequent stage involves the utilization of an fuzzy inference system (FIS) to prognosticate the output variable by relying on the input variable. The process of fuzzy inference involves the utilization of input variables and fuzzy rules to generate a crisp output value. Ultimately, the assessment of surface quality will be conducted through an analysis of the projected outcome. The aforementioned data can be employed to modify the machining parameters by enhancing the surface characteristics of the machined part. Through the establishment of a correlation between the machining process variables and the surface quality of the machined component, the system can facilitate the optimization of machining parameters by enhancing the surface quality. This particular aspect is demonstrated in Figures 13.15 and 13.16.

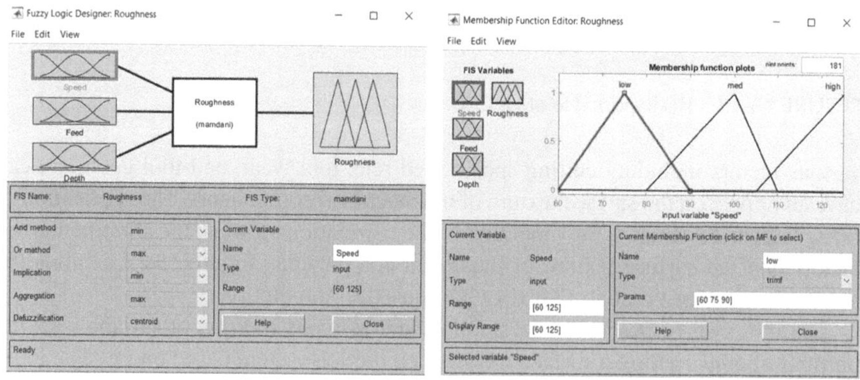

FIGURE 13.15 The input (speed, feed, depth) and predicting output (surface roughness) parameters feeding into fuzzy logic designer are exhibited, and (right) the triangular membership function is defined for process parameters.

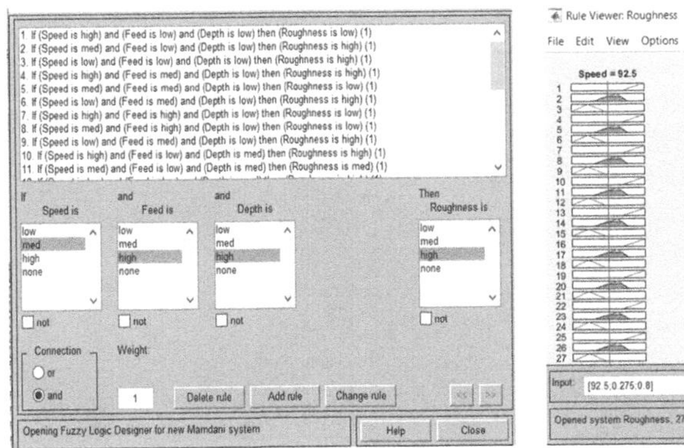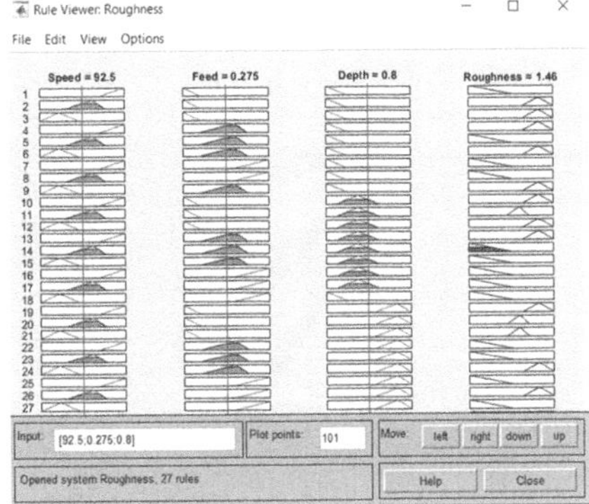

FIGURE 13.16 If–then rules are imposed for prediction, and (right) predicted surface roughness (1.46 μm) obtained is demonstrated for inputs of speed of 92.5 m/min, feed of 0.275 mm/rev, and depth of 0.8 mm using fuzzy logic toolbox.

13.3.1.3 ANFIS-based Surface Integrity Modeling

The ANFIS system is a type of hybrid intelligent system that integrates the advantageous features of both neural networks and FL. The basic framework of ANFIS techniques is shown in Figure 13.17. The utilization of ANFIS is deemed to be a useful approach in the prediction of the surface reliability of composites in the course of machining. The mechanical properties and quality of composite materials are significantly influenced by surface integrity, which is a crucial factor. ANFIS employs a methodology whereby the input parameters are subjected to fuzzification through the utilization of fuzzy sets, and subsequently, the output is produced by means of a fuzzy inference system. The ANFIS model's parameters are optimized through a hybrid learning algorithm that integrates backpropagation and least squares techniques. ANFIS can undergo a training process using a collection of input–output data pairs to acquire knowledge of the fundamental association between the input and output.

An instance of predicting the surface roughness of a composite material during machining can be achieved by utilizing input variables of the machining process, in order to determine the output variables. The acquisition of input–output data pairs is necessary through the machining of composite material under varying cutting conditions, followed by the measurement of resultant surface roughness. The ANFIS model has the capability to undergo training to forecast surface roughness by utilizing the input variables as per the available data. The predictive accuracy of the ANFIS model will be estimated through testing with validation data. Upon completion of the ANFIS training, the model is subsequently employed to forecast the surface roughness of the composite material during machining operations under novel cutting conditions. This model incentivizes manufacturers to optimize the machining process in order to attain the desired surface superiority and mechanical properties of the composite material. The application of ANFIS is deemed a potent approach for forecasting the surface integrity that transpires during the composite machining process. Through the utilization of ANFIS modeling

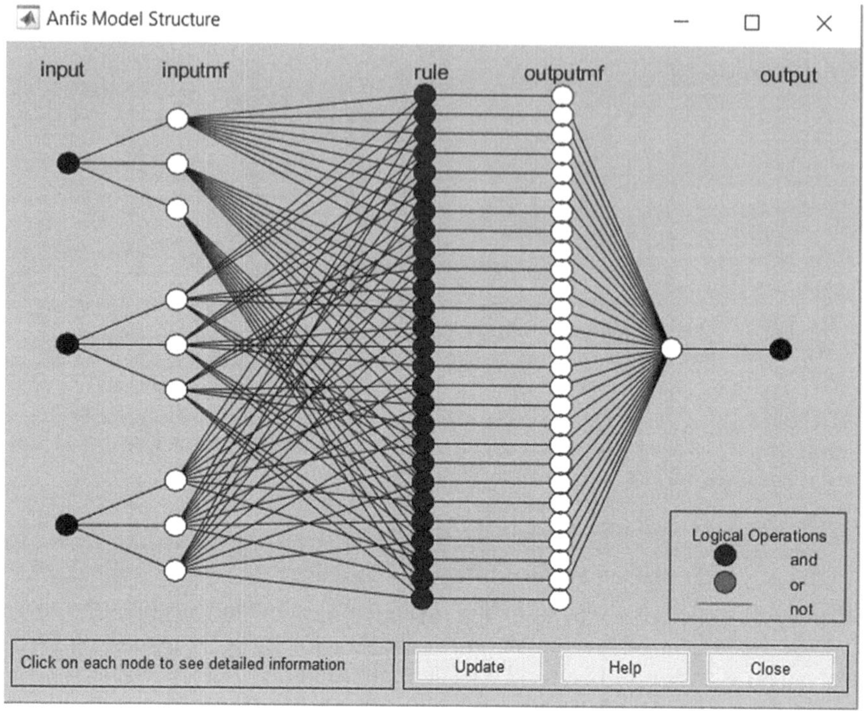

FIGURE 13.17 Structure of an ANFIS model.

techniques on input–output data pairs, manufacturers can effectively optimize the machining process in order to attain the desired mechanical properties and surface quality of composite materials. This particular aspect is demonstrated in Figures 13.18 to 13.20. Error data calculated through different approaches are tabulated in Table 13.3.

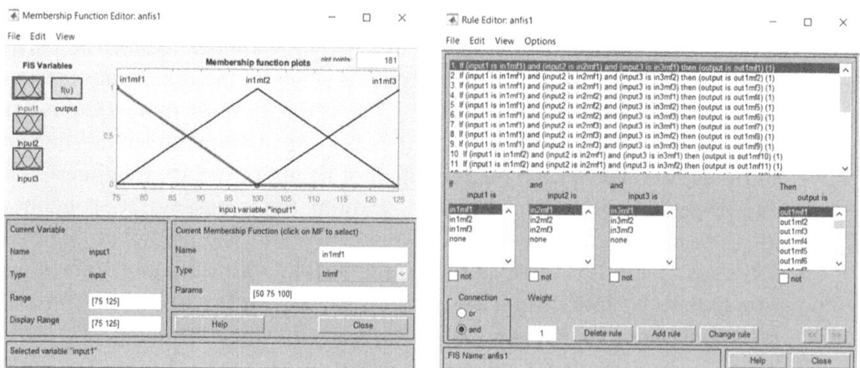

FIGURE 13.18 The triangular membership function is defined for process parameters with the membership function editor, and (right) if–then rules are imposed for prediction with a rule editor of the ANFIS toolbox.

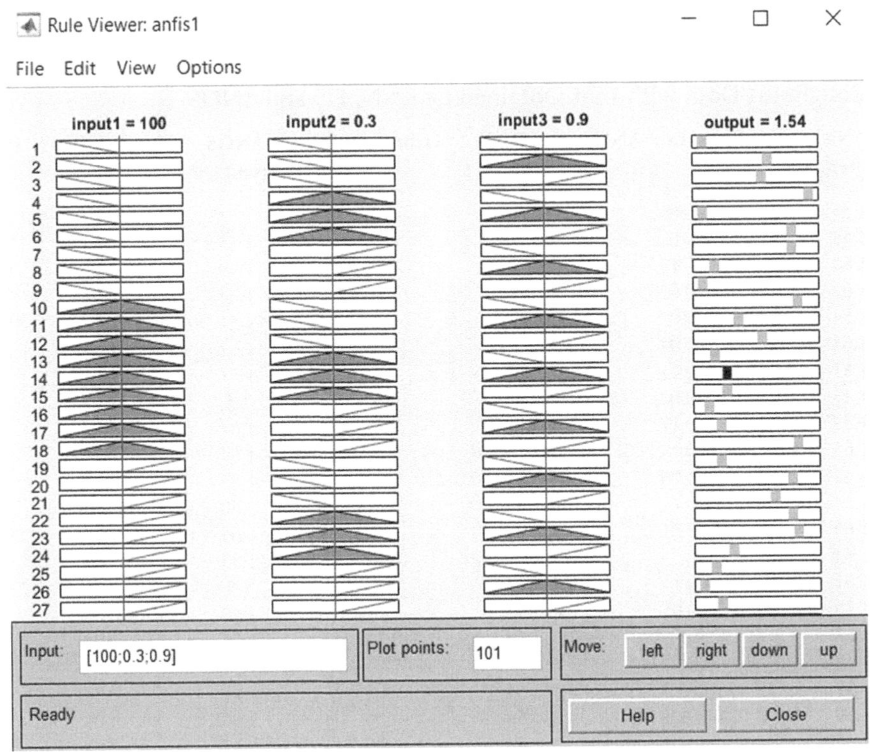

FIGURE 13.19 Predicted surface roughness (1.54 μm; blue box) obtained for inputs of speed of 100 m/min, feed of 0.3 mm/rev, and depth of 0.9 mm using the ANFIS toolbox.

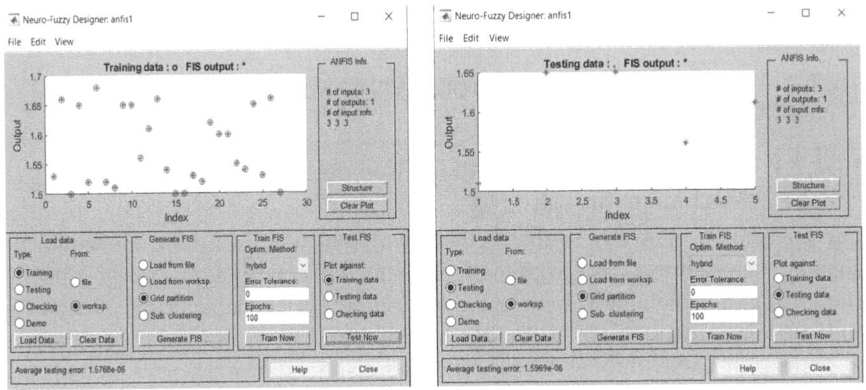

FIGURE 13.20 The training data used for training the ANFIS model and (right) testing data used to validate the ANFIS model are exhibited.

TABLE 13.3

Error Data Obtained Correspond to Experimentally Observed Roughness Data with That Obtained by ANN, FL, and ANFIS

ANN Predicted	Error=Exp.-ANN	FL Predicted	Error=Exp.-FL	ANFIS Predicted	Error=Exp.-ANFIS
1.59	−0.06	1.54	−0.01	1.52	0.01
1.68	−0.02	1.54	0.12	1.65	0.01
1.64	−0.14	1.54	−0.04	1.49	0.01
1.63	0.02	1.63	0.02	1.64	0.01
1.54	−0.02	1.46	0.06	1.51	0.01
1.64	0.04	1.46	0.22	1.67	0.01
1.51	0.01	1.46	0.06	1.51	0.01
1.49	0.02	1.46	0.05	1.5	0.01
1.54	0.11	1.54	0.11	1.64	0.01
1.65	0	1.54	0.11	1.64	0.01
1.62	−0.06	1.54	0.02	1.55	0.01
1.61	0	1.54	0.07	1.60	0.01
1.65	0.01	1.47	0.19	1.65	0.01
1.54	0	1.47	0.07	1.53	0.01
1.61	−0.11	1.63	−0.13	1.5	0
1.56	−0.06	1.47	0.03	1.49	0.01
1.53	0	1.63	−0.1	1.52	0.01
1.52	0	1.47	0.05	1.52	0
1.59	0.03	1.54	0.08	1.61	0.01
1.59	0.01	1.54	0.06	1.59	0.01
1.63	−0.03	1.54	0.06	1.59	0.01
1.61	−0.06	1.54	0.01	1.54	0.01
1.56	−0.02	1.54	0	1.53	0.01
1.6	0.05	1.54	0.11	1.64	0.01
1.62	−0.09	1.54	−0.01	1.53	0
1.6	0.06	1.54	0.12	1.65	0.01
1.51	−0.01	1.54	−0.04	1.49	0.01

13.3.2 AI-BASED MODELING FOR TOOL WEAR AND TOOL LIFE IN COMPOSITE MACHINING

13.3.2.1 Tool Wear and Tool Life Prediction during Composite Machining through ANN

The application of ANNs in the prediction of tool wear and tool life within the domain of machining. The procedure entails the collection of an extensive dataset comprising of machining parameters, composite material characteristics, tool wear, and tool longevity information. The aforementioned data are utilized for the purpose of training the ANN through a process of iterative modification of its weights and biases. This process continues until the ANN is capable of accurately predicting tool wear and tool life based on the input data. Upon completion of the training process, the ANN can be employed to forecast tool wear and tool life in novel machining conditions and composite materials.

13.3.2.2 FL-based Assessment of Tool Wear and Tool Life

FL-based forecasting of the wear of tools and their lifespan when machining composites entails the contemplation of machining parameters, orientation of fiber, matrix type, and density as input parameters. Fuzzy sets and rules were employed to establish and interrelate these variables. The aforementioned methodology enables the handling of uncertain data and furnishes a convenient means of evaluating tool wear and tool longevity in the context of composite machining.

13.3.2.3 Hybrid Neuro-fuzzy Inference System for Tool Wear and Tool Life Modeling

Tool wear is a significant concern in machining, leading to decreased performance, poor surface finish, and increased expenses. The efficacy of utilizing an ANFIS has been effective in the modeling of the interrelationship between tool wear, tool life, and input machining parameters in the context of machining of composite. Accurately forecasting the remaining lifespan of tools enables operators to strategically plan tool replacements at the optimal time, thereby reducing the likelihood of tool failure and improving the overall efficacy of the operation.

13.3.3 AI-BASED MODELING FOR MRR AND FEATURE ACCURACY IN COMPOSITE MACHINING

13.3.3.1 ANN-Driven Control of MRR and Accuracy

ANN AI tool is commonly used to predict the MRR and accuracy during the machining of composites. The ANN model is trained based on input–output pairs representing the relationship between input-output variables. The input variables, in this case, could be cutting speed, feed, depth, tool geometry (such as rake angle and clearance angle), tool material, work material (including fiber type and matrix type), and machining environment (coolant type, pressure, and temperature), while the output variables could be MRR and the accuracy of the machining process.

13.3.3.2 FL-based Control of MRR and Accuracy

FL-based prediction is a popular approach for predicting MRR and accuracy during the machining of composites. FL can help to predict the MRR and accuracy during machining by modeling the complicated relation between the process input and output variables. The input variables may include speed, feed, cutting depth, orientation of fiber, and tool geometry, while predictable responses are MRR and accuracy. Utilizing FL-based prediction has the potential to enhance the optimized machining parameters, as well as lead to enhanced quality and efficiency in the machining process for composites.

13.3.3.3 Neuro-Fuzzy-based Hybrid System for Controlling MRR and Accuracy

ANFIS has been employed in the domain of composite machining to forecast both the MRR and precision of the procedure. The ANFIS-based prediction of MRR and accuracy in composite machining is dependent on input variables such as cutting speed, feed rate, depth of cut, and the specific type of composite material being machined. The output responses comprise MRR and the precision of the procedure. To conclude, applying ANFIS presents a significant contribution to predicting MRR and accuracy in the context of composite machining. ANFIS can be utilized by engineers to optimize the machining process, thereby achieving optimal MRR and accuracy.

13.3.4 OTHER ML APPROACHES FOR SURFACE INTEGRITY PREDICTION

AI tools, including ANN, FL, and ANFIS, as well as other machine learning algorithms such as regression, random forest (RF), decision trees, and support vector machines (SVMs), are employed to examine the correlation between response variables in the prediction of machining parameters. By utilizing these models, it becomes feasible to enhance the machining process and attain prediction of MRR, surface roughness, and tool wear during composite machining.

- SVMs: One of the standard ML algorithms for surface integrity prediction. A nonlinear regression model can make accurate predictions by finding the best hyperplane that separates the data into different classes.
- RFs: RF is an ensemble learning technique that combines several decision trees to make predictions. It can handle nonlinear interactions between the input and output variables and make accurate predictions even with noisy data.

13.4 METAHEURISTIC OPTIMIZATION OF PROCESS PARAMETERS TO ENHANCE MACHINABILITY

The application of metaheuristic optimization was found to be a potent technique for optimizing process parameters with the aim of improving machinability during machining operations. The process of machining composites poses a significant challenge owing to their anisotropic and heterogeneous characteristics, which can potentially lead to complications such as delamination and fiber pullout. Attaining machined parts of superior quality with minimal defects is feasible through appropriate optimization of process parameters. Various metaheuristic optimization methods are currently utilized to optimize process parameters for the machining of composites. Some of the commonly used methods include genetic algorithm (GA), particle Swarm optimization (PSO), and others.

13.4.1 GA-DRIVEN OPTIMIZATION FOR PERFORMANCE ENHANCEMENT IN COMPOSITE MACHINING

GA is a robust metaheuristic optimization methodology that employs concepts derived from genetics and natural selection to explore potential solutions to a given problem (Figure 13.21). These models exhibit a high degree of suitability for optimization problems that involve a large number of variables or intricate, non-linear associations.

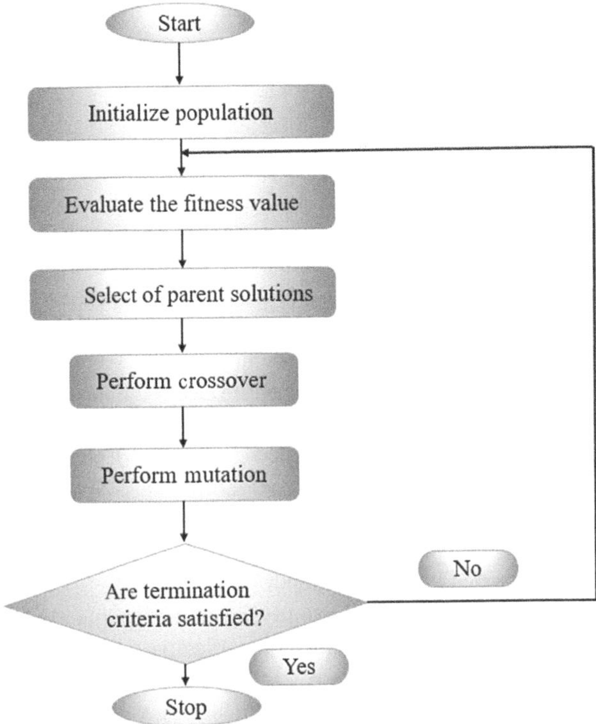

FIGURE 13.21   Flow chart of working principles of GA.

GA-driven optimization has been observed in the domain of composite machining materials, which are widely applied in various industries such as aerospace and automotive, owing to their favorable characteristics such as high strength-to-weight ratio. Moreover, their anisotropic and heterogeneous nature renders them challenging to machine. The optimization of the machining process for composite materials is a crucial aspect in guaranteeing the production of superior quality parts, while simultaneously minimizing material wastage and reducing expenses. The utilization of GA-driven optimization can facilitate the identification of the most favorable amalgamation of cutting parameters. This process can be primed for the attainment of optimal machining performance, which encompasses the optimization of the surface roughness and tool wear, as well as MRR, for a specific composite material. An example of implementing GA-driven optimization for composite machining is discussed next.

To optimize the drilling parameters for the purpose of creating holes in a composite material made of CFRP. The objective is to optimize the MRR while simultaneously minimizing Ra and TW. The optimization of the cutting process requires the consideration of three key variables, namely, the speed (V), feed (f), and depth (d). Establishing realistic values for each variable can be achieved by defining upper and lower bounds. We can define the fitness function as follows:

$$\text{Fitness} = -\text{MRR} + \frac{1}{\text{Ra}} + \frac{1}{\text{TW}}.$$

The negative sign for MRR is attributed to its maximization, whereas the remaining terms are positive and necessitate minimization. The reciprocal terms for Ra and TW guarantee that greater numerical values yield lower fitness scores. The steps involved in GA are as follows.

- Define objective functions: Objective functions are mathematical expressions that are used to quantify the performance of a system or process. They are used in optimization problems to determine the optimal values of decision variables that will maximize or minimize the objective function. The objective function serves as an indicator of the machining process's effectiveness, encompassing the parameters of MRR, surface roughness, and tool wear. The objective function receives the machining parameters as its input and generates the corresponding outputs of MRR, surface roughness, and tool wear.
- Define design variables: *Design variables* refers to the parameters or factors that are manipulated or controlled during the design process in order to achieve a desired outcome. These variables can include physical dimensions, material properties, machining conditions, and other relevant factors that impact the design. The design variables under consideration in this instance can be speed, feed, and depth.
- Initialize the population: A randomly initialized population represents a solution set of values evaluated using an objective function.
- Evaluate the fitness: The fitness of each population solution is determined by evaluating the objective function using design variables.
- Selection: Select the parental solution to be used for reproduction. The selection process can be based on tournament selection, proportional selection, or rank-based selection.
- Crossover: The process of crossover involves merging a specifically chosen solution in order to generate a fresh set of solutions. The process involves the exchange of segments of solution between specific pairs.
- Mutation: The process of mutation involves the random modification of a newly generated set of solutions, with the aim of introducing genetic diversity and preventing the occurrence of local optima.
- Evaluate the fitness of the new population: Assess the level of fitness exhibited by the recently formed population. Assess the suitability of the recently generated population by means of the objective function.
- Repeat the process: Iterate the sequence of selection, crossover, and mutation until the machining parameters that optimize the MRR and minimize the surface roughness and tool wear are achieved. Upon achieving convergence on an optimal set of cutting parameters, the GA can effectively utilize them to attain optimal performance during the machining process.

To summarize, the employment of GA-driven optimization has the potential to significantly improve performance in composite machining. Identifying the most effective combination of cutting parameters can lead to superior machining outcomes, decreased material waste, and cost savings.

13.4.2   PSO Implementation for Machinability Improvement during Composite Machining

PSO is an optimization algorithm motivated by the collective behavior of bird flocks or fish schools. It can be used to optimize widespread problems, including machining process parameters optimization for composite materials. Here is an example of how PSO can improve machinability during composite machining (Figure 13.22).

For optimizing the cutting parameters for milling a GFRP composite material. The goal is to minimize Ra and TW while maintaining a high MRR. The cutting parameter variables needed for optimization are the cutting speed (V), feed (f), and depth (d). Set the upper and lower bounds for each variable to ensure the values are realistic. The PSO algorithm maintains a population of candidate solutions called particles, each representing a potential set of cutting parameters. The particle moves around in search space, and the best solution guides their movements found so far (global best) and the best solution found by each particle (the personal best). Implementing a PSO algorithm can be carried out in the following way:

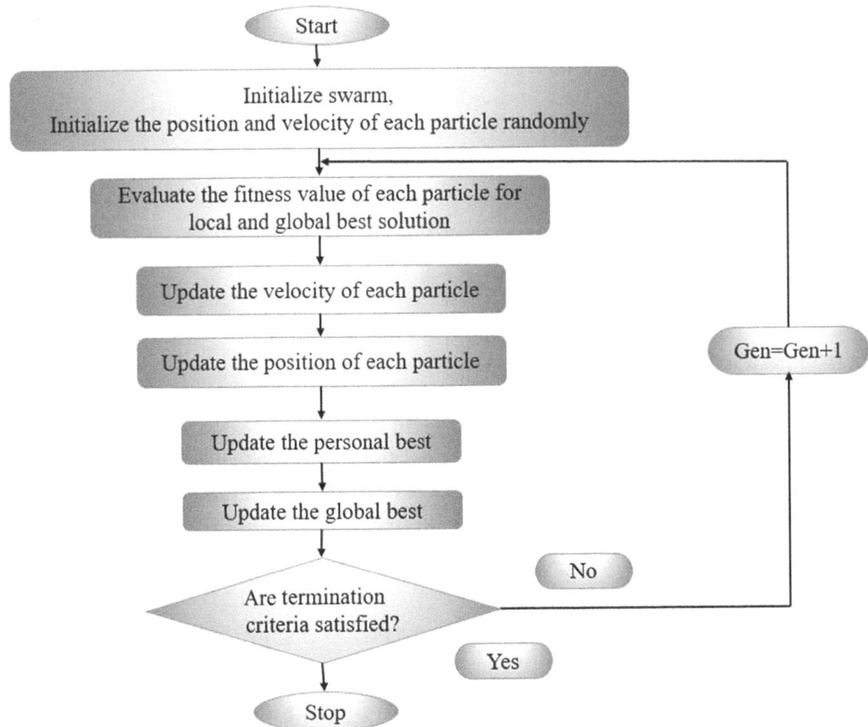

FIGURE 13.22   Flow chart of working principles of PSO.

- Initialization: Randomly generates a population of particles with initial positions and velocities in the search space.
- Evaluations: Calculate the fitness of each particle using the following fitness function.

$$\text{Fitness} = -\text{MRR} + \frac{1}{\text{Ra}} + \frac{1}{\text{TW}}$$

- Update personal best: For each particle, compare its current fitness with its personal best fitness. If the current fitness is better, update the personal best position and fitness.
- Update global best: Identify the particle with the best fitness among all particles, and update the global best position and fitness.
- Update velocities and positions: Update the velocities and positions of each particle based on the global best and personal best solutions. The updated equations are as follows:

$$v_{id}^{new} = w_i \times v_{id}^{old} + c_1 \times r_1 \times \left(p_{id} - x_{id} \right) + c_2 \times r_2 \times \left(p_{gd} - x_{id} \right),$$

where $d = 1, \ldots, D, i = 1, \ldots, E$;
v_{id}^{new} = velocity of the particle i in the iteration t;
v_{id}^{old} = velocity of the particle i in the iteration $(t - 1)$;
c_1 and c_2 = coefficients of acceleration;
w_i = weight of inertia;
r_1 and r_2 = random numbers [0, 1];
p_{id} = best position of each individual particle i; and
p_{gd} = best position of the entire swarm.

With the updated velocity, the position of each particle is updated with the following equation:

$$x_{id}^{new} = x_{id}^{oid} + v_{id}^{new},$$

where x_{id}^{new} = new position of the particle in iteration t and x_{id}^{oid} = position of the particle in iteration $(t - 1)$.

Repeat steps 2 to 5 until the optimized criteria are reached. Once the PSO algorithm has converged to an optimal set of cutting parameters, then these parameters can be used in the machining process to achieve the best possible performance. Minimizing surface roughness and tool wear while maintaining a high MRR can enhance the machinability of the composite material and improve the quality of the machined parts.

13.4.3 Integration of AI-driven Modeling and Optimization Tools

The employment of AI-based modeling and optimization techniques in composite machining has been shown to augment the machinability of these materials and enhance the quality of machined components. This topic presents an overview of the integration of AI-driven modeling and optimization techniques in the machining of composites (Figure 13.23).

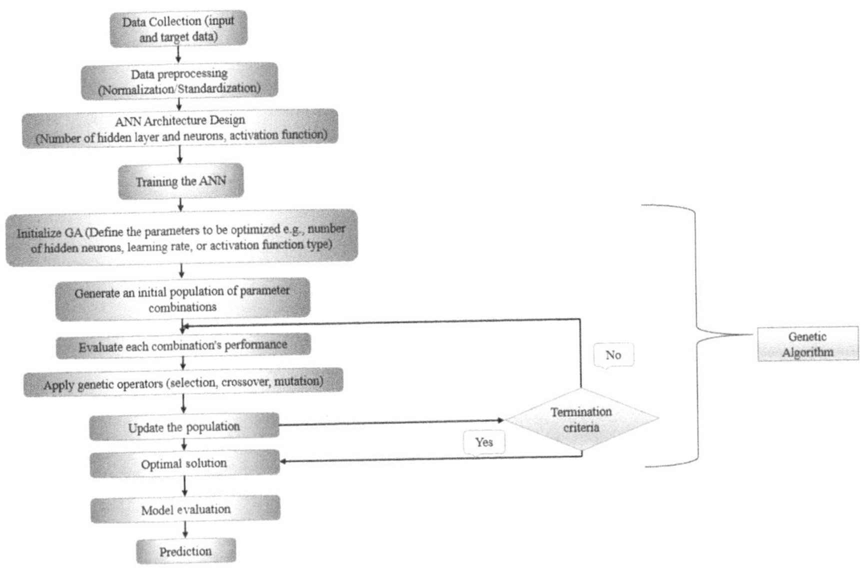

FIGURE 13.23   Flow chart of working principles of ANN–GA.

During the preliminary phase, an AI-driven approach incorporating neural networks, fuzzy logic, and other machine learning algorithms is utilized to develop a machining process model that is customized to a particular composite material. The predictive performance of the model can be improved by training it on data obtained from prior machining experiments or simulations. This training process enables the model to anticipate the material's reaction to various cutting conditions. This methodology facilitates the determination of the most suitable cutting parameters for attaining the intended machining efficacy.

Upon the development of the model, optimization techniques such as GAs, PSO, or similar methodologies are employed to ascertain the most optimal cutting parameters. The optimization tools take into account multiple factors such as material properties, tool wear, and surface finish requirements in order to optimize the machining process and attain optimal performance.

The objective of the proposed investigation is to construct an integrated neural network framework with optimization techniques that can forecast the machining performance of composite materials when subjected to diverse cutting conditions, and it yields the desired machining performance.

13.5   FUTURE SCOPE OF PRECISION MACHINING OF COMPOSITES

Existing research has demonstrated that the outcome of the machining process is influenced by numerous time-dependent factors, including tribological contact between the tool and the workpieces, tool wear, machine dynamics, and metallurgical

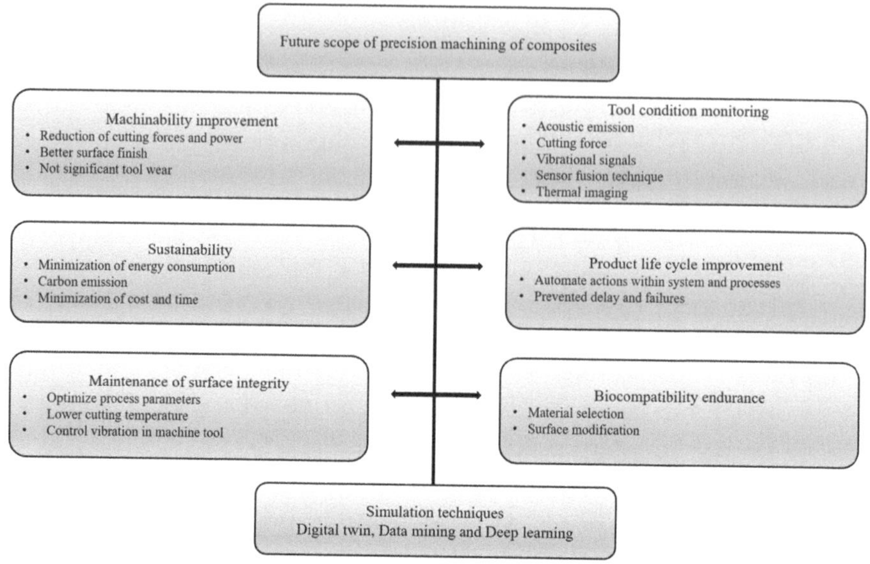

FIGURE 13.24   Framework for the future scope of precision machining.

imperfections in the workpiece. Predicting machining responses for multivariable systems becomes challenging due to the diverse nature of these factors. In the realm of precision and advanced manufacturing, the integration of sensors for real-time communication and the application of data analytics techniques, such as data mining and deep learning, align with the principles of Industry 4.0. This convergence, facilitated by advancements in sensor technology and the Industrial Internet of Things, has given rise to the concept of digital twins. Digital twin technology involves creating a comprehensive digital representation of a physical system, enabling valuable insights into the workpiece's condition during machining through physics-based simulation models and data-driven intelligence.

There is still a wide range of potential scopes for extending this research area to better analysis, real-time assessment, and improved monitoring of the machining system while working with composite workpieces (Figure 13.24). The following are some potential scopes:

Real-time surface roughness mapping: Development of a machine learning framework capable of virtually creating and real-time mapping of the machined surface considering mechanistic and dynamic aspects of the process, material aspects, and temporal changes in the tool condition

Virtual display of real-time surface integrity: Augmented reality or virtual reality technologies enable surfaces to be animated with high fidelity. The ability to visualize instantaneously facilitates expeditious analysis and evaluation. The convergence of virtual and physical realms results in the integration of precision and efficiency.

Tool condition monitoring: Real-time data acquisition and analysis ensure timely detection of tool wear or damage. ML algorithms identify patterns and deviations, enabling predictive maintenance strategies. Enhanced productivity and cost savings are achieved through optimized tool usage and reduced downtime.

MRR enhancement without sacrificing tool life cycle: Data-driven analytics optimize cutting parameters for improved efficiency. Adaptive control systems dynamically adjust machining conditions, reducing wear and enhancing performance. Real-time monitoring enables proactive maintenance, ensuring longevity without compromising productivity.

Sustainability assessment: Advanced data analytics unveil hidden patterns and optimize resource utilization. The machine learning algorithms modeled can identify sustainability risks and recommend mitigation strategies. Real-time monitoring enhances decision-making, enabling proactive sustainability measures for long-term viability.

Tool path generation: Advanced algorithms optimize cutting paths for enhanced efficiency and precision. Real-time data analysis enables adaptive tool paths, adjusting for changing conditions and material properties. Intelligent tool path generation reduces cycle time, minimizes tool wear, and improves overall machining performance.

Zero-defect manufacturing: Early defect detection and prevention are made possible by real-time data analysis with the help of ML algorithms modeled. Continuous progress is ensured by proactive monitoring, raising productivity and client satisfaction.

Biocompatibility assessment: Virtual simulations have the capability to visually represent the potential impact of machining on biocompatibility, thereby enabling proactive modifications prior to actual production.

13.6 CONCLUSION

This chapter reveals various obstacles that must be overcome in order to advance the implementation of AI within contemporary manufacturing systems. The present study focuses on the challenge of training AI models with limited training sets or samples. We propose potential solutions to this challenge by utilizing advanced information and communication technologies, such as edge computing, which can facilitate centralized or distributed approaches to training learning models. The present study investigates nascent patterns for forthcoming manufacturing educational models and outlines a conceptual framework that integrates human and machine intelligence. The aforementioned perspective is grounded on the premise that the human function within manufacturing systems will persistently progress towards a state of mutual existence and interdependence with mechanized manufacturing resources, as opposed to being solely responsible for operating machinery. Hence, this perspective posits that humans and intelligent machines are not adversaries in terms of fulfilling labor demands in forthcoming production systems. Rather, they are collaborators that facilitate the emergence of inventive and imaginative paradigms.

REFERENCES

1. Pedergnana, A. and Ollé, A. (2018). Building an experimental comparative reference collection for lithic micro-residue analysis based on a multi-analytical approach. *Journal of Archaeological Method and Theory*, 25, 117–154. https://doi.org/10.1007/s10816-017-9337-z

2. Wang, Z.J. and Liu, Z. (2023). Characterizing ancient materials. *Nature Materials*, 22(1), 144–145. https://doi.org/10.1038/s41563-022-01423-x

3. Saha, S., Deb, S. and Bandyopadhyay, P.P. (2020). An analytical approach to assess the variation of lubricant supply to the cutting tool during MQL assisted high speed micromilling. *Journal of Materials Processing Technology*, 285, 116783. https://doi.org/10.1016/j.jmatprotec.2020.116783

4. Teti, R. (2002). Machining of composite materials. *CIRP Annals*, 51(2), 611–634. https://doi.org/10.1016/S0007-8506(07)61703-X

5. Tortorella, G.L., Saurin, T.A., Hines, P., Antony, J. and Samson, D. (2023). Myths and facts of Industry 4.0. *International Journal of Production Economics*, 255, 108660. https://doi.org/10.1016/j.ijpe.2022.108660

6. Hu, W., Zhang, T., Deng, X., Liu, Z. and Tan, J. (2021). Digital twin: A state-of-the-art review of its enabling technologies, applications and challenges. *Journal of Intelligent Manufacturing and Special Equipment*, 2(1), 1–34. https://doi.org/10.1108/JIMSE-12-2020-010

7. Aheleroff, S., Xu, X., Zhong, R.Y. and Lu, Y. (2021). Digital twin as a service (DTaaS) in industry 4.0: an architecture reference model. *Advanced Engineering Informatics*, 47, 101225. https://doi.org/10.1016/j.aei.2020.101225

8. Zhou, G., Xu, C., Ma, Y., Wang, X.H., Feng, P.F. and Zhang, M. (2020). Prediction and control of surface roughness for the milling of Al/SiC metal matrix composites based on neural networks. *Advances in Manufacturing*, 8, 486–507. https://doi.org/10.1007/s40436-020-00326-x

9. Chandrasekaran, M. and Devarasiddappa, D. (2012). Development of predictive model for surface roughness in end milling of Al-SiCp metal matrix composites using fuzzy logic. *World Academy of Science, Engineering and Technology*, 6(7), 928–933.

10. Daniel, S.A.A., Pugazhenthi, R., Kumar, R. and Vijayananth, S. (2019). Multi objective prediction and optimization of control parameters in the milling of aluminium hybrid metal matrix composites using ANN and Taguchi-grey relational analysis. *Defence Technology*, 15(4), 545–556. https://doi.org/10.1016/j.dt.2019.01.001

11. Azmi, A.I. (2015). Monitoring of tool wear using measured machining forces and neuro-fuzzy modelling approaches during machining of GFRP composites. *Advances in Engineering Software*, 82, 53–64. https://doi.org/10.1016/j.advengsoft.2014.12.010

12. Boga, C. and Koroglu, T. (2021). Proper estimation of surface roughness using hybrid intelligence based on artificial neural network and genetic algorithm. *Journal of Manufacturing Processes*, 70, 560–569. https://doi.org/10.1016/j.jmapro.2021.08.062

13. Latha, B. and Senthilkumar, V.S. (2010). Modeling and analysis of surface roughness parameters in drilling GFRP composites using fuzzy logic. *Materials and manufacturing processes*, 25(8), 817–827. https://doi.org/10.1080/10426910903447261

14. Palanikumar, K., Latha, B., Senthilkumar, V.S. and Davim, J.P. (2013). Application of artificial neural network for the prediction of surface roughness in drilling GFRP composites. *Materials Science Forum* 766, 21–36. https://doi.org/10.4028/www.scientific.net/MSF.766.21

15. Kumar, P. (2022). Machine Learning in Drilling of GFRP Composite Using ANN. *Advancement in Materials Processing Technology: Select Proceedings of AMPT 2020* (pp. 157–169). Singapore: Springer Singapore.

16. Bagci, E. and Işık, B. (2006). Investigation of surface roughness in turning unidirectional GFRP composites by using RS methodology and ANN. *The International Journal of Advanced Manufacturing Technology, 31*, 10–17. https://doi.org/10.1007/s00170-005-0175-x

17. Karim, M.R., Dilwar, F. and Siddique, R.A. (2018). Predictive modeling of surface roughness in mql assisted turning of sic-al alloy composites using artificial neural network and adaptive neuro fuzzy inference system. *J Adv Res Manuf Mater Sci Met Eng, 5*(3), 12–28.

18. Rajasekaran, T., Palanikumar, K. and Vinayagam, B.K. (2011). Application of fuzzy logic for modeling surface roughness in turning CFRP composites using CBN tool. *Production Engineering, 5*, 191–199. https://doi.org/10.1007/s11740-011-0297-y

19. Chandrasekaran, M. and Devarasiddappa, D. (2014). Artificial neural network modeling for surface roughness prediction in cylindrical grinding of Al-SiCp metal matrix composites and ANOVA analysis. *Advances in Production Engineering & Management, 9*(2), 59–70. http://dx.doi.org/10.14743/apem2014.2.176

20. Gopan, V., Wins, K.L.D. and Surendran, A. (2018). Integrated ANN-GA approach for predictive modeling and optimization of grinding parameters with surface roughness as the response. *Materials Today: proceedings, 5*(5), 12133–12141. https://doi.org/10.1016/j.matpr.2018.02.191

21. Aurtherson, P.B., Sundaram, S., Shanawaz, A.M. and Prakash, M.S. (2011). Grinding process on AlSic composite material and optimization of surface roughness by ANFIS. *International Journal of Engineering and Technology, 3*(4), 425.

22. Bhuyan, R.K., Mohanty, S. and Routara, B.C. (2017). RSM and Fuzzy logic approaches for predicting the surface roughness during EDM of Al-SiC$_p$ MMC. *Materials Today: Proceedings, 4*(2), 1947–1956. https://doi.org/10.1016/j.matpr.2017.02.040

23. Dewangan, S., Gangopadhyay, S. and Biswas, C.K. (2015). Multi-response optimization of surface integrity characteristics of EDM process using grey-fuzzy logic-based hybrid approach. *Engineering Science and Technology, an International Journal, 18*(3), 361–368. https://doi.org/10.1016/j.jestch.2015.01.009

24. Hourmand, M., Sarhan, A.A., Farahany, S. and Sayuti, M. (2019). Microstructure characterization and maximization of the material removal rate in nano-powder mixed EDM of Al-Mg$_2$Si metal matrix composite—ANFIS and RSM approaches. *The International Journal of Advanced Manufacturing Technology, 101*, 2723–2737. https://doi.org/10.1007/s00170-018-3130-3

25. Chaki, S., Bathe, R.N., Ghosal, S. and Padmanabham, G. (2018). Multi-objective optimisation of pulsed Nd: YAG laser cutting process using integrated ANN–NSGAII model. *Journal of Intelligent Manufacturing, 29*, 175–190. https://doi.org/10.1007/s10845-015-1100-2

26. Patel, P., Sheth, S. and Patel, T. (2016). Experimental analysis and ANN modelling of HAZ in laser cutting of glass fibre reinforced plastic composites. *Procedia Technology, 23*, 406–413. https://doi.org/10.1016/j.protcy.2016.03.044

27. Najjar, I.M.R., Sadoun, A.M., Abd Elaziz, M., Abdallah, A.W., Fathy, A. and Elsheikh, A.H. (2022). Predicting kerf quality characteristics in laser cutting of basalt fibers reinforced polymer composites using neural network and chimp optimization. *Alexandria Engineering Journal, 61*(12), 11005–11018. https://doi.org/10.1016/j.aej.2022.04.032

28. Kumaran, S.T., Ko, T.J., Kurniawan, R., Li, C. and Uthayakumar, M. (2017). ANFIS modeling of surface roughness in abrasive waterjet machining of carbon fiber reinforced plastics. *Journal of Mechanical Science and Technology, 31*, 3949–3954. https://doi.org/10.1007/s12206-017-0741-9

29. Bhowmik, S. and Ray, A. (2019). Prediction of surface roughness quality of green abrasive water jet machining: a soft computing approach. *Journal of Intelligent Manufacturing*, 30(8), 2965–2979. https://doi.org/10.1007/s10845-015-1169-7

30. Shukla, M. and Tambe, P.B. (2010). Predictive modelling of surface roughness and kerf widths in abrasive water jet cutting of Kevlar composites using neural network. *International Journal of Machining and Machinability of Materials*, 8(1–2), 226–246. https://doi.org/10.1504/IJMMM.2010.034498

31. Popli, D., Sharma, P. and Verma, S. (2020). Artificial Neural Network Models for the Prediction of Metal Removal Rate in Rotary Ultrasonic Machining. In *Advances in Intelligent Manufacturing: Select Proceedings of ICFMMP 2019* (pp. 115–132). Springer Singapore.

32. Banerjee, B., Doloi, S., Das, S. and Dhupal, D. (2022). Parametric Optimization of MRR During Ultrasonic Machining Process. In *Advances in Modern Machining Processes: Proceedings of AIMTDR 2021* (pp. 257–270). Singapore: Springer Nature Singapore.

33. Saha, S., Deb, S. and Bandyopadhyay, P.P. (2021). Progressive wear based tool failure analysis during dry and MQL assisted sustainable micro-milling. *International Journal of Mechanical Sciences*, 212, 106844. https://doi.org/10.1016/j.ijmecsci.2021.106844

34. Saha, S., Kumar, A.S., Malayath, G., Deb, S. and Bandyopadhyay, P.P. (2023). Energy balance model to predict the critical edge radius for adhesion formation with tool wear during micro-milling. *Journal of Manufacturing Processes*, 93, 219–238. https://doi.org/10.1016/j.jmapro.2023.03.034

35. Saha, S., Ansary, S.I., Deb, S. and Bandyopadhyay, P.P. (2023). Influence of tool wear on chip-like burr formation during micro-milling, and image processing based measurement of inwardly-deflected burrs. *Wear, 530–531*, 205024. https://doi.org/10.1016/j.wear.2023.205024

36. Causes of heat generation in metal cutting and thermal aspects of metal machining https://www.mccholic.com/ (accessed on 24.05.2023)

37. Kehren, D., Simonow, B., Bäger, D., Dziurowitz, N., Wenzlaff, D., Thim, C., Neuhoff, J., Meyer-Plath, A. and Plitzko, S. (2019). Release of respirable fibrous dust from carbon fibers due to splitting along the fiber axis. *Aerosol and air quality research*, 19(10), 2185–2195. https://doi.org/10.4209/aaqr.2019.03.0149

38. Altin Karataş, M., Motorcu, A.R. and Gökkaya, H. (2021). Study on delamination factor and surface roughness in abrasive water jet drilling of carbon fiber-reinforced polymer composites with different fiber orientation angles. *Journal of the Brazilian Society of Mechanical Sciences and Engineering, 43*, 1–29. https://doi.org/10.1007/s40430-020-02741-4

39. Chan, K.C., Cheung, C.F., Ramesh, M.V., Lee, W.B. and To, S., (2001). A theoretical and experimental investigation of surface generation in diamond turning of an Al6061/SiCp metal matrix composite. *International Journal of Mechanical Sciences*, 43(9), 2047–2068.https://doi.org/10.1016/S0020-7403(01)00028-5

40. Sheikh-Ahmad, J., and Davim, J.P. (2012). Tool wear in machining processes for composites. In *Machining technology for composite materials* (pp. 116–153). Woodhead Publishing

41. Isbilir, O. and Ghassemieh, E. (2012). Finite element analysis of drilling of carbon fibre reinforced composites. *Applied Composite Materials*, 19, 637–656. https://doi.org/10.1007/s10443-011-9224-9

42. Ghafarizadeh, S., Chatelain, J.F. and Lebrun, G. (2016). Finite element analysis of surface milling of carbon fiber-reinforced composites. *The International Journal of Advanced Manufacturing Technology*, 87, 399–409. https://doi.org/10.1007/s00170-016-8482-y

43. Lotfi, M., Amini, S. and Aghaei, M. (2018). 3D FEM simulation of tool wear in ultrasonic assisted rotary turning. *Ultrasonics, 88,* 106–114. https://doi.org/10.1016/j.ultras.2018.03.013

44. Dandekar, C.R. and Shin, Y.C. (2013). Multi-scale modeling to predict sub-surface damage applied to laser-assisted machining of a particulate reinforced metal matrix composite. *Journal of Materials Processing Technology, 213*(2), 153–160. https://doi.org/10.1016/j.jmatprotec.2012.09.010

45. Kumaran, S.T., Ko, T.J., Uthayakumar, M. and Islam, M.M. (2017). Prediction of surface roughness in abrasive water jet machining of CFRP composites using regression analysis. *Journal of Alloys and Compounds, 724,* 1037–1045. https://doi.org/10.1016/j.jallcom.2017.07.108

46. Liu, C., Gao, L., Jiang, X., Xu, W., Liu, S. and Yang, T. (2020). Analytical modeling of subsurface damage depth in machining of SiCp/Al composites. *International Journal of Mechanical Sciences, 185,* 105874. https://doi.org/10.1016/j.ijmecsci.2020.105874

47. Saravanakumar, A., Sasikumar, P. and Nilavusri, N. (2016). Optimization of machining parameters using taguchi method for surface roughness. *J. Mater. Environ. Sci, 7*(5), 1556–1561.

48. Shetty, R., Pai, R.B., Rao, S.S. and Nayak, R. (2009). Taguchi's technique in machining of metal matrix composites. *Journal of the Brazilian Society of Mechanical Sciences and Engineering, 31,* 12–20. https://doi.org/10.1590/S1678-58782009000100003

49. Sahoo, A.K. and Pradhan, S. (2013). Modeling and optimization of Al/SiCp MMC machining using Taguchi approach. *Measurement, 46*(9), 3064–3072. https://doi.org/10.1016/j.measurement.2013.06.001

50. Alwarsamy, T., Abhinav, T. and Krishnakant, C.A. (2012). Surface roughness prediction by response surface methodology in milling of hybrid aluminium composites. *Procedia Engineering, 38,* 745–752. https://doi.org/10.1016/j.proeng.2012.06.094

51. Bhowmik, S. and Ray, A. (2016). Prediction and optimization of process parameters of green composites in AWJM process using response surface methodology. *The International Journal of Advanced Manufacturing Technology, 87,* 1359–1370. https://doi.org/10.1007/s00170-015-8281-x

52. Laghari, R.A., Li, J., Xie, Z. and Wang, S.Q. (2018). Modeling and optimization of tool wear and surface roughness in turning of Al/SiCp using response surface methodology. *3D Research, 9,* 1–13. https://doi.org/10.1007/s13319-018-0199-2

14 Grinding of Thermal Spray Ceramic Coating

Rounak Pal, Dipankar Chetia, Alfa Bisoi, and Simanchal Kar

14.1 INTRODUCTION

Thermally sprayed ceramic coatings consist primarily of cermets and nonmetallic that are deposited using ceramics, metals, and alloy powders. They can be either crystalline, amorphous, or both in nature. The use of ceramics is increasing primarily due to their excellent properties like fracture toughness, hardness, chemical inertness, high-temperature stability, high thermal strength, and wear resistance [1]. Ceramic layers in the form of splats form the 'lamellar layer' of plasma-sprayed ceramic coatings. Owing to their unique properties, plasma-sprayed ceramic coatings are used in different applications, such as turbine components, automobile parts, and others [2]. These applications call for high-precision details with flawless surface and dimensional tolerances which can be attained by finishing procedures like grinding and polishing. High grinding speeds provide thinner maximum undeformed chip thickness values, which minimize surface damage, enhance surface integrity, and lower wheel wear [3]. Grinding ceramic poses multiple challenges, including increased wheel wear, susceptibility to microfractures and surface cracks, and compromised surface quality. During ceramic grinding, the ratio of force, precisely the normal to tangential force, is approximately 2–3 times greater than steel [4]. There are few investigations on ceramic coatings including tungsten carbide, alumina, chromia, and alumina–titania coatings showing the effect of grinding on such surfaces [5]. Kar et al. have investigated the high-speed grinding of plasma-sprayed ceramic coatings using monolayer electroplated diamond grinding wheels [6]. Ceramics possess bond arrangements that can be classified as either ionic or covalent, sometimes even as a combination of both. This characteristic of ceramics results in the restricted mobility of atoms within the material, rendering them susceptible to cracking [7]. Compared to metals, ceramic materials have a distinct characteristic: their limited ability to undergo significant deformation before fracturing due to restricted dislocation movement. However, fresh dislocations are introduced continuously during the grinding process due to shear stresses. These dislocations accumulate at specific internal microstructural barriers such as second-phase particles, grain boundaries, micro-twins, and others. The accumulation of these dislocations sets off a crack or void initiation at the respective barrier [8]. At higher wheel speeds, they have seen reduced grinding forces, tensile residual stress, and grinding damage on the ground surface. At low grinding speeds, material removal occurs mainly through

DOI: 10.1201/9781032665375-14

the micro-brittle fracture. Regardless of the grinding parameters, the grinding chips maintain their irregular and blocky shape, exhibiting no change. No thermal-induced residual stresses were found on the ground surfaces, indicating that the ceramic coatings retain their physical properties during grinding.

Unlike metal grinding, grinding ceramic materials offers two modes of removal: brittle and ductile [5]. Pore defects and low bonding strength are common issues associated with plasma-sprayed ceramics, making them more susceptible to crushing compared to sintered ceramics. Consequently, studies suggest that micro-pulverization is the primary method of removing plasma-sprayed coatings [9]. The critical depth of cut determines the grinding characteristics of ceramic coating. The ceramic grinding begins with ductile removal when the uncut chip thickness falls below a threshold value. Grinding damage is commonly associated with two basic fracture systems: median cracks (also known as radial cracks) and lateral cracks. Median or radial fractures contribute predominantly to strength deterioration, whereas lateral cracks play an essential role in material loss and surface creation [10].

14.2 THERMAL SPRAYING: PLASMA SPRAYING

The thermal spray procedure is a surface engineering technique that deposits solid particles on a substrate in molten and semi-molten phases [2]. The plasma spray operation is used to cover a variety of materials. Melting or incomplete melting of feedstock substance, commonly in a powdered or wired form, is followed by spraying the molten material onto a substrate with a spray cannon. This technique is used for various industrial applications because of the finished coating's high degree of adhesion, wear durability, and corrosion resistance.

Assisted by a plasma produced by an electric arc, the feedstock material gets melted and sprayed over the substrate surface in plasma spraying as shown in Figure 14.1. A gas, often argon, and sometimes nitrogen is ionized and passed through a high-voltage electrical field to produce the plasma arc. Vacuum and atmospheric plasma spraying are the two most often utilized plasma spraying methods.

The choice of thermal spray technique is influenced by the material being sprayed, the outer layer of the base, and the intended coating properties. Each strategy has distinct advantages and disadvantages, so selecting the optimal one necessitates a thorough examination of the process and its parameters.

14.3 PLASMA SPRAY PROCESS PROCEDURE

Plasma spray coating can be deposited on any solid material with some preprocessing to improve its surface roughness. A typical example is a mild steel substrate, ground on both sides and cleaned in acetone before the coating process is illustrated to explain the coating deposition. The intended surface for coating was grit blasted using SiC particles of approximately 325 mesh size. The roughened surface was subsequently coated using a Sulzer Metco 9MB plasma gun on a computer numerically controlled (CNC) bench. For all the coatings, the primary gases were argon and nitrogen, while the secondary plasma gas was hydrogen. Furthermore, nitrogen was used as a carrier gas to carry powders from the powdered substance feeder to the

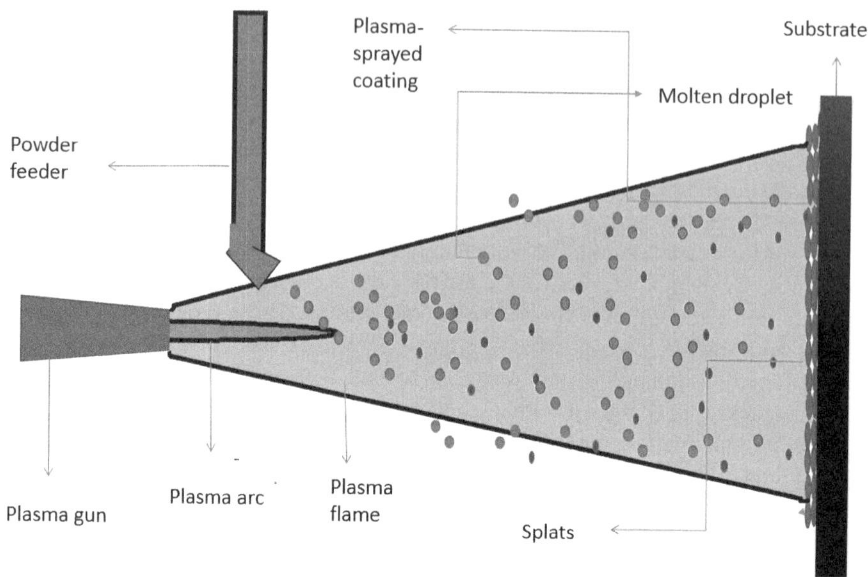

FIGURE 14.1 Coating of a substrate with the help of the plasma spray process.

TABLE 14.1

Coating Parameters Considered for a Typical Experiment based on the Ideal Parameters Provided by the APS Manufacturer

Sl No	Parameters	Value
1	Stand of distance (SOD) (mm)	100, 120, 150
2	Diameter of nozzle (mm)	7
3	Arc current (A)	450, 500, 600
4	Arc voltage (V)	65–74
5	Powder injection angle	90°
6	Hydrogen secondary gas flow rate (slpm)	10
7	Primary secondary gas flow rate (slpm)	80
8	Flow rate of powder (kg/hr)	1.5
9	Substrate material	Mild steel
10	Preheating of the substrate (°C)	200–250

cannon. The process parameters utilized for applying the coatings are provided in Table 14.1. The influence of arc current and standoff distance is studied for a given coating system. During the coating process, all coatings were sprayed at a spray angle of 90°. Two auxiliary air jets were employed to enhance the deposition process and ensure optimal coating quality. The parallel positioning of these jets to the plasma flame serves a dual function: cooling the substrate and removing any powder particles from the coating surface that have not melted or are inadequately bonded. Bond coats of a thickness of approximately 100–150 mm were applied before depositing the ceramic top coat, which had a thickness ranging from 250 to 350 mm. To enhance adhesion between ceramic topcoats and metal substrates, a bond coat is applied to minimize the impact of thermal expansion coefficient mismatch. Figure 14.2 depicts a scanning electron microscopy (SEM) view of the mild steel cross section and its top surface and bond coating. Then, sample specimens were collected for analysis. Appropriate samples were cut employing a low-speed diamond cutter to 10 × 10 × 10-mm dimensions. After obtaining the specimens, implant them in a resin bond. This involves placing the samples in a resin material to provide support and stability during the subsequent polishing process. The embedded samples were prepared for SEM examination through cross-sectional polishing. SiC abrasive papers with coarse grits were used progressively, followed by finer grit sizes. This step helps remove any roughness or irregularities from the sample's surface. After initial polishing with abrasive papers, diamond paste polishing was applied to further refine the sample's surface. The diamond pastes were applied with progressively finer grit sizes, similar to the SiC abrasive papers, to achieve a smoother finish. When the cross-sectional polishing is finished, an SEM image is used to evaluate the polished area's cross sections. The Zeiss EVO 60 SEM from Germany was employed in this case. Figures 14.3 and 14.4 illustrates the SEM and optical images of a typical alumina coating cross section and top surface after deposition, respectively.

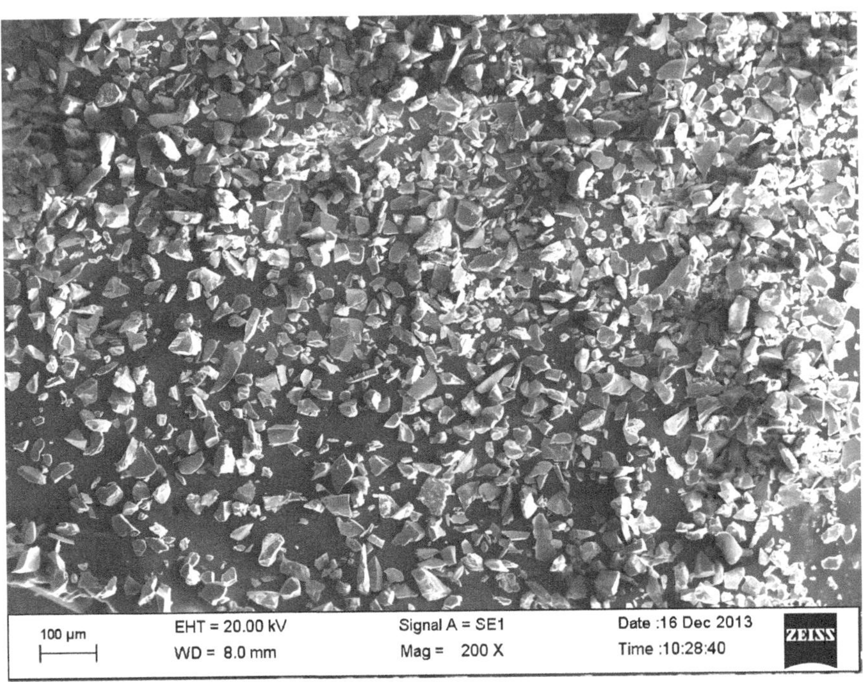

FIGURE 14.2   Cross-sectional view of an alumina powder with crushed particles.

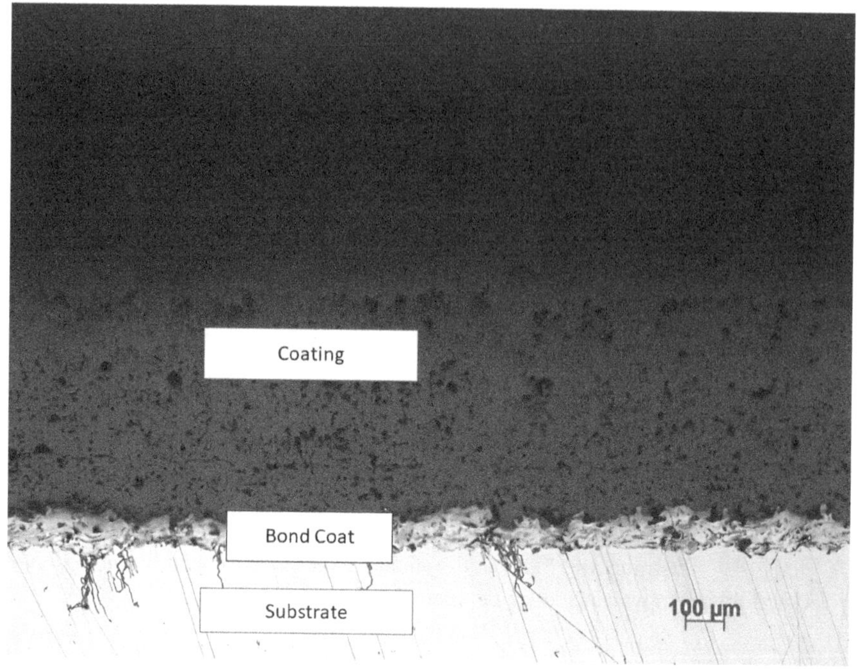

FIGURE 14.3 A SEM image of alumina coating with NiCrAlY bond coat.

FIGURE 14.4 Optical view of an alumina coating at 10× magnification.

14.4 THERMAL SPRAY PROPERTIES

14.4.1 YOUNG'S MODULUS

The tensile stress to tensile strain ratio relates to Young's modulus, a material property that shows how easily the material may expand and deform. The depth-sensing indentation technique is used to calculate Young's modulus. A pyramidal or spherical indenter is often used with a depth-sensing indentation method to determine Young's modulus on a substrate material. The following equation is used primarily for determining the Young's modulus on coating surfaces [11]:

$$E^{SI} = \frac{1}{\beta} \frac{\sqrt{\pi}}{2} \frac{1}{\sqrt{A}} \frac{1}{C}, \tag{14.1}$$

where E^{SI} is the reduced modulus of the specimen and indenter; β refers to the correction factor, which takes the indenter's geometry into account; A refers to the contact area at the maximum load of the indentation, and C refers to the compliance. The reduced modulus of the specimen E^* is evaluated by

$$\frac{1}{E^{SI}} = \frac{1}{E^*} + \frac{1}{E_i^*}. \tag{14.2}$$

using the relation

$$E^* = \frac{E}{1-v^2} \text{ and } E_i^* = \frac{E_i}{1-v_i^2}, \tag{14.3}$$

where E and E_i symbolize the Young's modulus of the specimen's and indenter's respective Poisson's ratios v and v_i and E_i^* implies the indenter's reduced modulus.

14.4.2 FRACTURE TOUGHNESS

The resistance of a material to propagating pre-existing faults is characterized as fracture toughness (K_{IC}). The indentation technique is used to measure fracture toughness in plasma-sprayed coatings. Evan and Charles (1976) created a mathematical equation to calculate ceramic components' fracture toughness (K_{IC}). This equation uses the fracture length and the diagonal length of a Vickers indentation, as illustrated in [12]:

$$K_{IC} = 0.15 \left(\frac{c}{a} \right)^{-3/2} \frac{H \sqrt{a}}{\varnothing}, \tag{14.4}$$

where k is the correction factor 3.2 for ceramics, c is the sum of the length of the crack and half of the Vickers diagonal in the direction of the crack, $\varnothing$ is a constraint factor approximately equal to 3, H is the Vickers hardness (MPa), and a = half of Vickers diagonal length measured in the direction of crack. The critical stresses for radial and transverse fractures in ceramics and ceramic coatings may be determined by employing the following ceramic materials expressions [12]:

(a) Critical load for radial crack initiation:

$$PR* = 54.5\left(\frac{\alpha}{\eta^2 \gamma^6}\right)\left(\frac{K_{IC}^4}{H^3}\right) \tag{14.5}$$

(b) Critical load for transverse or lateral crack initiation:

$$PL* = x\left(\frac{K_{IC}^4}{H^3}\right) f\left(\frac{E}{H}\right), \tag{14.6}$$

where, α, η, and γ are constant.

For Vickers indenter, $\alpha = 2/\text{p}$, $\eta \approx 1$ and $\gamma = 0.2$. x is a dimensionless constant, f (E/H) is an attenuation function equivalent to 2×10^5[11].

14.4.3　Maximum Uncut Chip Thickness (h_m) and Critical Depth of Cut (d_c)

One of the most significant aspects determining material removal during ceramic grinding is the critical depth of cut (d_c). The material removal mode is primarily governed by the ratio of the magnitude of d_c and h_m. The value of critical depth, that is, d_c, is obtained by the following equation [13]:

$$d_C = 0.15\left(\frac{E}{H}\right)\left(\frac{K_{IC}}{H}\right)^2, \tag{14.7}$$

where E represents the Young's modulus, H represents hardness, and K_{IC} represents fracture toughness. Similarly, the maximum uncut chip thickness (h_m) may be calculated using the following equation [13]:

$$h_m = \left(\frac{3}{c\tan\theta}\frac{v_w}{v_s}\sqrt{\frac{a_i}{d_s}}\right)^2, \tag{14.8}$$

where V_w defines table speed, V_s defines grinding wheel speed, c defines active grid concentration, down feed is defined by a_i, while the diameter of the grinding wheel is defined by d_s. It is observed that when the value of d_c is lower than h_m, than shear mode happens; otherwise, micro-brittle fracture is predominant.

14.4.4　Hardness

Hardness is a material's mechanical characteristic that demonstrates its resistance to indentation, scratching, and deformation. It evaluates how well a material can endure localized stress without permanently deforming or failing. Brinell, Rockwell, and Vickers hardness tests are the three most popular techniques for determining hardness. These techniques entail employing a hard indenter to apply a standardized load to the material's surface and then measuring the depth or size of the indentation for specific indenters, resulting in the determination of the material hardness. Figure 14.5a depicts a typical indentation curve, whereas Figure 14.5b depicts deformation details in loading and unloading situations [14].

14.4.5 TENSILE ADHESION STRENGTH

During a typical spray process, the coating substance is melted and accelerated by a plasma jet during plasma spray coating, and then it is sprayed onto the substrate, where it solidifies and adheres to the surface. For a sustainable layer, the tensile adhesion plays a significant role as it indicates the tensile adhesion of the coating with the substrate. Various factors, like the coating's composition, shape, and microstructure, as well as the features of the substrate and the spray process parameters, all collectively affect the coating's adhesion strength to the substrate.

A high-tensile adhesion strength is desirable in plasma spray coating because it indicates a firm connection between the coating and the substrate, assuring the coated component's performance and endurance under various operating conditions.

14.4.6 FRICTION AND WEAR

Friction and wear are two distinctive properties that are closely related. Coating friction and wear affect the longevity and performance of the underlying substrate. Plasma spray coatings are frequently employed to increase the wear resistance of materials; however, under some circumstances, the coatings themselves may also be prone to wear and deterioration. Numerous variables, including the substrate's and the coating's surface abrasiveness and topography, the coating's chemical makeup and microstructure, and the environment's temperature, humidity, and pressure, can affect friction in plasma spray coatings. In some applications, plasma-sprayed coatings are applied to reduce friction, such as high entropy alloy (HEA) coatings as in anilox rolls, and in some cases, they are applied to increase the friction of, for example, brake liners.

14.5 FINISHING CERAMIC COATING

Coating surface finish can significantly impact the coating's characteristics and performance. The following are a few popular ways for surface finishing coatings [2]:

(a) Use of post-remelting method using high energy beams: This process involves precise surface finishing, using various advanced machining processes such as electron beam machining (EBM), laser beam machining (LBM), and ion beam machining (IBM). Each of these techniques offers distinct advantages and can be utilized to attain the desired surface finish. All the previously mentioned advance-machining processes can be used to

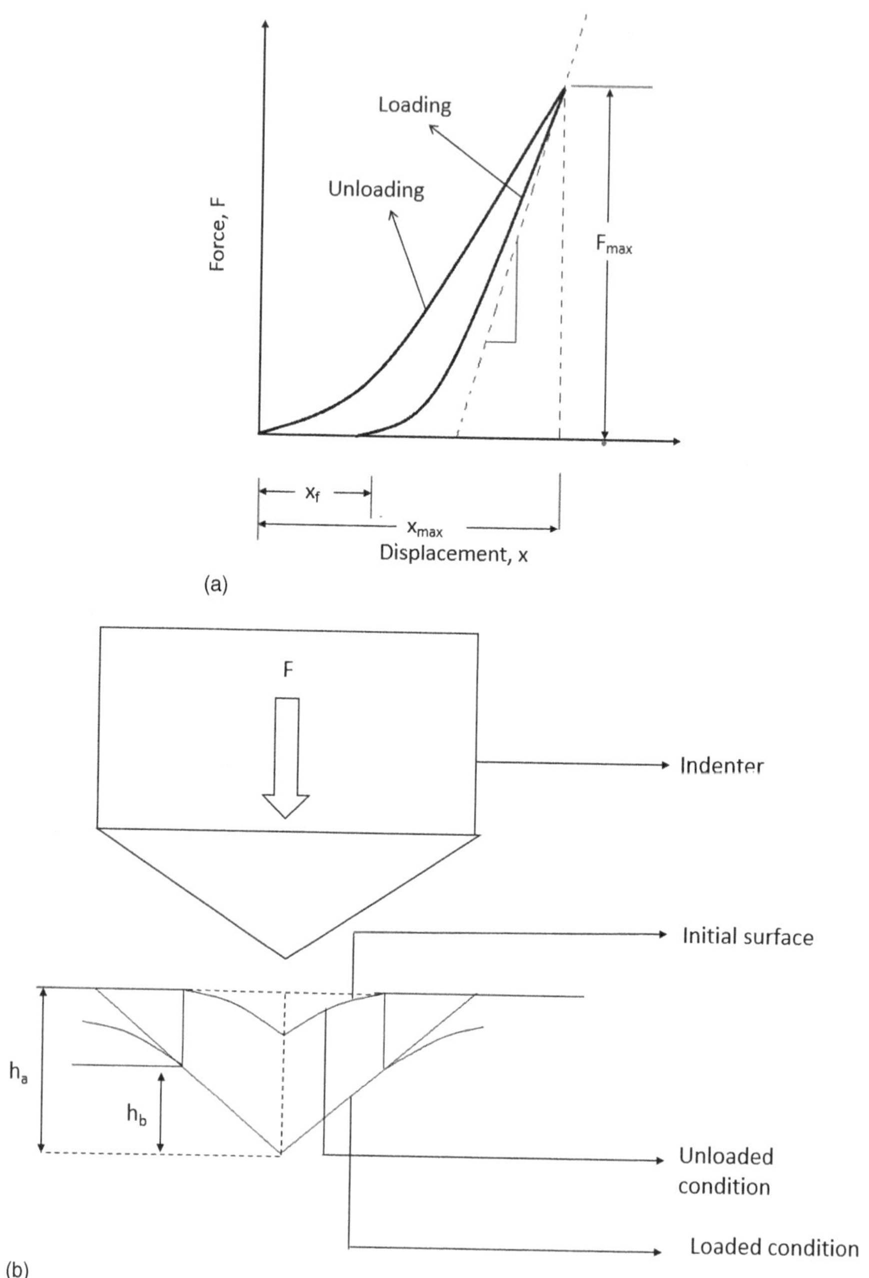

FIGURE 14.5 (a) A typical loading-unloading curve observed during depth sensing indentation on a ceramic (b) Schematic illustration of surface deformation features under loaded and unloaded situations.

Reproduced from the book Oliver, W.C., and G.M. Pharr, 1992 [14].

achieve very precise surface finishes on coatings, with a high degree of control over the shape and depth of the surface features. However, the process is costly and induces thermal residual stresses that are detrimental to coating properties.

(b) Another low-cost process is the grinding of the coating surface. To achieve a smooth and even surface on a ceramic coating, grinding can be employed to eliminate any rough spots or imperfections. This process involves the use of specialized grinding wheels, such as diamond or cubic boron nitride wheels, which are specifically designed for the grinding operation on ceramic coatings.

(c) Use of advanced micromachining processes: Polishing operations like lapping, honing, magnetorehelogical fluid (MRF)–assisted polishing, and hybrid grinding are a few rea commendable methods that are being studied for such typical applications. Micromachining of coating surface does have some limitations, including the need for specialized equipment, relatively slow process speed, and high cost. Depending on the coating thickness and composition, it may also be unsuitable for some types of coatings. Presently, super-abrasive grinding is evaluated as a low-cost alternative to get the desired surface finish in coatings.

14.5.1 SURFACE GRINDING THE CERAMIC COATING

The surface quality of thermally sprayed coatings is frequently harsh and requires further treatment to attain ideal results. To ensure a superior surface finish, coated components undergo post-finishing processes. These additional steps are essential for enhancing the surface quality of the coatings, resulting in improved overall performance and appearance. Grinding is the most popular method for performing surface finish operations on coatings. Ceramic coatings are ground with diamond and cubic boron nitride (CBN) wheels. Silicon carbide wheels, by comparison, are ideal for grinding carbide-reinforced metallic or metal matrix composites [2].

Chen et al. [14] investigated when grinding alumina with a monolayer brazed diamond disc at varying grinding velocities, the grindability characteristics may be analyzed in terms of numerous parameters. Several parameters were considered, including, but not limited to, grinding forces, force ratio, average grinding force per grain, specific grinding energy, surface morphology, grinding temperature, and energy partitioning. At first, it was discovered that when the peripheral wheel speed rose, the undeformed chip thickness decreased noticeably. As a result of the reduced chip thickness, the grinding forces were reduced. When the down feed and work speed were increased, the undeformed chip thickness rose as well, resulting in greater grinding forces. Additionally, a noteworthy observation was made regarding the difference between down grinding (grinding in the opposite direction of the wheel rotation) and up grinding (grinding in the same direction as the wheel rotation). It was found that down grinding resulted in reduced grinding forces, attributed to the variations in chip thickness. The study also revealed that smaller undeformed chip thickness promoted more ductile flow and fewer microfractures during grinding. However, this condition resulted in a higher specific grinding energy consumption. The grinding temperature was another important parameter, which increased with higher table and wheel speeds. An increase in specific grinding energy was identified as the primary cause of this temperature rise.

Inasaki [7] talked about significant variations in grind-ability between ceramic samples and metal specimens regarding crystal structure, material removal process, and grinding caused damages. Different ceramics were compared, including alumina, silicon nitride, zirconia, and silicon carbide. These ceramics were ground using two different methods: pendulum grinding and creep feed grinding. Notably, when low work speeds were used, creep feed grinding had a lower grinding ratio. The grinding ratio quantifies the link between material removal and wheel wear rates. This grinding process also displayed minimal grinding stress. Pendulum grinding, by comparison, had a distinct pattern. The grinding circumstances, like metal grinding, affect the surface roughness. In particular, a smaller infeed, lower work speed, and greater grinding speed resulted in a better surface polish. It is worth mentioning, however, that infeed's influence was less prominent in pendulum grinding. Rather than the specific process settings, the roughness levels were shown to be more responsive to the features of the workpiece material.

14.6 CASE STUDY: GRINDING THERMALLY SPRAYED ALUMINA COATING

Thermally sprayed coatings have higher surface roughness than other surface coatings, such as physical vapor deposition, chemical vapor deposition, and electroplated coatings. The primary reason for this roughness can be traced to the deposition process's intrinsic features. In thermal spraying, the surface is coated by depositing micron-sized particles, some of which remain unmelted or semi-molten and become embedded within the coating. Due to this roughness, many of the components coated through thermal spraying necessitate post-finishing treatments. For example, the paper industry utilizes press rolls coated with alumina, necessitating a high-quality surface finish [15].

14.6.1 BACKGROUND

The remarkable hardness, excellent chemical stability, and extraordinary wear resistance of plasma-sprayed ceramic coatings make them desirable. Nonetheless, these coatings have intrinsic flaws that limit their performance, such as brittleness, limited heat conductivity, and the tendency to retain residual stress. Consequently, when subjected to grinding, ceramic coatings are inclined to develop cracks and delamination. Moreover, their high hardness makes it challenging to achieve a satisfactory finish through conventional machining or grinding methods. These issues significantly restrict the industrial applicability of ceramic coatings. Hence, there is a pressing need to enhance the grindability of these coatings to maximize their potential within an industrial setting.

In order to discuss the tribological effect of grinding, we can take the example of grinding an alumina coating. In Figure 14.6, the scanning electron microscopy of the grinding surface is shown. The grinding process involves the development of two types of forces: normal and tangential grinding forces. These forces can be equivalently

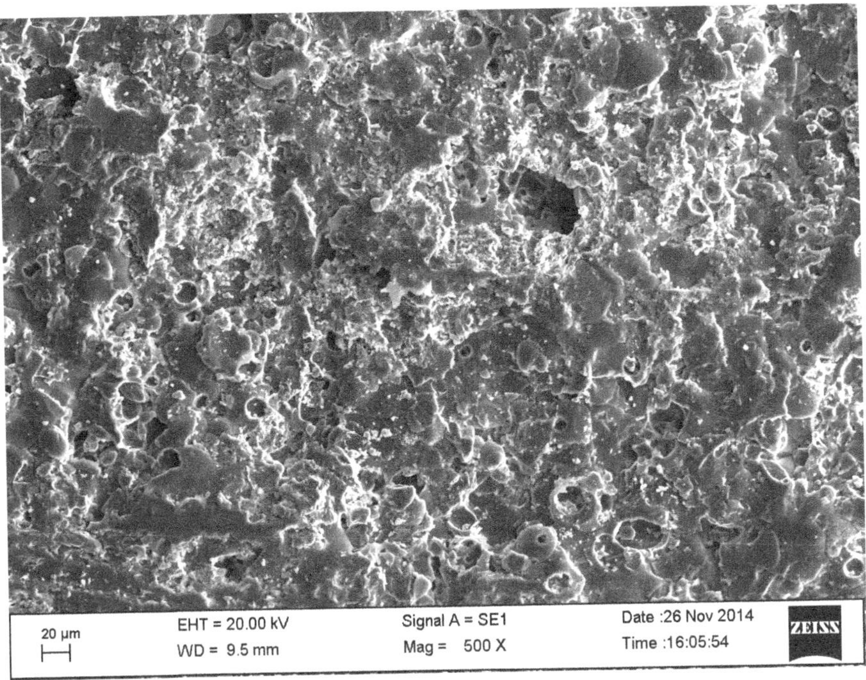

FIGURE 14.6 SEM picture of an alumina-coated grinding surface.

represented as the vertical force (F_V) and the horizontal force (F_H). The forces developed in ceramic grinding involve several steps. Initially, brittle fracture is the predominant mechanism of material loss, complemented in some cases by shear deformation. As a result, additional interactions, such as rubbing and ploughing, occur. The following is a presentation of the reduced force model [11]:

$$F_T = \left(K_1 + K_2 \ln \frac{v_s^{1.5}}{a_i^{0.25} v_w^{0.5}} \right) \frac{v_w a_i}{v_s} b_w + b_w A \left(\alpha + \frac{4\beta \rho_o v_w}{d_e v_s} \right) (d_e a_i)^{\frac{1}{2}} \qquad (14.9)$$

$$F_N = \left(K_3 + K_4 \ln \frac{v_s^{1.5}}{a_i^{0.25} v_w^{0.5}} \right) \frac{v_w a_i}{v_s} b_w + \frac{4 b_w A \rho_o v_w}{v_s} \left(\frac{a_i}{d_e} \right)^{\frac{1}{2}}, \qquad (14.10)$$

where K_4 is $\mu_{dynamic}$ K_2, K_3 is μ_{static} K_1, and $K_1 \left(v_w a \dfrac{b}{v_s} \right)$ refers to the static tangential force while $k_2 \left(\ln \dfrac{v_s^{1.5}}{a_i^{0.25} v_w^{0.5}} \right)$ refers to the dynamic tangential chip formation force.

14.6.2 Tribological Aspects and Utility of Grinding of Ceramic Coatings

Considering the tribological aspect, it is observed that plasma-sprayed ceramic coatings are porous in nature and this nature affects their thermal conductivity, thermal-cycling resistance, and strength characteristics. One notable drawback of the spray process is the formation of a coarse surface texture on the coating. A smooth surface is required in many dynamic components. To achieve a smooth coating surface on engine piston rings coated with ceramic material, a machining operation needs to be incorporated into the overall production procedure. This ensures that the rings can slide smoothly on the surface and maintain a reliable seal between the cylinder bore surface and the ring, effectively containing the high-pressure gases generated during engine operation [12].

Ceramic coating removal in the context of grinding is comparable to grinding a bulk ceramic material in that it requires material removal by surface microfracturing. Shear deformation, by comparison, is the principal mechanism of material removal while grinding metallic alloys. However, brittle fracture and plastic deformation can be observed in ceramic grinding, while brittle microfracture is the primary mechanism. The grinding process involves high grinding forces, resulting in a high grinding force ratio ($F_N F_T$). The chips produced during grinding exhibit a blocky fracture pattern. During the spraying process and subsequent grinding of plasma-coated ceramic samples, the ground surface can undergo changes in residual stresses and the formation of microcracks. In this context, conventional wheels are inadequate because their hardness is similar to the coating, leading to excessive wheel wear. To overcome this obstacle, super abrasive wheels like as cBN and diamond wheels are used for grinding. These sophisticated wheels outperform their traditional counterparts in several ways, including improved strength, remarkable wear resistance, and high hardness. By employing cBN and diamond wheels, we can effectively address the issue of wheel wear and enhance grinding efficiency in this particular scenario. By utilizing these super abrasive wheels, the problem of excessive wheel wear can be overcome, leading to more efficient and effective grinding of plasma-coated ceramic samples.

Thermally sprayed coatings have a rough surface due to the deposition process, which distinguishes them from other surface coatings such as physical vapor deposition, chemical vapor deposition, and electroplated coatings. Micron-sized particles are deposited during this process and certain particles that are unmelted or partly molten become lodged inside the coating, adding to its irregular texture. As a result, many components that are coated with thermally sprayed coatings require post-finishing operations to achieve the desired surface quality. The following are a few published reports from journals that have shown the effect of machining ceramic coatings.

Lacalle et al. [16] have investigated the presence of coolant, and the machinability of thick Ni–Al coatings using CBN tools has been examined. However, the cohesion of the coating has proven to be weaker compared to conventional materials. During the turning process, microcracking has been observed as the coating fails. Furthermore, the constant friction between the machined surface and the cutting tool has resulted in abrasive wear, particularly on the tool's secondary flank.

Liu and Zhang [17] showed that using a cast iron bond wheel, grinding can be effectively performed by horizontal spindle surface grinding on alumina–titania and tungsten carbide–cobalt coatings. The presence of significant surface and subsurface damage was reported. Unfortunately, the grinding process led to the formation of numerous lateral cracks and median cracks in both sample coatings.

Ramazani et al. [18] have compared two types of feedstock powders, namely, nanostructured and conventional, which were used to produce Al_2O_3–TiO_2 coatings via air plasma spraying. These coatings' mechanical characteristics and grindability were assessed. During the grinding process, it was discovered that the material removal rate of the nanostructured coatings was greater than that of the traditional coating. Furthermore, major changes in fracture patterns were detected between the two coatings. A few big cracks perpendicular to the grinding direction were found in the traditional coating, but the nanostructured coating had numerous thin cracks in no particular direction. It is worth noting that the nanostructured coating displayed characteristic semi-molten phases, which acted as effective crack arresters.

Liu et al. [19] examined the wheel parameters on ceramic surface. The goal of this investigation was to see how different grinding settings affect the surface quality of thermally sprayed nanostructured n-WC/12Co and n-Al_2O_3/13TiO_2 coatings placed on a low carbon steel substrate. Although the surface criteria for a good quality surface encompasses both mechanical as well as metallurgical properties, the researchers were more interested in the effect of table speed and down feed on the surface roughness of the ground n-WC/12Co coating. The results showed that when the table speed and down feed were increased, so did the surface roughness of the ground n-WC/12Co coating.

As a result of the preceding literature, it is clear that ceramic coating surfaces are grindable, however this results in the production of many surface and subsurface fissures. As a result, coating characteristics and grinding settings have a large impact on ground topology.

14.6.3 Grinding Experiment

Precision grindability studies were conducted using HMT ALEX NH 500 in plunge surface grinding mode with coated samples (prepared in Section 14.3). The grinding process utilized a monolayer diamond grinding wheel. Table 14.2 contains detailed information on the wheel specs and grinding settings. A work-holding device was attached to a force dynamometer to properly measure the forces involved in grinding. A high-resolution piezoelectric-type dynamometer, combined with two charge amplifiers and a data-collecting system, was used to quantify normal and tangential grinding forces.

TABLE 14.2

An Example of Common Grinding Parameters Utilized on an HMT Surface Grinding Machine during the Finishing Operation

Grinding Machine	Grinding Parameters	Type/Values
HMT ALEX NH 500	Wheel specification (Diamond grits) Down feed	Single-layer galvanic-bonded Grit size: 126 μm 14 μm, 20 μm, 26 μm, 32 μm, 38 μm
	Table speed	6 m/min, 12 m/min
	Wheel speed	32 m/s

14.6.4 RESULT

Successful experiments were conducted using the above parameters and the ground sample was further investigated in terms of surface characterization, grinding forces, and grinding energy. Figure 14.6 depicts a scanning electron microscopic image of the alumina coating's grinding surface.

Furthermore, it was discovered that when the table speed was set to $V_w = 6$ m/min, the specific grinding energy initially increased with a certain down feed value. However, beyond that specific down feed point, the specific grinding energy started

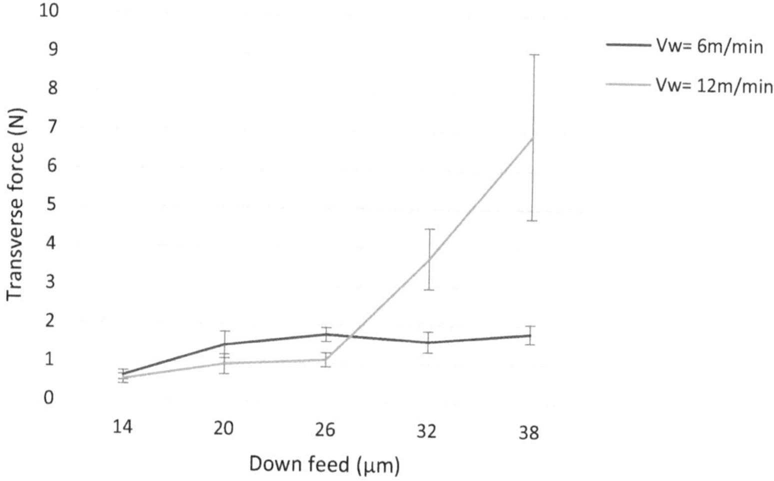

FIGURE 14.7 Transverse force versus down feed.

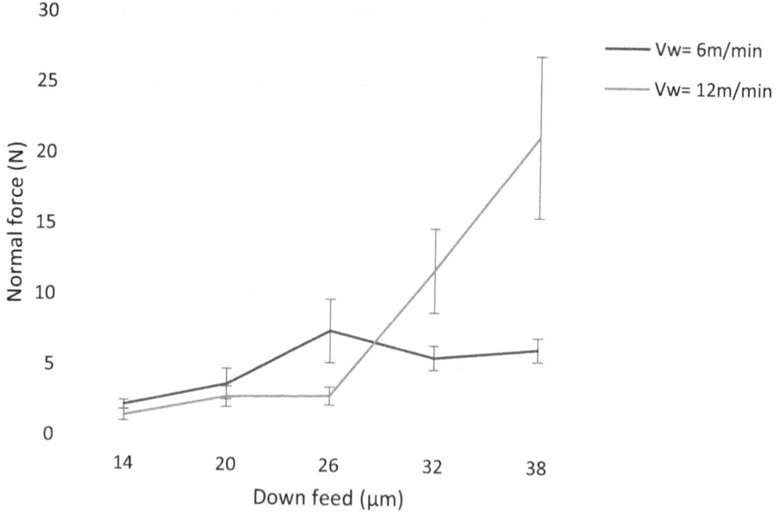

FIGURE 14.8 Normal force versus down feed.

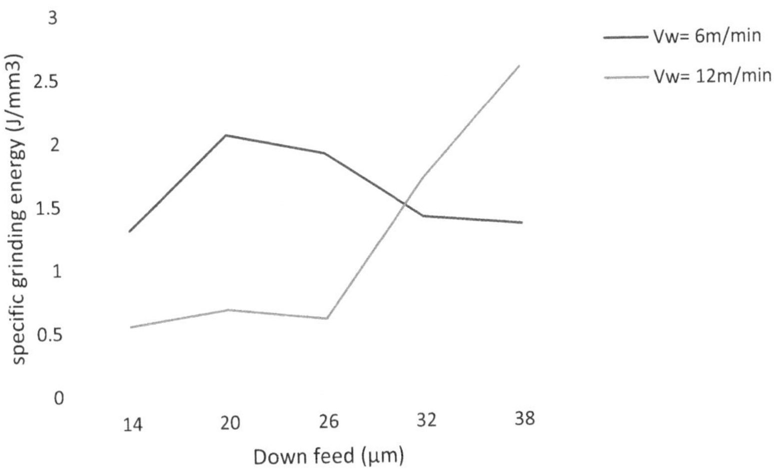

FIGURE 14.9   Specific grinding energy versus down feed.

to decrease. Similarly, when the table speed was set to V_w = 12 m/min, the specific grinding energy displayed a linear increase as the down feed was adjusted to a certain value. The most common mechanism of material removal is micro-brittle fracture, as seen in Figure 14.9, where an erratic departure in specific grinding energy is evident as down feed increases. A similar phenomenon was reported by Liu and Zhang [20].

14.7   SUMMARY

Ceramics are known for their hardness and brittleness and possess poor thermal conductivity and low fracture toughness. Grinding ceramic materials poses a significant challenge due to several factors. First, it leads to increased wheel wear. Additionally, ceramics are prone to microfractures and surface cracks during grinding, further exacerbating the difficulty. Consequently, these issues culminate in unsatisfactory surface integrity. The grinding process in the case of ceramic coating is determined by the critical depth of cut, which is strongly connected to material properties such as elastic modulus, coating, hardness, and fracture toughness. The fracture toughness

TABLE 14.3
Forces When Table Speed = 6 m/min

Down Feed	F_T	σ_{FT}	F_N	σ_{FN}
14 µm	0.6485	0.121718	2.234	0.321619
20 µm	1.454	0.344622	3.751	1.070798
26 µm	1.7505	0.182133	7.5865	2.247229
32 µm	1.5785	0.274008	5.735	0.867194
38 µm	1.8055	0.243027	6.381	0.845854

TABLE 14.4

Forces When Table Speed = 12 m/min

Down Feed	F_T	σ_{FT}	F_N	σ_{FN}
14 µm	0.547	0.125901	1.4855	0.396903
20 µm	0.953	0.256517	2.86	0.732393
26 µm	1.093	0.183415	2.972	0.632642
32 µm	3.722	0.783579	11.928	2.966523
38 µm	6.9145	2.146465	21.4625	5.72987

is primarily obtained through the mathematical formulation given in Equation 14.4 which is applicable only for ceramic coatings. Since it is a property of coating hardness and crack propagation. The hardness of a material plays a crucial role in influencing the mode of chip formation during machining. According to studies, micro-cutting happens when the maximum uncut chip thickness falls below the essential depth of cut. A micro-brittle fracture is possible if the maximum uncut chip thickness exceeds the critical depth of cut. Furthermore, particle size, particle velocity, plasma temperature, splat melting temperature, and other ambient parameters all have a major impact on coating surface roughness. Defects like porosity and unmelted or partially melted splats contribute to surface roughness. In sintered ceramics, such defects are less; hence, the mechanical properties are found to be one order larger than coated ceramics if the sintering is done optimally. However, in thermal spray, it is impossible to avoid these defects owing to the process limitations. The use of ceramic materials in grinding processes exhibits distinct properties. Compared to sintered ceramics, ceramic coatings result in significantly lower grinding forces. Due to the decreased hardness and fracture toughness of these materials, micro brittle fracture is the most common form of material removal. This leads to an extremely low critical depth of cut, making achieving a maximum uncut chip thickness smaller than this critical depth of cut impossible. However, customized abrasive processes like lapping, polishing, and MRF-assisted polishing are a few such operations that may result in a higher surface finish.

14.8　FUTURE SCOPE

The present work offers several potential avenues for extension, a few of them are as follows:

1. Investigation of grinding wheel wear on ceramic coatings and the interaction between the coating and grit during the grinding of thermally sprayed ceramic coatings. This involves investigating the grindability of thermally sprayed carbides and biocompatible coatings. The investigation will look at the material removal process, residual stress, fracture toughness, and surface finish.

2. The investigation on the effect of grinding fluid on high-speed and precision grinding of fragile ceramic materials will concentrate on optimizing coolant

pressure utilizing the least amount of lubricating method. The purpose of this study is to better understand the impacts of grinding fluid on the grinding process, especially in terms of increasing the performance of high-speed and precision grinding while dealing with fragile ceramic materials.

3. The effect of high grinding speed and low down feed is truly challenging and highly essential for micro/nano finishing of ceramic surfaces. This work can be extended to study the process using ductile regime grinding. The use of MRF and its application to nano-finishing ceramic material are a few current research areas that influenced by the outcomes of this study.

REFERENCES

[1] C.B. Carter, M.G. Norton, *Ceramic Materials*, Springer New York, New York, NY, 2013. https://doi.org/10.1007/978-1-4614-3523-5

[2] L. Pawlowski, Technology of thermally sprayed anilox rolls: State of art, problems, and perspectives, *J Therm Spray Technol.* 5 (1996) 317–334. https://doi.org/10.1007/BF02645884

[3] L. Yin, H. Huang, K. Ramesh, T. Huang, High speed versus conventional grinding in high removal rate machining of alumina and alumina–titania, *Int J Mach Tools Manuf.* 45 (2005) 897–907. https://doi.org/10.1016/j.ijmachtools.2004.10.016

[4] K. Kitajima, G.Q. Cai, N. Kurnagai, Y. Tanaka, H.W. Zheng, Study on mechanism of ceramics grinding, *CIRP Annals.* 41 (1992) 367–371. https://doi.org/10.1016/S0007-8506(07)61224-4

[5] S. Kar, P.P. Bandyopadhyay, S. Paul, Precision superabrasive grinding of plasma sprayed ceramic coatings, *Ceram Int.* 42 (2016) 19302–19319. https://doi.org/10.1016/j.ceramint.2016.09.100

[6] S. Kar, P.P. Bandyopadhyay, S. Paul, High speed and precision grinding of plasma sprayed oxide ceramic coatings, *Ceram Int.* 43 (2017) 15316–15331. https://doi.org/10.1016/j.ceramint.2017.08.071

[7] I. Inasaki, Grinding of Hard and Brittle Materials, *CIRP Annals.* 36 (1987) 463–471. https://doi.org/10.1016/S0007-8506(07)60748-3

[8] Zhang B, Study on Grindability of Thermo-Spray Coated nanophase Materials, 2003.

[9] S. Kar, S. Kumar, P.P. Bandyopadhyay, S. Paul, Grinding of hard and brittle ceramic coatings: Force analysis, *J Eur Ceram Soc.* 40 (2020) 1453–1461. https://doi.org/10.1016/j.jeurceramsoc.2019.12.058

[10] S. Agarwal, P.V. Rao, Experimental investigation of surface/subsurface damage formation and material removal mechanisms in SiC grinding, *Int J Mach Tools Manuf.* 48 (2008) 698–710. https://doi.org/10.1016/j.ijmachtools.2007.10.013

[11] W.C. Oliver, G.M. Pharr, Measurement of hardness and elastic modulus by instrumented indentation: Advances in understanding and refinements to methodology, *J Mater Res.* 19 (2004) 3–20. https://doi.org/10.1557/jmr.2004.19.1.3

[12] S. Kar, P.P. Bandyopadhyay, S. Paul, Effect of arc–current and spray distance on elastic modulus and fracture toughness of plasma-sprayed chromium oxide coatings, *Friction.* 6 (2018) 387–394. https://doi.org/10.1007/s40544-017-0166-6

[13] A.S. Kumar, S. Kar, P.P. Bandyopadhyay, S. Paul, Grinding of ceramics – sintered ceramics versus ceramic coatings, *Adv Mater Process Technol.* 4 (2018) 538–547. https://doi.org/10.1080/2374068X.2018.1479820

[14] W.C. Oliver, G.M. Pharr, An improved technique for determining hardness and elastic modulus using load and displacement sensing indentation experiments, *J Mater Res.* 7 (1992) 1564–1583. https://doi.org/10.1557/JMR.1992.1564

[15] Q. Chen, W.G. Mao, Y.C. Zhou, C. Lu, Effect of Young's modulus evolution on residual stress measurement of thermal barrier coatings by X-ray diffraction, *Appl Surf Sci.* 256 (2010) 7311–7315. https://doi.org/10.1016/j.apsusc.2010.05.071

[16] J.R. Davis, *Handbook of Thermal Spray Technology*, 2004.

[17] L.N.L. de Lacalle, A. Gutiérrez, A. Lamikiz, M.H. Fernandes, J.A. Sánchez, Turning of Thick Thermal Spray Coatings, *J Therm Spray Technol.* 10 (2001) 249–254. https://doi.org/10.1361/105996301770349349

[18] X. Liu, B. Zhang, Effects of grinding process on residual stresses in nanostructured ceramic coatings, *J Mater Sci.* (2002) 3229–3239. https://doi.org/10.1023/A:1016174731658

[19] M. Ramazani, J. Khalil-Allafi, R. Mozaffarinia, Grindability evaluation, and fatigue and wear behavior of conventional and nanostructured Al2O3-13 Wt.% TiO2 air plasma sprayed coatings, *J Therm Spray Technol.* 19 (2010) 611–619. https://doi.org/10.1007/s11666-009-9459-2

[20] X. Liu, B. Zhang, Grinding of nanostructural ceramic coatings: damage evaluation, *Int J Mach Tools Manuf.* 43 (2003) 161–167. https://doi.org/10.1016/S0890-6955(02)00157-8

Index

Pages in *italics* refer to figures and pages in **bold** refer to tables